Nuclear Power Plant Emergencies in the USA

Dean Kyne

Nuclear Power Plant Emergencies in the USA

Managing Risks, Demographics and Response

Springer

Dean Kyne
Department of Sociology and Anthropology
The University of Texas Rio Grande Valley
Edinburg, TX
USA

ISBN 978-3-319-84376-6 ISBN 978-3-319-50343-1 (eBook)
DOI 10.1007/978-3-319-50343-1

Printed on acid-free paper

This Springer imprint is published by Springer Nature
The registered company is Springer International Publishing AG
The registered company address is: Gewerbestrasse 11, 6330 Cham, Switzerland

To my parents, my father U Moe Kyaw and my late mother Daw Thein Thein Tin.

To my family, my wife Herina Miyamoto, and my two sons, Marc and Will.

Preface

There are currently 99 commercial reactors, operating at 61 nuclear power plant sites in the USA. Nuclear power plants (NPPs) are associated with potential core meltdown accidents, which could ultimately lead to catastrophic events. The three main nuclear disasters with the most significant historic and catastrophic impacts are as follows: Three Mile Island in the USA, which occurred in 1979; Chernobyl, Ukraine, in 1986; and Fukushima Daiichi, Japan, in 2011. The Fukushima Daiichi accident, the most recent disaster of the three, demonstrates that the risks of nuclear core meltdown accidents could be magnified by a natural disaster event such as an earthquake, tsunami, or similar natural events.

The immediate impacts from nuclear power core meltdown accidents is the release of high-level radioactive materials into the air which could be carried away for several hundreds of miles away, depending on the geographical location of an NPP, the size of the population living in areas around the facility, wind speed and directions, and weather conditions on the day the event is taking place. The Chernobyl accident caused hundreds of thousands of people to evacuate, and additionally, millions of people were exposed to radioactive radiation. The larger areas near the nuclear power plants were radiation-contaminated, and some areas were permanently restricted for habitation due to high level of radiative contamination. According to Perrow (1992), accidents related to NPPs are normal because the operating systems are coupled, and failure in one of the systems will have impact on other parts of the system, leading to the failure of an entire nuclear system. Given that accidents are considered to be normal for nuclear power plants, the risks of such accidents are inevitable.

There are three options to manage such inevitable risks: shutting down all NPPs, transforming inevitable risks to evitable ones, and effectively managing future nuclear power emergencies. The first option of shutting down all NPPs is not a currently feasible option, for a number of political and economic reasons. The second option involves the Nuclear Regulatory Commission (NRC), which has initiated reactor oversight framework which emphasise on three performance areas: reactor, radiation, and safeguards. However, the agency faces challenges of human errors and the nature of the inevitable accident (Perrow's Normal Accident Theory) in its efforts to transform the inevitable risks. The last option presents opportunities to minimize the negative impacts associated with the NPPs.

Managing the nuclear power emergencies is significantly different from managing those of other disasters such as fire or flood emergencies. For example, in the case of a fire emergency, as soon as the 911 call center alerts the fire department, firemen are dispatched to the place the fire is taking place, usually within minutes of the initial call. However, managing nuclear power emergencies requires certain procedures and protocol to go through, prior to an initial announcement of evacuation could take place. To effectively manage nuclear power emergencies, it is necessary to understand the problems in the previous nuclear power accidents, as well as demographic data of the populations prone to high-level radiation doses, living in areas surrounding the NPPs. Critically important in nuclear disaster management plans are the places the radioactive plume will disperse, and the individuals under the plume path at risks of exposure to the high-level radiation dose, and the challenges faced in evacuating individuals living in the areas with high-level radiation doses.

The development of commercial NPPs in the USA, the impacts of the three historic nuclear power core meltdown accidents, and the problems associated with response and evacuation are discussed in Chap. 1. The impacts of the three nuclear core meltdown accidents are analyzed, coupled with addressing the problems with response and evacuation, from a disaster and emergency management point of view. Chapter 2 captures the geographical locations of the 61 nuclear power plants within the USA and the communities exposed to the potential risks of core meltdown accident associated with the NPPs. In an event of a nuclear emergency, it is vital to carry out evacuation activities immediately so that the people living around the NPPs could be protected from the potential high-level doses of radiation. Chapter 3 examines the current radiological emergency plan and carefully investigates the process and potential problems that could lead to undermining the effectiveness of immediate response and evacuation. To evacuate people, it is imperative to know where the radioactive plume will go, given the weather conditions on the day the event takes place. Chapter 4 demonstrates utilization of powerful computer code, namely Radiological Assessment Systems for Consequence Analysis (RASCAL) to estimate the places the radioactive plume could be carried away by the given weather conditions during a nuclear core meltdown accident. The chapter provides two simulation exercises at two NPPs, namely the Palo Verde Nuclear Generating Station in Arizona, and the Indian Point Nuclear Generating Station in New York. The simulation exercises utilizing the RASCAL computer code with step-by-step procedures provides fundamental understanding and special technical skills needed to carry out a part of the effective nuclear power emergency management process. Chapter 5 examines the issues related to the nuclear power emergency plan in place. The discussion focuses on issues in terms of policy, priorities, process, participation, evacuation, and recovery. Chapter 6 proposes the three options for minimizing the risks associated with NPPs, suggesting the elimination of all NPPs in operation in USA, transforming inevitable risks to evitable risks, and transforming the current radiological plan into an effective emergency management plan.

This book would not have been made possible without the comments from my former academic advisors, for whom I have the outmost respect for, including Prof. Bob Bolin, Arizona State University; Prof. Christopher Boone, Arizona State University; and Prof. Dawid Pijawaka, Arizona State University. All three have been instrumental and extremely helpful in shaping my research skills and strengthening my work in nuclear emergency management. I would also like to express immense gratitude to two anonymous reviewers who provided me with very helpful comments and feedback in my work. Additionally, many thanks go to the instructors of the RASCAL course, Lou Brandon, PMT and RASCAL Program Manager, Nuclear Regulatory Commission (NRC), George F. Athey, of Athey Consulting, and James Van Ramsdell, of the Pacific Northwest National Laboratory. My heartfelt appreciation goes to Edlira Kuka, who edited my entire book manuscript, in multiple editing phases and providing invaluable feedback from inception to completion. Significant gratitude and appreciation is also extended to Dr. Robert Doe, Editor, Springer International Publishing, and Ms. Naomi Portnoy, project production coordinator, for their kind support throughout the publishing process. This book is considered a first-of-its-kind in providing a comprehensive understating of how to effectively manage nuclear power emergencies in the USA.

Edinburg, TX, USA Dean Kyne

Contents

About the Author

Dean Kyne is assistant professor of environmental sociology at the Department of Sociology and Anthropology, College of Liberal Arts, at the University of Texas Rio Grande Valley.[1] Being an environmental social scientist and geographer, he has conducted research studies covering the issues related to nuclear power plants, disaster and emergency management, environmental justice, social capital index, and siting controversial facilities.

[1]He graduated with a Ph.D. in Environmental Social Science from Arizona State University.

Chapter 1
The US Commercial Nuclear Power Plants and Their Potential Risks

1.1 U.S. Nuclear Power Development

In the history of nuclear power development, there are three major events that precede as the first commercial nuclear power plant was developed and it became fully operational in USA (Table 1.1). The three development events are as follows: discovery of fission, discovery of self-sustaining chain reaction, and the political vision of peaceful applications of atoms.

1.1.1 Discovery of Fission

The first preceding event before the birth of US commercial nuclear power plants was the discovery of fission in 1993 (US DOE 1995). To understand fission, one requires the knowledge of "atoms." This can be traced back to about 2,400 years ago when a pre-Socratic Greek philosopher claimed that the world is made up of small and invisible substances called "atoms" (Ferguson 2011). Fission can be defined as follows:

> The splitting of an atom, which releases a considerable amount of energy (usually in the form of heat) that can be used to produce electricity. Fission may be spontaneous, but is usually caused by the nucleus of an atom becoming unstable (or "heavy") after capturing or absorbing a neutron. During fission, the heavy nucleus splits into roughly equal parts, producing the nuclei of at least two lighter elements. In addition to energy, this reaction usually releases gamma radiation and two or more daughter neutrons (US NRC 2016b).

Fission was discovered by Enrico Fermi, a physicist who conducted an experiment on neutrons in 1934 (US DOE 1995). His findings revealed that the neutrons could split many different types of atoms. 4 years later, two German scientists, namely Otto Hahn and Fritz Strassman, added their new discovery to the findings of Fermi. Their findings were that after firing with a source containing *radium* and *beryllium* into *uranium* (atomic number 92), there are even lighter elements such as

D. Kyne, *Nuclear Power Plant Emergencies in the USA*,
DOI 10.1007/978 3-319-50343-1_1

Table 1.1 Chronology of nuclear research and development

1940s	
1942	The first self-sustaining nuclear chain reaction occurs at the University of Chicago
1945	The U.S. Army's Manhattan Engineer District (MED) tests the first atomic bomb at Alamogordo, New Mexico, under the code name Manhattan Project
1945	The atomic bomb, nicknamed Little Boy is dropped on Hiroshima, Japan on August 6. 3 days later, another bomb, Fat Man, is dropped on Nagasaki, Japan. Japan surrenders on August 15 of 1945, ending World War II
1946	The Atomic Energy Act of 1946 creates the Atomic Energy Commission (AEC) to control nuclear energy development and explore peaceful uses of nuclear energy
1947	The AEC first investigates the possibility of peaceful uses of atomic energy, issuing a report the following year
1949	The AEC announces the selection of a site in Idaho for the National Reactor Testing Station
1950s	
1951	In Arco, Idaho, Experimental Breeder Reactor I produces electric power from nuclear energy for the first time, lighting four light bulbs
1952	Keel for the Navy's first nuclear submarine, Nautilus, is laid at Groton, Connecticut
1953	Nautilus starts its nuclear power units for the first time
1953	President Eisenhower delivers his "Atoms for Peace" speech before the United Nations. He calls for greater international cooperation in the development of atomic energy for peaceful purposes
1954	President Eisenhower signs The Atomic Energy Act of 1954, the first major amendment of the original Atomic Energy Act, giving the civilian nuclear power program further access to nuclear technology
1955	The AEC announces the Power Demonstration Reactor Program. Under the program, AEC and industry will cooperate in constructing and operating experimental nuclear power reactors
1955	Arco, Idaho, population 1,000, becomes the first town powered by a nuclear reactor, the experimental boiling water reactor BORAX III
1955	Geneva, Switzerland, hosts the first United Nations International Conference on the Peaceful Uses of Atomic Energy
1957	The first power from a civilian nuclear unit is generated by the Sodium Reactor Experiment at Santa Susana, California. The unit provided power until 1966
1957	The Price-Anderson Act provides financial protection to the public and AEC licensees and contractors if a major accident occurs at a nuclear reactor
1957	The United Nations creates the International Atomic Energy Agency (IAEA) in Vienna, Austria, to promote the peaceful use of nuclear energy and prevent the spread of nuclear weapons around the world
1958	The world's first large-scale nuclear reactor begins operation in Shippingport, Pennsylvania. The plant reaches full power three weeks later and supplies electricity to the Pittsburgh area
1958	Construction begins on the world's first nuclear-powered merchant ship, the N.S. Savannah, in Camden, New Jersey. The ship is launched on July 21, 1959
1959	Dresden-1 Nuclear Power Station in Illinois, the first U.S. nuclear plant built entirely without government funding, achieves a self-sustaining nuclear reaction
1960s	
1960	The third U.S. nuclear reactor, Yankee Rowe Nuclear Power Station, achieves a self-sustaining nuclear reactor
1960	Small nuclear-power generators are first used in remote areas to power weather stations and to light buoys for sea navigation
1961	The U.S. Navy commissions the world's largest ship, the U.S.S. Enterprise. It is a nuclear-powered aircraft carrier with the ability to operate at speeds up to 30 knots for distances up to 400,000 miles (740,800 km) without refueling

(continued)

Table 1.1 (continued)

1964	President Lyndon B. Johnson signs the Private Ownership of Special Nuclear Materials Act, which allows the nuclear power industry to own the fuel for its units. After June 30, 1973, private ownership of uranium fuel is mandatory
1963	Jersey Central Power & Light Company announces its commitment for the Oyster Creek nuclear reactor, the first nuclear plant ordered as an economic alternative to fossil-fuels
1964	Three nuclear-powered surface ships, the Enterprise, Long Beach, and Bainbridge, complete "Operation Sea Orbit," a cruise around the world
1965	The first nuclear reactor in space (SNAP-10A) is launched by the United States. SNAP stands for Systems for Nuclear Auxiliary Power
1970s	
1970	The United States, United Kingdom, Soviet Union, and 45 other nations ratify the Treaty for Non-Proliferation of Nuclear Weapons
1971	Twenty-two commercial nuclear reactors are in full operation in the United States. They produce 2.4% of U.S. electricity at this time
1973	U.S. utilities order 41nuclear reactors, a 1-year record
1974	The first 1,000-MW-electric nuclear reactor goes into service—Commonwealth Edison's Zion 1 Plant
1974	The Energy Reorganization Act of 1974 divides AEC functions between two new agencies—the Energy Research and Development Administration (ERDA), to carry out research and development, and the Nuclear Regulatory Commission (NRC), to regulate nuclear power
1977	President Jimmy Carter announces the United States will defer indefinitely plans for reprocessing spent nuclear fuel
1977	President Carter signs the Department of Energy Organization Act, which transfers ERDA functions to the new Department of Energy (DOE)
1977	DOE begins operations
1979	The worst accident in U.S. commercial reactor history occurs at the Three Mile Island nuclear power station near Harrisburg, Pennsylvania. The accident is caused by a loss of coolant from the reactor core due to a combination of mechanical malfunction and human error. No one is injured, and no overexposure to radiation results from the accident. Later in the year, the NRC imposes stricter reactor safety regulations and more rigid inspection procedures to improve the safety of reactor operations
1979	Seventy-two licensed reactors generate 12% of the electricity produced commercially in the United States
1980s	
1980	DOE initiates the Three Mile Island research and development program to develop technology for disassembling and de-fueling the damaged reactor. The program will continue for 10 years and make significant advances in developing new nuclear safety technology
1982	After 25 years of service, the Shippingport Power Station is shut down. Decommissioning would be completed in 1989
1983	The Nuclear Waste Policy Act (NWPA) establishes a program to site a repository for the disposal of high-level radioactive waste, including spent fuel from nuclear reactors. It also establishes fees for owners and generators of radioactive waste and spent fuel, who pay the costs of the program
1983	Nuclear power generates more electricity than natural gas
1984	The atom overtakes hydropower to become the second largest source of electricity, after coal. Eighty-three nuclear power reactors provide about 14% of the electricity produced in the United States
1985	The Institute of Nuclear Power Operations forms a national academy to accredit every nuclear reactor's training program

(continued)

Table 1.1 (continued)

1986	The Perry Power Plant in Ohio becomes the 100th U.S. nuclear reactor in operation.
1986	April 26. Operator error causes two explosions at the Chernobyl No. 4 nuclear reactor in the former Soviet Union. The reactor has an inadequate containment building, and large amounts of radiation escape. A plant of such design would not be licensed in the United States
1987	December 22. The Nuclear Waste Policy Act (NWPA) is amended. Congress directs DOE to study only the potential of the Yucca Mountain, Nevada, site for disposal of high-level radioactive waste
1988	U.S. electricity demand increases 50% higher than in 1973
1989	109 nuclear reactors provide 19% of the electricity used in the U.S.; 46 units have entered service during the decade.
1989	The NRC proposes a plan for reactor design certification, early site permits, and combined construction and operating licenses
1990s	
1990	DOE launches a joint initiative to improve operational safety practices at civilian nuclear reactors in the former Soviet Union
1990	America's 110 nuclear reactors set a record for the amount of electricity generated, surpassing all fuel sources combined in 1956
1990	The final shipment of damaged fuel from the Three Mile Island nuclear plant arrives at a DOE facility in Idaho for research and interim storage. This ends DOE's 10-year Three Mile Island research and development program
1991	111 nuclear reactors operate in the U.S. with a combined capacity of 99,673 MW. They produce almost 22% of the electricity generated commercially in the U.S.
1992	110 nuclear reactors account for nearly 22% of all electricity used in the U.S.
1992	DOE signs a cooperative agreement with the nuclear industry to co-fund the development of standard designs for advanced light-water reactors
1992	The Energy Policy Act of 1992 is signed into law. The Act makes several important changes in the licensing process for nuclear reactors
1992	The 50th anniversary of the historic Fermi experiment is observed worldwide
1993	The U.S. nuclear utility consortium, the Advanced Reactor Corporation (ARC), signs a contract with Westinghouse Electric Corporation to perform engineering work for an advanced, standardized 600-MW pressurized water reactor. Funding for this next-generation plant comes from ARC, Westinghouse, and DOE
1993	The U.S. nuclear utility consortium, ARC, signs a contract with General Electric Company for cost-shared, detailed engineering of a standardized design for a large, advanced nuclear reactor. The engineering is being funded under a joint program among utilities, General Electric, and DOE

Source US DOE (1995)

barium (atomic number 56), in the leftover materials which weigh about half the atomic mass of uranium. However, they faced a puzzle that when they added all messes of fission products, the sum did not equal before the weights of messes before. Their big question was where did the lost mass go? Another scientist, Lise Meitner, who worked with Niels Bohr and his nephew, Otto R. Frisch in Copenhagen, used Einstein's theory to provide the answer to the big question of the previous two scientists. Lise finally provided that the lost mass was due to occurrence of fission (US DOE 1995) which was a significant discovery leading to the future of nuclear work.

1.1.2 Self-sustaining Chain Reaction

The second development in nuclear discoveries was the achievement of self-sustaining chain reaction (US DOE 1995). It was in 1929 that Niels Bohr who worked with Hahn–Strassmann–Meitner met Einstein and shared his discoveries. Bohr also met with Fermi in a conference in the USA and shared his excitement about a potential to sustain nuclear reaction. At that time, many scientists around the world began to believe that the potential for a self-sustaining nuclear reaction could be realized. In 1941, Fermi and Leo Szilard suggested a possible design that could help sustain the fission chain reaction process; taking uranium and placing it in a stack of graphite to make a cubelike frame of fissionable materials (US DOE 1995), they began their work in constructing the world's first nuclear reactor, which took place at Chicago University, in 1942. Their first reactor is known as Chicago Pile-1 consisting of control rods, made of *cadmium*, *uranium*, and *graphite*. *Cadmium* is a metallic element which absorbs neutrons resulting in slowing down of the chain reactions. In the reactor, the fission is controlled by cadmium. When the *cadmium* rod was pulled out in the pile, it provided more neutrons available to split atoms that resulted in speeding up the chain reactions. It was exactly 3:25 pm on December 2, 1942, that Fermi and his group successfully transformed their theories into realities, leading the world to enter into the nuclear age.

1.1.3 Peaceful Applications of Atoms

The third development in nuclear energy was the study and examination for peaceful applications of nuclear power (US DOE 1995). Prior to World War II, there were many scientists who made great efforts to harness atoms in order to produce nuclear weapons. Producing nuclear weapons was secretly under the code name Manhattan Project at that time. On the other hand, there were many other scientists who worked on developing breeder reactors. A turning point was reached when the use of atomic bombs began. In August 1945, atomic bombs were dropped in Japan, in Hiroshima and Nagasaki. People from different walks of life including politicians, journalists, scientists, and business leaders saw the horrific negative side of using atomic power for destruction and war. The events changed the attitudes of people around the world, and many began condemning the use of atoms for destructive purposes. Meanwhile, in December 1945, Alvin M. Weinberg added another excitement to the world through his claim on peaceful use of atoms. He claimed the potentials of nuclear power to benefit all mankind in a report to the US Senate's Special Committee. His assertion was that "atomic power can cure as well as kill. It can fertilize and enrich a region as well as devastate it. It can widen man's horizons as well as force him back into the cave (US NRC 2016c)." Based on Alvin's claim, in 1946, the US government encouraged the development of nuclear energy for peaceful civilian purposes.

The official launch was made when the government established the Atomic Energy Commission (AEC), under the Atomic Energy Act of 1946. Under the act, the AEC was responsible for regulating nuclear power. The government's intention to develop atomic energy became more intense when the President Eisenhower made a popular public speech, in December 1953. In delivering his "Atoms for Peace" speech, Eisenhower stated that "this greatest of all destructive forces can be developed into a great boon, for the benefit of all mankind (US NRC 2016c)." He highlighted his plan to transform nuclear fission materials—the subject of much public fear—into a resource with a peaceful end use (Eisenhower 1953). 8 years after the establishment of the AEC, Congress replaced the existing law with the Atomic Energy Act of 1954, which allowed the agency to develop, for the first time, a commercial nuclear power plant (US NRC 2016a). The new act authorized the AEC to promote the use of nuclear power and to regulate its safety. One notion was that the AEC was playing a dual role; on the one hand, the agency was required to promote the nuclear energy sector, and on the other, it was responsible to regulate the industry. 4 years after the famous Atoms for Peace speech, on December 18, 1957, the world's first commercial nuclear power plant—built on the Ohio River 25 miles northeast of Pittsburgh—began generating nuclear power (US NRC 2011) out of the Shippingport Atomic Power Station. The AEC received a lot of criticisms for the competing dual roles it played as promoter and regulator of atomic power. These dual roles were competing with each other, and the agency was criticized for being unable to impose strict rules and regulations on the use of nuclear energy. According to the NRC (US NRC 2016c), during the 1960s, the AEC was criticized for its weak actions on several important areas, such as protecting the public from radiation, establishing strong radiation protection standards, establishing guidelines for reactor safety, siting plants, and protecting environments. Taking those criticisms into account, in 1974, Congress decided to abolish the AEC and created the Energy Reorganization Act of 1974. Under the new act, the Nuclear Regulatory Commission (NRC) was established (US NRC 2016c); the agency officially began its operations on January 19, 1975. According to the NRC (US NRC 2016c), the agency's regulatory activities emphasized on the following:

(1) reactor safety oversight and reactor license renewal of existing plants;
(2) materials safety oversight and materials licensing for a variety of purposes;
(3) waste management of both high-level and low-level wastes; and
(4) evaluating new applications for nuclear plants.

1.1.4 Nuclear Power Reactor with Fission Technology

Nuclear power plants are not different from other power plants that use heat to produce electricity. Referring to US EIA (2015), a nuclear power plant could be simply described as follows: Nuclear power reactors were developed after the discoveries of the self-sustaining fission reactions because the fission reaction is used as a source of the heat for nuclear power plants. In a nuclear fission reaction, the atoms are split into smaller atoms that release energy. Nuclear fission takes place in the uranium fuel which is placed in the reactor of a power plant to heat up the water. The uranium fuel rod consists of ceramic pellets which are placed in 12-foot metal fuel. Each ceramic pellet could produce the similar energy amount produced by 150 gallons of oil. A fuel assembly may include hundreds of fuel rods, and a reactor core contains many fuel assemblies. When the water in the reactor is heated by nuclear fission reactions, taking place in the fuel rods, the water is transformed into steam, which rotates the turbine blades, making generators and thus producing electricity. The water is cooled down and sent back to the reactor to be reused. According to the NRC (US NRC 2010), there are two types of US commercial nuclear reactors, namely boiling water reactors (BWR) (Fig. 1.1a) and pressurized water reactors (PWR) (Fig. 1.1b).

Referring to the NRC (2010), both nuclear power reactors contain a core which is placed inside the reactor where fuel assembles, and bundles of hundreds of fuel rods are housed in order to produce heat. The heat is generated by using the nuclear fission reactions in fuel rods. The core of the nuclear reactor is indicated "1" in both reactor types. The reactor obtains heat energy from nuclear fission reactions, and the next step is to transform the heat to steam. In the BWR type (Fig. 1.1a), the stream–water mixture is generated by a flow of very pure water (reactor coolant), through the core of the reactor where the heat is absorbed (indicated as "2"), then the stream-water mixture passes through separators and dryers (indicated as "3"). In the PWR type (Fig. 1.1b), the stream generation is achieved by two steps: First, reactor coolant is pressurized, and then, the pressurized coolant carries the heat to the stream generator (indicated as "2"). The heat vaporizes the water in a secondary loop that generates steam (indicated as "3"). The differences between the PWR and BWR types are that the steam is generated in the core of a reactor in the BWR, whereas the steam is produced in a separate chamber of steam generator in PWR

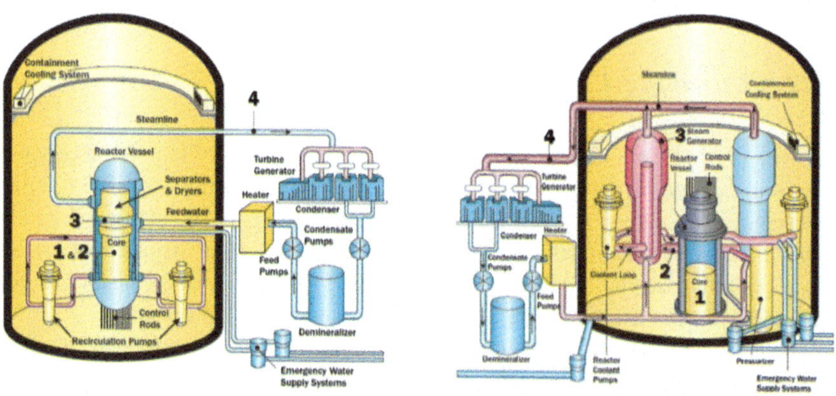

(**a**) Typical Boiling-Water Reactor (**b**) Typical Pressurized-Water Reactor

Fig. 1.1 Types of US commercial nuclear power reactors. *Source* US NRC (2010)

type of reactor. The generated steam in both types of reactors is directed to turbine generators, in order to drive electric generators to produce electricity (indicated as "4"). In both reactors, unused steam is recycled by placing it back to the condenser and then reheating it, by flowing through the reactor. The BWR type uses more fuel rods than the PWR type does. The first one normally carries between 370 and 800 fuel assemblies, whereas the latter does between 150 and 200 fuel assemblies.

The most important and challenging task is to control the fission reaction in the core in order to sustain the amount of heat that is appropriate to generate the steam. If the fission reaction is out of control, the core containing nuclear fuel assemblies could become overheated, leading to a core-meltdown. In the BWR-type reactor, the fuel assemblies are cooled by water circulated using electrically powered pumps, which receive electricity from the electrical grid. In the PWR, the fuel assemblies are cooled by circulating water pumped with electrical pumps. A backup pumping system is installed using electricity, produced by onsite diesel generators. Another cooling system that further cools the containment is also installed. The dome-shaped containment structure is built with iron and concrete walls of about 4 ft in thickness. In case of a core-meltdown accident, the radioactive materials are contained inside, preventing from polluting the environment.

1.2 U.S. Commercial Nuclear Power Plants in Operation

Currently, the USA has brought online a fleet of commercial nuclear reactors, located at 61 sites scattered across 30 states. Before the current commercial reactors could begin their operation, the owners of the nuclear power plants are required to apply for a license, even before they begin to build a site. The following three sections discuss why a license is required; when a license could be renewed; and the current nuclear reactors currently in operation.

1.2.1 Operating License Requirement

All owners of US commercial nuclear power plants are legally required to obtain a license to operate or build from the US NRC. According to the US NRC (2016d), the agency was authorized for licensing and regulating the operation of nuclear power plants. As mentioned earlier, the AEC also authorizes issuance of nuclear power plant licenses before the US NRC was formed in 1974. In the history of both the AEC and the NRC, they are authorized to issue nuclear licenses.

The US NRC explains conditions that require those seeking licenses to submit an application for a license to operate. There are three conditions that may warrant the license requirement which are as follows:

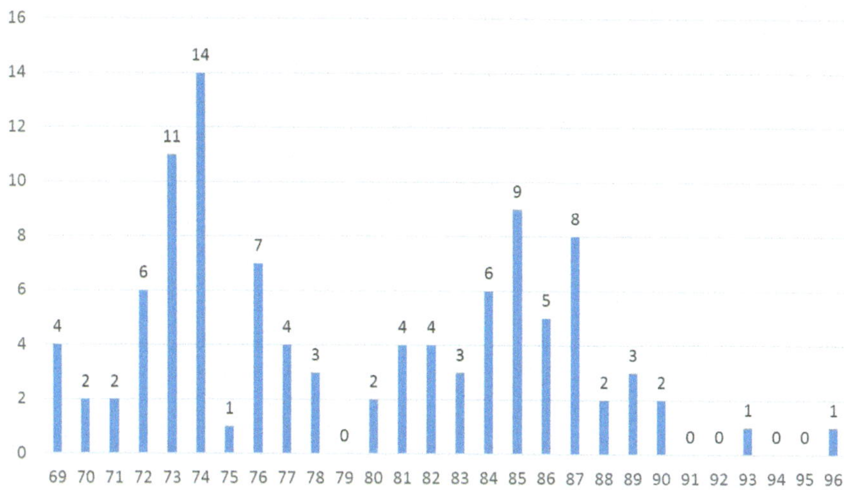

Fig. 1.2 US commercial nuclear power reactor operating licenses—issued by year, Data. *Source* US NRC (2010), US NRC (2016a)

(1) to construct, operate, and decommission commercial reactors and fuel cycle facilities.
(2) to possess, use, process, export and import nuclear materials and waste, and handle certain aspects of their transportation.
(3) to site, design, construct, operate, and close waste disposal sites (US NRC 2016d)

As of August 2009, both the ACE and the NRC have issued a total of 104 US commercial nuclear reactors which are located at 65 nuclear power plant sites in 31 states, to 26 various operating companies (Fig. 1.2). The reactors are produced by four different reactor vendors, and there are 80 different designs (US NRC 2010). In 2013, three nuclear power plants with a total of four reactors were permanently shut down. They are Crystal River Power Plant in Florida (one reactor) in February; the Kewaunee Power Plant in Wisconsin (one reactor) in April; and the San Onofre Power Plant in California (two reactors) in June. In 2014, the Vermont Yankee Nuclear Power Plant in Vermont (one reactor) was also permanently shut down in December (US EIA 2016). 1973 is seen as the year in which the USA has seen a peak of construction and operating of nuclear power plants (Fig. 1.2). After 1973, the demand for nuclear power plants slowed down until in the mid-1980s, where the construction resumed gradually. However, there were no licenses related to nuclear power plant facilities issued after 1996.

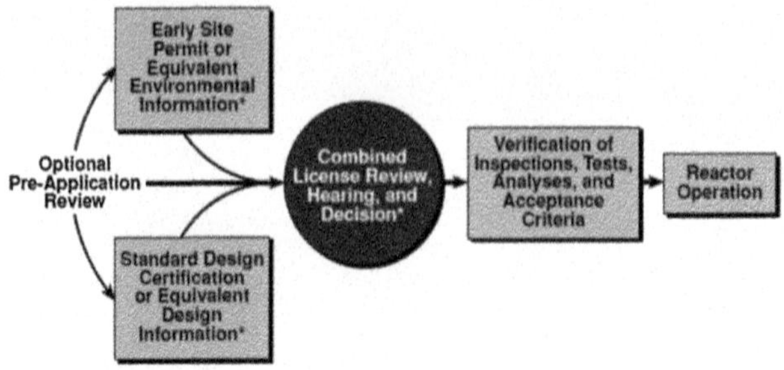

*A combined license application can reference an early site permit, a standard design certification, both, or neither. If an application does not reference an early site permit and/or a standard design certification, the applicant must provide an equivalent level of information in the combined license application.

Fig. 1.3 Relationships between combined licenses, early site permits, and standard design certifications. *Source* US NRC (2004)

1.2.2 Operating License Process

In the past, nuclear power plants were licensed under a two-step licensing process. This process required both a construction permit and an operating license separately. In 1989, the NRC established an alternative licensing process that essentially combines a construction permit and an operating license with certain conditions, into a single license (Fig. 1.3). Under either process, before an applicant can build and operate a nuclear power plant, it must obtain the approval from the NRC (US NRC 2004).

During the licensing process, one of the key elements is "public involvement." One of the reasons to have a public process is to encourage the public to participate in the environmental decision-making process and provide feedback. This key element could be related to the National Environmental Policy Act (NEPA). In 1970, NEPA went into effect and the Environmental Protection Agency (EPA) was established, with its mission to incorporate environmental protection policy into all federally sponsored projects. According to the EPA, the NEPA could be summarized as follows:

> The NEPA is one of the first laws ever written that establishes the broad national framework for protecting our environment. NEPA's basic policy is to assure that all branches of government give proper consideration to the environment prior to undertaking any major federal action that significantly affects the environment (US EPA 2016).

NEPA required that all federal agencies include Environmental Assessments (EAs) and Environmental Impact Statements (EISs) from carrying out the projects for airports, buildings, military complexes, highways, parkland purchases, and other federal activities (US EPA 2016). The US EPA is the agency which assures that all federal activities incorporate the NEPA requirements by advocating environmental justice, which could be defined as:

> Fair treatment and meaningful involvement of all people regardless of race, color, national origin, or income with respect to the development, implementation, and enforcement of environmental laws, regulations, and policies. *Fair treatment means* that no group of people should bear a disproportionate share of the negative environmental consequences resulting from industrial, governmental and commercial operations or policies. *Meaningful Involvement* means that: people have an opportunity to participate in decisions about activities that may affect their environment and/or health; the public's contribution can influence the regulatory agency's decision; their concerns will be considered in the decision making process; and the decision makers seek out and facilitate the involvement of those potentially affected (US EPA 2016)

In the combined license process, there are opportunities for the public to participate in environmental decision making (Fig. 1.4). At least two opportunities for public involvement are mandatory, namely public meetings on environmental review and mandatory hearings on the combined license application process. The NRC conducts its licensing activities based on the current existing regulations. The two existing regulations that are regarded as the most relevant to the design, siting, construction, and operation of new commercial nuclear power facilities are 10 CFR Part 51, "Environmental Protection Regulations for Domestic Licensing and Related Regulatory Functions," and 10 CFR Part 52, "Licenses, Certifications, and Approvals for Nuclear Power Plants." The combined licenses have been awarded to Fermi 3, South Texas Project Unit 3, South Texas Project Unit 4, V.C. Summer Unit 2, V.C. Summer Unit 3, Vogtle Unit 3, and Vogtle Unit 4.

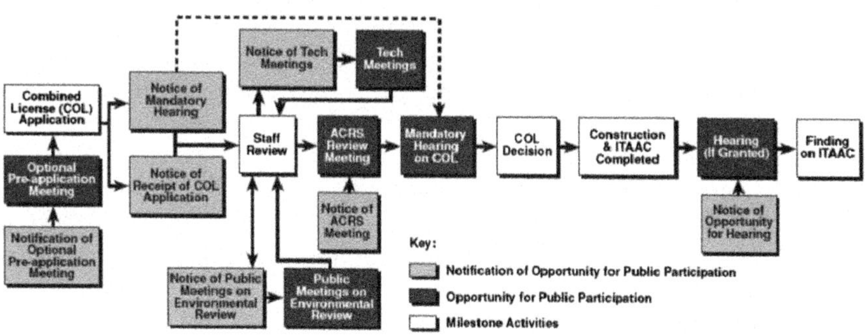

Fig. 1.4 Opportunities for public involvement during the review of combined licenses. *Source* US NRC (2004)

1.2.3 License Renewal

After the amendment of the Atomic Energy Act of 1954, the NRC was authorized to issue a 40-year initial license for commercial nuclear power reactors. The act also allows the NRC to renew the licenses for another 20 years at a time. After 40 years of use, some of the essential parts of a reactor, which are engineered with an expected use of 40 years, will expire and the license holders must replace and maintain the parts in good condition, before renewing their existing licenses. As of June 2015, the NRC has renewed the 74 currently operating reactors located at 43 plant sites to another 20 years (US NRC 2015a). The regulations governing the renewal process are specified in 10 CFR Part 54, "Requirements for Renewal of Operating Licenses for Nuclear Power Plants," while the NRC's environmental review requirements for license renewal are found in 10 CFR Part 51, "Environmental Protection Regulations for Domestic Licensing and Related Regulatory Functions (US NRC 2015a)."

1.2.4 Current Reactors in Operation

As of January 27, 2016, there are 99 US commercial nuclear reactors which are currently in operation located at 61 power plant sites in 30 states (Fig. 1.5). Many of the reactors are located in the east, northeast, and southwest of the nation. Among the 61 nuclear power plant sites, 35 of these plants have two or more reactors. The Palo Verde Power Plant in Arizona, the largest nuclear power plant, which contains three reactors, has the largest combined net summer generating capacity of 3,937 megawatts (MW) in 2014. The Fort Calhoun in Nebraska was the smallest single reactor with the smallest net summer capacity of 479 megawatts (MW) in 2014 (US EIA 2016). According to the EIA (2015), the USA is ranked as the highest capacity for generation of nuclear energy, among 30 countries generating electricity with nuclear power plants.

Of the 30 countries in the world that have commercial nuclear power plants, the USA has the most nuclear capacity and generation suppling 20% of its total energy usage. France, the country with the second highest nuclear capacity, relies on nuclear power for nearly 75% of its total electricity. Fourteen other countries generate more than 20% of their electricity from nuclear power (US EIA 2015).

As noted earlier, as of June 2015, the NRC has renewed the 74 currently operating reactors located at 43 plant sites to another 20 years (US NRC 2015a). It could be implied that the current operators are aging. There are about 35 reactors out of a total 99 (35%) that are over 40 years old (Fig. 1.6). The primary license provided a 40-year operation span; many of them were initially designed to operate for a 60-year span. The owners who would like to operate beyond the 40 years of nuclear reactor life spans are required to replace parts that were expired from what is their expected life of use, before they could renew their licenses. There are issues associated with aging of reactors which can cause safety concerns. Copper (2013)

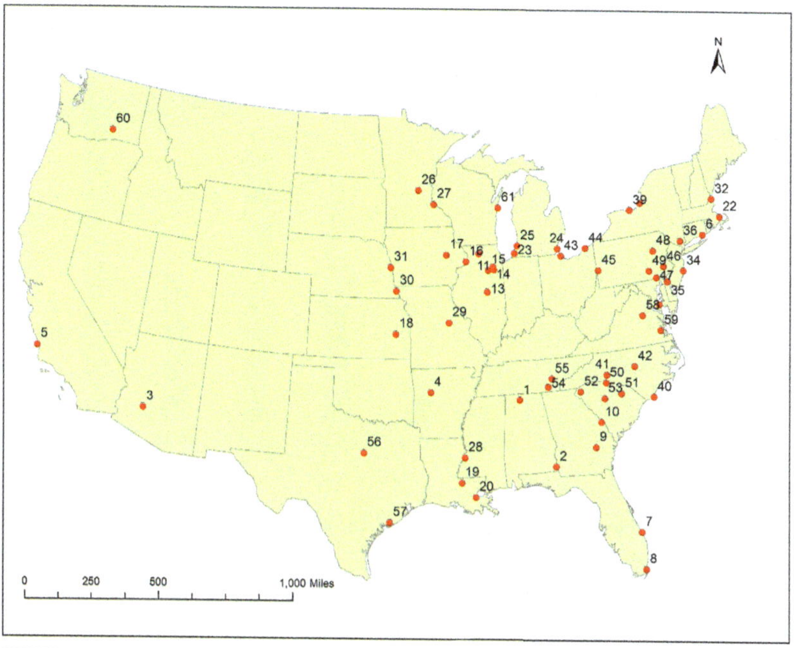

Index	Plant Name	State	Index	Plant Name	State
1	Browns Ferry Nuclear Plant	Alabama	31	Fort Calhoun Station	Nebraska
2	Joseph M. Farley Nuclear Plant	Alabama	32	Seabrook Station	New Hampshire
3	Palo Verde Nuclear Generating Station	Arizona	33	Hope Creek Generating Station	New Jersey
4	Arkansas Nuclear One	Arkansas	34	Oyster Creek Nuclear Generating Station	New Jersey
5	Diablo Canyon Nuclear Power Plant	California	35	Salem Nuclear Generating Station	New Jersey
6	Millstone Power Station	Connecticut	36	Indian Point Nuclear Generating	New York
7	St. Lucie Plant	Florida	37	James A. FitzPatrick Nuclear Power Plant	New York
8	Turkey Point Nuclear Generating	Florida	38	Nine Mile Point Nuclear Station	New York
9	Edwin I. Hatch Nuclear Plant	Georgia	39	R.E. Ginna Nuclear Power Plant	New York
10	Vogtle Electric Generating Plant	Georgia	40	Brunswick Steam Electric Plant	North Carolina
11	Braidwood Station	Illinois	41	McGuire Nuclear Station	North Carolina
12	Byron Station	Illinois	42	Shearon Harris Nuclear Power Plant	North Carolina
13	Clinton Power Station	Illinois	43	Davis-Besse Nuclear Power Station	Ohio
14	Dresden Nuclear Power Station	Illinois	44	Perry Nuclear Power Plant	Ohio
15	LaSalle County Station	Illinois	45	Beaver Valley Power Station	Pennsylvania
16	Quad Cities Nuclear Power Station	Illinois	46	Limerick Generating Station	Pennsylvania
17	Duane Arnold Energy Center	Iowa	47	Peach Bottom Atomic Power Station	Pennsylvania
18	Wolf Creek Generating Station	Kansas	48	Susquehanna Steam Electric Station	Pennsylvania
19	River Bend Station	Louisiana	49	Three Mile Island Nuclear Station	Pennsylvania
20	Waterford Steam Electric Station	Louisiana	50	Catawba Nuclear Station	South Carolina
21	Calvert Cliffs Nuclear Power Plant	Maryland	51	H. B. Robinson Steam Electric Plant	South Carolina
22	Pilgrim Nuclear Power Station	Massachusetts	52	Oconee Nuclear Station	South Carolina
23	Donald C. Cook Nuclear Power Plant	Michigan	53	Virgil C. Summer Nuclear Station	South Carolina
24	Fermi	Michigan	54	Sequoyah Nuclear Plant	Tennessee
25	Palisades Nuclear Plant	Michigan	55	Watts Bar Nuclear Plant	Tennessee
26	Monticello Nuclear Generating Plant	Minnesota	56	Comanche Peak Steam Electric Station	Texas
27	Prairie Island Nuclear Generating Plant	Minnesota	57	South Texas Project	Texas
28	Grand Gulf Nuclear Station	Mississippi	58	North Anna Power Station	Virginia
29	Callaway Plant	Missouri	59	Surry Nuclear Power Station	Virginia
30	Cooper Nuclear Station	Nebraska	60	Columbia Generating Station	Washington
			61	Point Beach Nuclear Plant	Wisconsin

Fig. 1.5 Locations of current US commercial nuclear power plants, Data. *Source* US NRC (2016a)

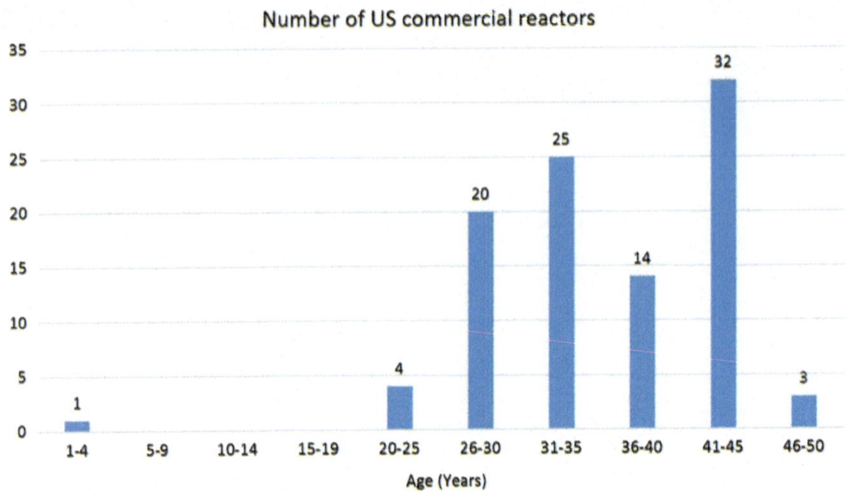

Fig. 1.6 Age of currently operating US commercial reactors, Data. *Source* The US NRC (2016a)

made a one notion that the construction proposals make a projection of a nuclear power plant life to be up to 60 years, but in reality, about one-sixth of the total US reactors retired prior to their projected expectancy of 60 years. In other words, they even retired before expiration of their initial license which was provided for up to 40 years.

There are at least three reasons why owners of current reactors made a decision to abandon the aging reactors (Cooper 2013). The following reasons are evident:

(1) Expensive repairs;
(2) Uneconomic to keep running; and
(3) Some of them have permanent damages beyond repair.

Costly repairs became major reason for abandonment of two nuclear power plants: Duke Energy's Crystal River in Florida and Edison International's San Onofre in California. Both of the plants needed multibillion-dollar repairs (Crooks 2016). According to the NRC, the Crystal River Unit 3 Nuclear Generating Plant with a capacity of producing 2,609 MW and a pressurized water reactor was issued a license to operate from December of 1976 to February 20, 2013. On September 26, 2009, when the plant was doing a refueling, the owner decided to replace the stream generators. To do so, a large hole must be created in the concrete dome, prior to replacing the generator. After replacing the generator, the hole must be restored again. However, the owner could not fix the repair and finally decided to permanently shut down the Crystal River Unit 3 reactor, after 33 years of operation. Similarly, San Onofre, in California, retired after 30 years of operation, in 2013, after observing steam tube degradation (US NRC 2015b). The repair needs were complex especially for the replacement of 2000 tubes, and finally, the owner decided to take the reactor down and end operations.

Further, some reactors are simply not economical to operate. A good case study to review is the Kewaunee Power Plant in Wisconsin, with a capacity of 556 MW, which was closed on May 7, 2013, after 39 years of operation, solely for uneconomic reasons (Dotson 2014). Dotson noted that the plant owner confirmed that the decision for permanently shutting down the plant was due to economic reasons and the cost of operation, which faced no regulatory reason or public opposition. Dominion, a Virginia-based power company, owns three plants, namely North Anna and Surry in Virginia and Millstone in Connecticut. In July 2005, the company acquired Kewaunee Power Plant, which was jointly owned by Wisconsin Public Service Corporation and Alliant for $129 million. The Kewaunee acted as a supplier for electricity to two companies in the past. The two companies did not renew their contracts. The company could not buy another power plant to make up the loss with Kewaunee and to make it economical to continue the plant running. The company put the plant on sale in April 2011, but it could not find any buyers. In October 2012, it decided to close the plant (Dotson 2014).

Some reactors received local opposition to discontinue their operations. On March 29, 1979, the Three Mile Island Unit 2 reactor in Middletown, PA, experienced a partial core-meltdown experience (US NRC 2013c). It was regarded as one of the most serious accidents in US commercial power plant history, and the plant was permanently shut down after being in operation since December 30, 1978.

The current fleet of nuclear power plants which are aging provide lessons and cautionary observations for the future of nuclear reactor maintenance and safe operations. They demonstrate some hindrances that could limit their full potentials. A study conducted by Massachusetts Institute of Technology (MIT 2003) found that there are four major problems within the nuclear industry. These four problems center on costs of operations, safety, proliferation, and risks.

Nuclear power plants are found to have higher overall lifetime costs than other source of energy in the long term. For example, the study shows that when compared to another source of energy, natural gas, with combined cycle turbine technology (CCGT) and coal, at least the nuclear power has an absence of a carbon tax or an equivalent "cap and trade" mechanism for reducing carbon emissions. Further, other issues exist, associated with safety, health, and environmental impacts. Nuclear power is associated with the issues that receive attention and concern from the public. In many cases, the public demonstrated their strong opposition against the use of nuclear power for safety reasons. The nuclear power accidents over time, Three Mile Island in 1979, Chernobyl reactor accidents in 1986 in Russia and Fukushima accident in 2011 in Japan, bring up safety issues associated with nuclear power reactors. In addition, nuclear fuel waste also poses safety concerns, especially while in the fuel pool, next to the nuclear reactor, which poses additional safety risks. During transportation from the nuclear power plant to the storage sites, fuel waste also poses risks and safety concerns. Nuclear power plants also pose the concerns for nuclear proliferation. Currently, nuclear power technology has separated to various counties. With regard to national security, there is a concern that nuclear power plants could be misused to produce uranium for nuclear weapons. The production of uranium for nuclear weapons is possible in the fuel cycles that include the reprocessing of spent fuel to separate plutonium and uranium enrichment technologies. Lastly, nuclear power is naturally associated with the management of nuclear waste as it involves high radioactive waste; thus, public safety is a major concern. Nuclear waste must be systematically deposited in a way that does not pose any danger to the environment and human health. Yucca Mountain, Nevada, has been designated to have high-level nuclear waste deposal, but there is no final decision reached yet with regard to this site. The project faced strong local opposition to use the place for high-level nuclear permanent waste deposal. In short, to prosper from nuclear power in the future, all key stakeholders must address the four unresolved problems collectively in a way that responds to the posed threats to the environment and exposure to communities.

1.3 U.S. Commercial Nuclear Power Plants and Their Potential Risks

Nuclear power is not a risk-free technology, and as earlier discussions depict, nuclear technology harnesses the most powerful destructive material in the world, uranium atoms, for producing heat energy. Human knowledge and ability to completely harness uranium are limited. The limitation does not warrant a development of a technology that could get rid of any potential risks. Nuclear power plants expose (1) core-meltdown risks, (2) contamination risks, and (3) radiation risks.

1.3.1 Core-Meltdown Risks

Since May 26, 1958, when Shippingport came online, nuclear power plants in the USA have been in operation for more than 59 years. Nuclear power plants house dangerous materials of many assemblies of fuel rods in the core of a nuclear reactor, which are always contained in an average of 4-ft-thick iron and concrete dome-shaped structures. When the system that cools down the nuclear fuel rods is disturbed, the fuel rods could melt down and containment might be breached. Further, highly radioactive toxic materials could be carried away by the winds and people living beneath the radioactive plume pathway might be exposed to it. In the history of nuclear power, it has been demonstrated that a destructive potential could go beyond the human capacity to manage and respond appropriately to such an event.

1.3.1.1 Three Mile Island (TMI) Accident

Preceding events: The first nuclear accident occurred at the Three Mile Island (TMI) reactor Unit 2, just 26 years after Shippingport was in operation. The plant experienced a major core-meltdown accident due to a disruption in the cooling system of the fuel rods, causing the cooling system shutdown, resulting in uncovering the fuel assemblies in the core and overheating and melting the fuel rods. It is noted that failure of the cooling systems was not the only contributing factor to the accident. The accident was caused by personnel error, deficiencies in design, and failures of parts and components in the system (US NRC 2013c).

The following description demonstrates the accident-related elements, including the sequence of events preceding the accident at the Unit 2 of the TMI power plant. It was on Wednesday, March 28, 1979, the main feed-water pump that normally sends water to the steam generators experienced a failure at about 4:00 am (Fig. 1.7, Event 1). The cause of the failure was unknown, and it was speculated that the failure could be due to shortage in power supply or a mechanical issue. In a standard operation, the steam generators utilize steam from the nuclear reactor, and thus, they reduce the heat inside the reactors. The major consequence of the shutdown of the steam generators is that the heat increases in the nuclear reactor causing an increase in pressure in the pressurizer. To reduce the pressure, the pilot-operated relief valve (a valve located at the top of the pressurizer) was opened (Fig. 1.7, Event 2). The valve was constructed to close by itself when the pressure is released and it reaches to a proper level. However, the valve was stuck open, resulting in cooling water in the reactor being released. In the control room, the operators noticed the water being released from the core reactor for a couple of

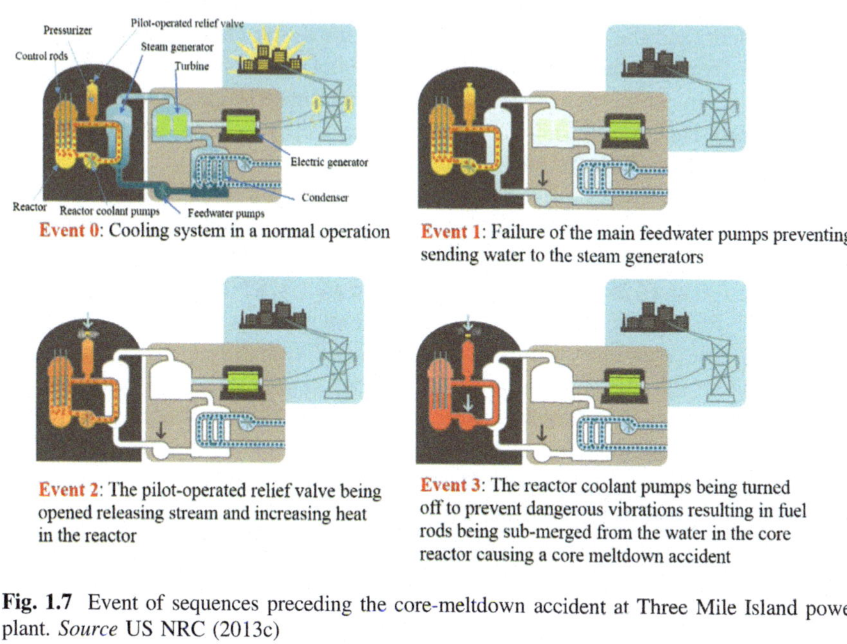

Fig. 1.7 Event of sequences preceding the core-meltdown accident at Three Mile Island power plant. *Source* US NRC (2013c)

reasons; the instruments did not provide sufficient information on water level in the reactor, and there was no specific instrument that monitors the water level in the reactor. At that time, the staff made a critical assumption; the high water level in the pressurizer keeps the core covered with water. The flashing warning lights after alarms running in the control room did not make the staff realize a loss-of-coolant accident the plant was experiencing. To respond to the warning, the staff moved their figures to push buttons on the control panel in order to carry out a series of tasks to prevent from further damage. First, they turned off the reactor coolant pumps in order to prevent vibration caused by escaping water through pilot-operated relief valve which was stuck open. Second, they reduced the amount of emergency cooling water being pumped into the primary system in order to prevent the pressurizer from filling up completely. These actions worsen the conditions causing reactor overheated. Overheating of the nuclear fuel caused zirconium cladding, which are long metal tubes holding the fuel pellets inside the nuclear reactor ruptured resulting in melting the fuel pellets (Fig. 1.7, Event 3) (US NRC 2013c).

Managing the nuclear emergency: The following chronological events (Table 1.2) in TMI nuclear power accident describe how the management of the TMI and concerned authorities carried out activities to respond and manage the nuclear power emergency, a core-meltdown accident which is regarded as the worst kind of nuclear accident. About 4 h after the incident was initiated, the plant management finally decided to inform the NRC's regional office located in King of Prussia, PA. A moment later, the NRC Headquarters in Washington, D.C., opened the Operation Center in Bethesda in MD. An official notification was reached to the White House, about 5 h the initiation of the accident took place. 7 h later after the first initial incident, management evacuated non-essential workers off the premise of the plant. The air sampling to monitor radioactivity took place after 8 h after the initial incident, and the sampling was recorded above the plant (US NRC 2013c).

Impacts from the nuclear emergency: The NRC and other agencies including the US Environmental Protection Agency, the Department of Health, Education and Welfare (now Health and Human Services), the Department of Energy, and the Commonwealth of Pennsylvania conducted detailed studies of the accident's radiological consequences (US NRC 2013c). It was estimated that there were about 2 million people who lived around the troubled TMI Unit 2 that might have received about 1 millirem above the usual background dose which was estimated to be between 100 and 125 millirem per year for the area (US NRC 2013c). According to the US NRC (2013c), the environmental samples taken from air, water, milk, vegetation, soil, and foodstuffs collected revealed low levels of radionuclides. A study conducted after 3 years the TMI accident took place shows that there is a modest post-accident increase in cancer near TMI, but the increase could not be explained by the radiation emissions form the accident (Hatch et al. 1991). 9 years later after, another study was conducted and findings reveal that there were higher-than-expected thyroid cancer incidences in two counties, namely Lancaster and York with high exposure to radiation to TMI accident (Levin 2008). After

Table 1.2 Chronological events in TMI nuclear power accident

March 28, 1979	
4:00 am	Shutdown of the main water-feed pumps sending water to stream generator
	Pilot-operated relief valve on top of the pressurizer stuck open
	Reactor coolant pumps were turned off
	Nuclear fuel overheated and caused zirconium cladding, an event of melting metal tubes holding the fuel pellets inside the nuclear reactor
7:45 am	The TMI's plant management informed the NRC's Regional Office in King of Prussia, PA
8:00 am	The Operation Center in Bethesda, MD was opened by the NRC Headquarters in Washington, D.C.
9:15 am	Notification was officially sent to the White House
11:00 am	Evacuation of all non-essential personnel off the plant's premises was carried out
12:00 am	The TMI, General Public Utilities Nuclear, and the Department of Energy took air-sample to monitor radioactivity in the atmosphere above the plant
Evening	The plant's operators believed that the core was adequately cooled down and the stabilization of the reactor was achieved
March 30, 1979	
Morning	A significant amount of radioactive materials was released from the plant's auxiliary building to relive pressure on primary system
	An advisory announcement was made by the governor of Pennsylvania, Richard L. Thornburgh for pregnant women and pre-school-age children within a 5-mile radius of the plant to leave the area
March 31, 1979	
	The authorities were concerned about possible hydrogen burst from hydrogen accumulated by chemical reactions in the melting fuel in the pressure vessel in the dome. It was anticipated that in the worst scenario, the hydrogen bubble might burn or explore leading toward rupturing the reactor vessel and core falling on the floor of containment building. If this happens, the containment building might be breached. This possible hydrogen burst created great anxiety among government authorities and the population on March 31, 1979
April 1, 1979	
	The authorities realized that the possibility of the bubble burst was very slim or non-existence because there were no oxygen in the pressure vessel that the bubble needs to burn or explode. The utility was able to reduce the size of bubbles in the pressure vessel

Source US NRC (2013c)

30 years of the TMI accident, Levin and his colleagues conducted another study and their findings revealed that thyroid cancer incident was higher, but the direct correlations with the radiation exposure from the accident could not be validated (Levin et al. 2013). They suggest that chronic low-level radiation exposure from nuclear accidents and routine emissions from power plants should be examined in the future studies.

Analysis of nuclear emergency management: The incident was originated due to machine failure or shortage of electricity, which could not be determined given that this failure was coupled with another system failure, that of a pressurizer's valve. When these machine failures take place, a failure of pilot-operated valve that was designed to close by itself did not work as intended. This design failure coupled with machine failure did not provide information to the operators timely. The operators were not able to make effective and timely decision to correct situations due to lack of information. The human errors made the situations worse and the accident took place. Fortunately, the event came to an end without requiring an evacuation on March 28, 1979 (US NRC 2013c). It has been estimated that more than two million people were exposed to radiation (US NRC 2013c), but the exact level of the doses they were exposed to was not then able to be determined with any precision. One notion is that the emergency management plan was developed effectively. The advisory evaluation was provided to residents living within a 5 mile radius of the plant, including preschool-aged children and pregnant women, who were evacuated but only after 2 days had passed since the accident occurred (Three Mile Island Alert 2013).

1.3.1.2 Chernobyl Accident

Preceding events: On April 26, 1986, the second demonstration of NPPs' powerful destructive potential occurred at the Chernobyl NPP in Ukraine, 7 years after the event at TMI. The Chernobyl plant experienced an accident that breached its reactor containment vessel and released radioactive effluents into the environment (US NRC 2013b). The accident took place at the reactor Unit 4 due to a sudden surge of power during a safety system test related to a routine maintenance outage. On April 25, 1986, the Unit 4 reactor, the Soviet-designed RBMK (*reaktor bolshoy moshchnosty kanalny*, high-power channel reactor), a pressurized water-cooled reactor which includes individual fuel channels and uses graphite as its moderator (Fig. 1.8), was scheduled for a routine maintenance shutdown. The plant intended to take a test if pumps that cool down the core were supplied with sufficient electricity generated by slowing operation management of the plant. This was intended as a test to assure that in the event of power outage from the station power, the pumps that cool down the core receive sufficient power supply from the slowing turbine generators, prior to the pumps are supplied with the power generated from the diesel emergency power supply (World Nuclear Association 2016a).

The test conducted last year showed unsatisfactory results due to a rapid decline in the power supply from the slowing turbine engine. Therefore, before the scheduled maintenance shutdown, management intended to undertake the test again. The test was considered as non-nuclear-related and sufficient information exchange with the in charge of nuclear safety of the nuclear reactor did not take place. The test required preconditions for the reactor before it is shutdown. One of them was to shut down the emergency core cooling system (ECCS) in order to proceed with the planned test. However, there was a risk associated with the

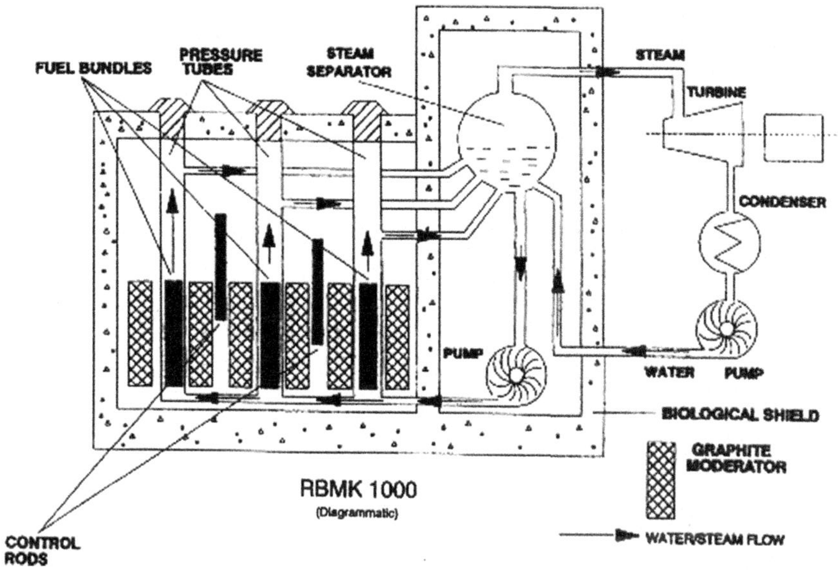

Fig. 1.8 Diagram of the RBMK-1000 reactor design. *Source* World Nuclear Association (2016a)

isolation of the ECCS because if the core experienced an incident, the system might be able to reduce the impacts. In fact, the ECCS could not be shut down because the reactor was operating at about half power. It was required to be switched off. Next, the required condition for the reactor was to stabilize at 700–1000 MW, but it was at 30 MW at 00:28 on April 26; but it could restore back to about 200 MW at 01:03 a.m. (Table 1.3). It was obvious that the operators underestimated that risks are associated with setting the preconditions for the test. As planned, the power needed to increase from 200 to 500 MW, but operational errors caused the power to lower down to 30 MW at 00:28 on 26 April. At 01:03, the reactor came into stabilized state at 200 MW and the test began. After 20 min, the signals of the power excursion rate emergency protection system were displayed and the reactor was operating at 530 MW. Since that time, there was a continued increase in the power that leads to rupture in fuel elements which increased steam generation resulting in increase in power. At 1:24, the fuel control rods were no longer functional and reached their lower limit stop switches; it leads to power being switched off for clutch mechanisms. Finally, the stem explosion took place first. This event was followed by another event after a couple of seconds, another explosion stemming from the buildup of hydrogen from zirconium–steam reaction (World Nuclear Association 2016a).

Managing the nuclear emergency: When the explosions took place, the plume that carries radioactive materials, smoke, debris from the core, and the building went to up to 1 km high into the air. On that day, the northwesterly winds carried the radioactive plume further away, leaving the heavier debris down near the plant.

Table 1.3 Chronological events in Chernobyl nuclear power accident

April 25, 1986	
01:06	The scheduled shutdown of the reactor started. Gradual lowering of the power level began
03:47	Lowering of reactor power halted at 1600 MW (thermal)
14:00	The emergency core cooling system (ECCS) was isolated (part of the test procedure) to prevent it from interrupting the test later. The fact that the ECCS was isolated did not contribute to the accident; however, had it been available it might have reduced the impact slightly The power was due to be lowered further; however, the controller of the electricity grid in Kiev requested the reactor operator to keep supplying electricity to enable demand to be met. Consequently, the reactor power level was maintained at 1600 MW and the experiment was delayed. Without this delay, the test would have been conducted during the day shift
23:10	Power reduction recommenced
24:00	Shift change
April 26, 1986	
00:05	Power level had been decreased to 720 MW and continued to be reduced. Although *INSAG-1* stated that operation below 700 MW was forbidden, sustained operation of the reactor below this level was not proscribed
00:28	With the power level at about 500 MW, control was transferred from the local to the automatic regulating system. The operator might have failed to give the 'hold power at required level' signal or the regulating system failed to respond to this signal. This led to an unexpected fall in power, which rapidly dropped to 30 MW
00:43:27	Turbogenerator trip signal blocked in accordance with operational and test procedures. *INSAG-1* incorrectly reported this event occurring at 01:23:04 and stated: "This trip would have saved the reactor." However, it is more likely that disabling this trip only delayed the onset of the accident by 39 s
01:00	The reactor power had risen to 200 MW and stabilized. Although the operators may not have known it, the required operating reactivity margin (ORM) of 15 rods had been violated. The decision was made to carry out the turbogenerator rundown tests at a power level of about 200 MW
01:03	A standby main circulation pump was switched into the left hand cooling circuit in order to increase the water flow to the core (part of the test procedure)
01:07	An additional cooling pump was switched into the right hand cooling circuit (part of the test procedure). Operation of additional pumps removed heat from the core more quickly leading to decreased reactivity, necessitating further absorber rod removal to prevent power levels falling. The pumps delivered excessive flow to the point where they exceeded their allowed limits. Increased core flow led to problems with the level in the steam drum
01:19 (approx.)	The steam drum level was still near the emergency level. To compensate, the operator increased feed water flow. This raised the drum level, but further reduced reactivity to the system. The automatic control rods went up to the upper tie plate to compensate but further withdrawal of manual rods was required to maintain the reactivity balance. System pressure began to fall and, to stabilize pressure, the steam turbine bypass valve was shut off. Since the

(continued)

Table 1.3 (continued)

	operators were having trouble with the pressure and level control, they deactivated the automatic trip systems to the steam drum around this time
01:22:30	Calculations performed after the accident found that the ORM at this point proved to be equal to eight control rods. Operating policy required that a minimum ORM of 15 control rods be inserted in the reactor at all times
01:23 (approx.)	Reactor parameters stabilized. The unit shift supervisors considered that preparations for the tests had been completed and, having switched on the oscilloscope, gave the order to close the emergency stop valves
April 26, 1986 (Test)	
01:23:04	Turbine feed valves closed to start turbine coasting. This was the beginning of the actual test. According to Annex I of *INSAG-7*, for the following approximately 30 s of rundown of the four coolant pumps, "the parameters of the unit were controlled, remained within the limits expected for the operating conditions concerned, and did not require any intervention on the part of the personnel"
01:23:40	The emergency button (AZ-5) was pressed by the operator. Control rods started to enter the core, increasing the reactivity at the bottom of the core
01:23:43	Power excursion rate emergency protection system signals on; power exceeded 530 MW
01:23:46	Disconnection of the first pair of main circulating pumps (MCPs) being 'run down', followed immediately by disconnection of the second pair
01:23:47	Sharp reduction in the flow rates of the MCPs not involved in the rundown test and unreliable readings in the MCPs involved in the test; sharp increase of pressure in the steam separator drums; sharp increase in the water level in the steam separator drums
01:23:48	Restoration of flow rates of MCPs not involved in the rundown test to values close to the initial ones; restoration of flow rates to 15% below the initial rate for the MCPs on the left side which were being run down; restoration of flow rates to 10% below the initial rate for one of the other MCPs involved in the test and unreliable readings for the other one; further increase of pressure in the steam separator drums and of water level in the steam separator drums; triggering of fast acting systems for dumping of steam to condensers
01:23:49	Emergency protection signal 'Pressure increase in reactor space (rupture of a fuel channel)'; 'No voltage—48 V' signal (no power supply to the servodrive mechanisms of the EPS); 'Failure of the actuators of automatic power controllers Nos 1 and 2' signals
01:24	From a note in the chief reactor control engineer's operating log: "01:24: Severe shocks; the RCPS rods stopped moving before they reached the lower limit stop switches; power switch of clutch mechanisms is off"

Source World Nuclear Association (2009)

About 4 min after the explosion, the very first task carried out by the 14 firemen at the site was to put out the fire associated with the explosion on the roof of adjacent turbine hall. Finally, a total of 69 firemen which consisted of the first group of 14 firemen who were joined by other firemen in nearby areas placed the fire under control at 5:00 pm. While the explosion had taken place at the Unit 4 reactor, the other reactor Units 1, 2, and 3 were still in operation. After two and a half hours of the explosion taking place, the plant operators could successfully shut down at 5:00 a.m. The other two, Units 1 and 2, were shut down the next morning on April 27. The firemen used both water and foam spray for flammable materials such as diesel oil, stored gas, and chemicals to put out the fire. There was one notable human error in carrying out the response activities which were feeding the water with emergency water pumps into the reactor core. After a day of injecting 200–300 tons per hour into the reactor core, the plant realized that the water was going into Units 1 and 2, and then, the water feed was stopped (World Nuclear Association 2016a).

2 days after the accident (on April 28), the second task for a massive accident management was carried out. The main purpose of the task was to combat the fire and the radioactive releases. The task was performed by 1800 helicopter flights which dumped about 5000 tons (40 tons of boron carbide, 2400 tons of lead, 1800 tons of sand and clay, and 800 tons of dolomite) over the troubled reactor. In carrying out the task, there was another notion of human error and safety caution taken into account of effective planning. At first time, the helicopters stabilized over the reactor and dumped the materials, but the radiation level was very high. Thus, the plan was changed to dump the materials while the helicopters travelled over the reactor. When the compounds were not dumped over the reactor, they became compounds that insulated heat and increased the temperature (World Nuclear Association 2016a).

A week later (on May 5), another plan to cool down the reactor temperature was implemented. The plan included to feed cold nitrogen to the areas surrounding the reactor. A day after injecting the cold liquid to the reactor, the temperature was noticeably decreased. It took about 15 days and required about 400 workers to construct a built-in cooling system beneath the reactor underground (World Nuclear Association 2016a). After 8 months the explosion took place (December 14), a "sarcophagus," a concrete roof, was constructed over the troubled reactor. That sarcophagus was constructed with about 300,000 tons of concrete, and 6,000 tons of metal was estimated to last for about 30 years (The Chernobyl Gallery 2016). The sarcophagus contained about 5,500 tons of radioactive sand, lead, and boric acid and about 220 tons of uranium and other unstable isotopes (Wendle 2016). The radiation level was very high inside (10,000 roentgens per hour) (500 roentgens over 5 h is a lethal dose). It is impossible to repair the sarcophagus due to high-level radiation inside, and a new containment called "New Safe Containment" was constructed (The Chernobyl Gallery 2016). The shape of the New Safety Containment looks like an outsized aircraft hangar with measurements of 360 ft high, 540 ft long, and 850 ft wide with a total estimated cost of $1.7 billion (Wendle 2016).

Impacts from the nuclear emergency: The Chernobyl nuclear power plant accident caused an evaluation of about 116,000 people from areas surrounding the nuclear power plant in the year the accident took place. In the following years, there were about 220,000 people who were evacuated from three countries: Belarus, the Russian Federation, and Ukraine (Fig. 1.9) (UNSCEAR 2008). The authorities sealed off the areas within 18 miles of the nuclear power plant. However, individuals involved in continued investigations were allowed to continue their

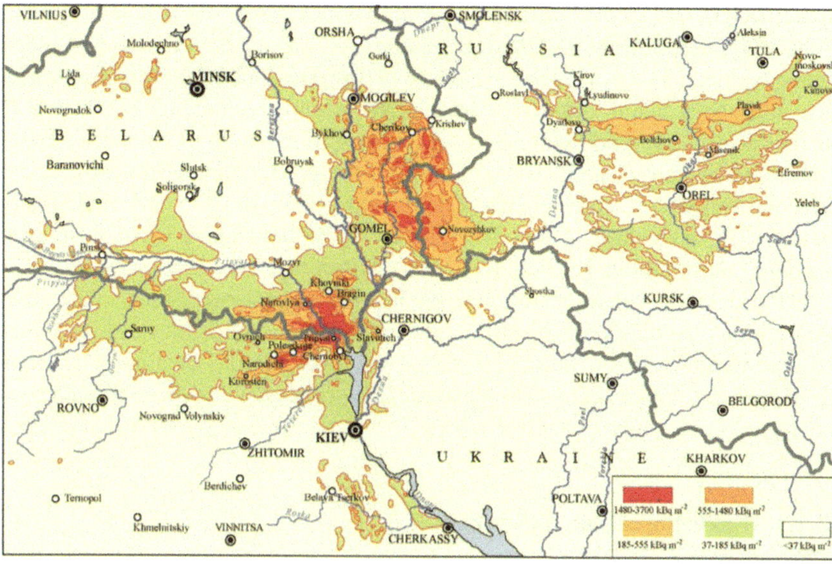

Fig. 1.9 Surface ground deposition of cesium 137 released in the Chernobyl accident. *Source* UNSCEAR (2008)

businesses (US NRC 2013b). The radioactive materials and radionuclides from the nuclear power accident were carried by the winds to far distances and they contaminated the larger areas in other countries in the northern hemisphere. People living in the areas surrounding the nuclear power plants in Belarus, the Russian Federation, and Ukraine included more than 6,000 individuals who were diagnosed with thyroid cancer up to the year 2005 (UNSCEAR 2008). The individuals consist of both children and adults. There were more than five million people who exposed to radiation after the nuclear power accident. Most of them were exposed to the radioactive materials during the evacuation. When the accident took place in the early morning of April 26, 1986, there were about 600 workers who were working on the site, 134 of them received high doses (0.8–1 Gy) and 28 of them died during the first 3 months after the accident and 19 died between 1987 and 2004. In addition, there were 530,000 individuals who were registered as recovery operation workers and exposed to doses between 0.02 and 0.5 Gy between 1986 and 1990 (UNSCEAR 2008). Those individuals were anticipated to experience cancer or other cancer-related diseases. However, in general, two decades after the accident, the increase in cancer incidents and mortality rates did not provide any clues to relate the cause of death to be radiation exposure after the nuclear power accident. It was observed that the incidence of leukemia in the general population did not increase to a cause of radiation exposure (UNSCEAR 2008).

Analysis of nuclear emergency management: The Chernobyl accident was ranked as a level 7, the highest level classified as a major event, according to the International Nuclear Event Scale. The accident was caused mainly by design failure, lack of sufficient coordination among various departments, underestimation of the risks, ignoring the standard procedure, lack of proper training, and human errors. Further, the emergency management plan was not effectively formulated. First, after the accident took place, there were not sufficient firemen available at the site, but later it was reinforced with other firemen from nearby stations and second, 2 days later, when it attempted to extinguish the fire and demonstrated a lack of sufficient knowledge and detailed plan to extinguish the fire and radionuclide releases. The high dose of radiation did not allow helicopters to stabilize while dumping sands and other materials. It was obvious that the radiation dose was not projected and the results were not incorporated into the emergency plan. Containing the radiation materials at the plant site took about 8 months after the accident. It was not a permanent solution, and in 30 years later, an arch-type containment structure that costed 2.1 billion euro was replaced. This costly containment structure was anticipated to last another 100-year period.

1.3.1.3 Fukushima Daiichi Accident

Preceding events: On March 11, 2011, an earthquake with a 9.0 magnitude which was epic entered in about 231 miles northeast of Tokyo struck Japan's Honshu Island coast. There were a total of fourteen reactors, Mark I type reactors (Fig. 1.10)

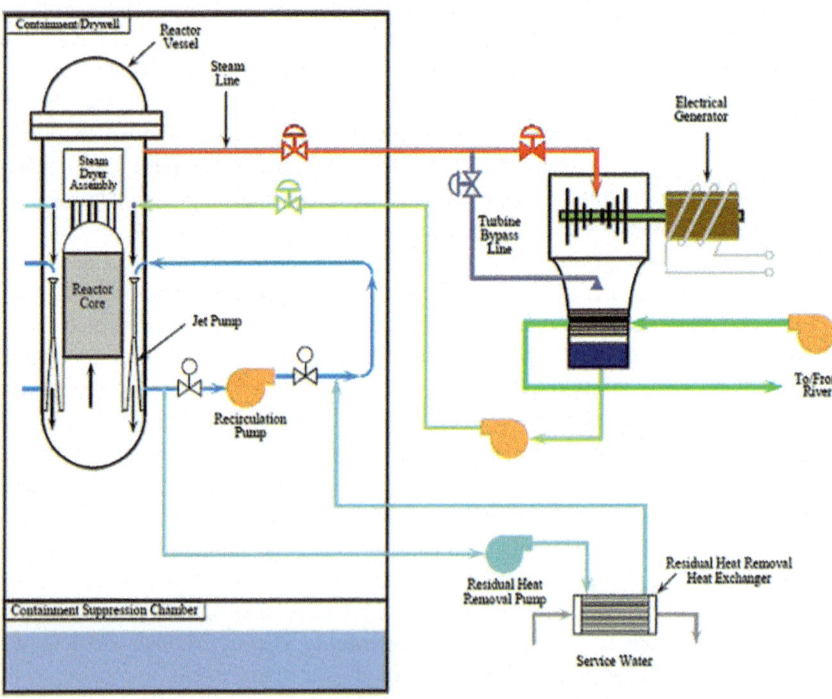

Fig. 1.10 Diagram of the Fukushima Daiichi reactor: GE boiling water reactors (BWR). *Source* World Nuclear Association (2012)

at four nuclear power plants, namely Fukushima Daiichi, Fukushima Daini, Onagawa, and Tokai. The eleven reactors that were in operation were as follows: Units 1, 2, and 3 at Fukushima Daiichi, Units 1, 2, 3, and 4 at Fukushima Daini, Units 1, 2, and 3 at Tohoku's Onagawa and Japco's Tokai, whereas Units 4, 5, and 6 at Fukushima Daiichi were under regular maintenance (World Nuclear Association 2016b).

When the earthquake struck, the eight reactors out of eleven were automatically shut down by design. The residual heat removal (RHR) system cooling pumps started to function to cool down the reactor within about 4 days, whereas the other three reactors at the Fukushima Daiichi plant site were flooded when about 50-ft-high tsunami waves generated by the powerful earthquake that struck the coast. At 2:42 pm, after an hour earthquake, the electricity supply to Units 1, 2, and 3 were disturbed when they were flooded by the 50-ft-high tsunami waves (Table 1.4). In addition, the flood caused damage to 12 of 13 backup generators as well as the heat exchangers which dump the reactor heat and decay heat from the reactor to the sea. As a result, the cooling system of the three reactors was dysfunctional and the temperature in the reactor increased. Then, the hydrogen formation took place inside the reactors due to zirconium cladding with the steam after the fuels submerged from the water inside. The emergency core cooling system (ECCS) was activated and firemen used pipes to inject the seawater into the core reactors. The Prime Minister's instructions were to cease the seawater injection into the reactor Unit 1, but the plant management did not follow. Gradually, the heat and pressure increased inside the reactors. The hydrogen explosion took place in Units 1, 2, and 4 which damaged the buildings and released radioactive materials from Units 1 and 2 (World Nuclear Association 2016b). Another concern was the used fuels stored in all six fuel pools located next to the reactors. When used fuels were removed from the reactors, they were stored in the fuel pools. Later, they were transferred into the central fuel pool located in the Unit 4. At the time of accident in 2011, about 60% of

Table 1.4 Event sequence following earthquake at Fukushima Daiichi accident

	Unit 1	Unit 2	Unit 3
Loss of AC power	+51 min	+54 min	+52 min
Loss of cooling	+1 h	+70 h	+36 h
Water level down to top of fuel[a]	+3 h	+74 h	+42 h
Core damage starts[a]	+4 h	+77 h	+44 h
Reactor pressure vessel damage[a]	+11 h	uncertain	uncertain
Fire pumps with freshwater	+15 h		+43 h
Hydrogen explosion (not confirmed for Unit 2)	+25 h service floor	+87 h suppression chamber	+68 h service floor
Off-site electrical supply	+11–15 days		
Freshwater cooling	+14–15 days		

Source World Nuclear Association (2016b). Timing from 14:46 on 11 March, 2011
Note [a]According to 2012 Modular Accident Analysis Program (MAAP)

the used fuels from the reactors at Fukushima Daiichi were stored in the central fuel pool. There were 6375 fuel assemblies out of the total capacity of 6840 in the central pool. The temperature increased to 73 °C, but later the temperature was placed under control and there was no damage in the fuel pool (World Nuclear Association 2016b).

Impacts from the nuclear emergency: The third and most recent NPP-related accident occurred, 25 years after the Chernobyl accident, when an underwater earthquake triggered a tsunami that hit the Japanese coastline, causing massive damage at the Fukushima Daiichi NPP (US NRC 2013f). This catastrophe has been regarded as a nuclear disaster of a degree humankind has never experienced (Cosmic Convergence 2013). The accident has proven to be far worse in its effects than the previous two nuclear power-related accidents. The Japanese government in its initial response evacuated approximately 160,000 people living within a 12–19 mile radius of the Fukushima plant (Morris-Suzuki et al. 2012), but the impacted areas have proven to be much wider than the evacuated areas. The earthquake was coupled with tsunami wave. In mid-December 2011, the plant management declared a cold shutdown condition after nitrogen was injected into all three containment vessels and pressure vessels leading to a minimal level of radioactive releases (World Nuclear Association 2016b). After the reactor core-meltdown accident, about 78,000 people living in areas within a 12 mile radius around the nuclear power plant were evacuated (Fig. 1.10). Later, about 62,000 people living in areas between 12 and 19 miles were evacuated by sheltering in their homes. In April 2011, areas in the northwest part of the plant were designated as deliberate evacuation areas because the high levels of radioactive materials were observed on the ground and people living in the areas could expose to high-level radiation. About 10,000 people living in the areas were evacuated because evacuation could reduce radiation by up to a factor of 10. However, evacuation came with undesired outcomes including evacuation-related deaths (World Nuclear Association 2016b). It was estimated that there were about 300,000 people evacuated from their homes in areas located surrounding the nuclear power plant according to Red Cross figures (Smith 2013). According to the Mainichi Shimbun survey, the deaths related to displacement amounted to about 1,600, whereas about 16,000 people were killed as a result of earthquake and tsunami (Smith 2013). Among the 1,600 deaths, 1,599 of these deaths were in the Fukushima Prefecture where the nuclear power plant accident took place (Smith 2013). According to the United Nations Scientific Committee of the Effects of Atomic Radiation report, a large portion of the atmospheric releases were carried away by the winds to the Pacific Ocean and direct radioactive liquid releases were discharged into surrounding sea until the reporting was made in May 2013 (UNSCEAR 2008) (Fig. 1.11).

Analysis of nuclear emergency management: First, the failure to design the plant to potential tsunami risk magnified the plant's vulnerability to natural disasters. The Fukushima Daiichi plant was built 10 meters above the sea level. When the plant was flooded by the tsunami, the plant was below about 5 meters of seawater. The Fukushima Daiichi plant was built under 13 meters above the sea level, and the impact of tsunami was less than the Daiichi plant. Second, the plant

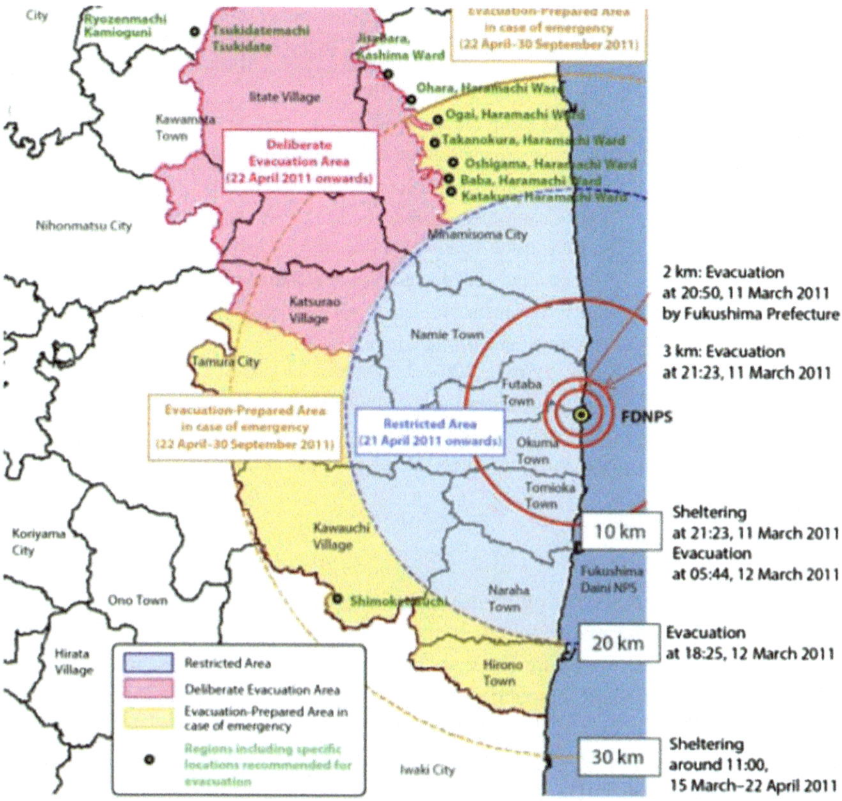

Fig. 1.11 Evacuation areas surrounding the Fukushima Daiichi nuclear power plant. *Source* UNSCEAR (2014)

management did not follow the instruction by the Office of the Prime Minister to not feed the seawater into the reactor cores. Third, 6 h after the plant accident initiated, the government started the evacuation of people living in areas within 2 mile (3 km) radius from the plant. The evacuation took place after 3 h core-meltdown started. This delay in evacuation could cause residents living near by the plant unnecessarily expose to the radiation. Fourth, the next day, the government made an announcement for evacuation of people living within 6 mile radius and 12 mile radius. Fifth, 4 days after the first accident initiated, a further evacuation to shelter inside was advised to residents living in areas between 12 mile (20 km) radius and 19 mile (30 km) radius surrounding the plant (Table 1.5).

These accidents at NPPs are normally preceded by an event or a series of events that trigger or contribute to the damage done to the reactor core. When brought under control, events with the potential to have escalated into major accidents—that is, those of the type capable of causing reactor core damage—are described as near-miss events. The NRC issues an annual report detailing all near-miss events

Table 1.5 Timeline of evacuation orders during the Fukushima Daiichi accident

Date	Time	Activity
March 11, 2011	14:46	JST The earthquake occurred.
	15:42	TEPCO made the first emergency report to the government.
	19:03	The government announced a nuclear emergency.
	20:50	The Fukushima Prefecture Office ordered 2 km radius evacuation.
	21:23	The government ordered 2-mile (3-km) evacuation and to keep staying inside buildings in the area of 2–6 mile (3–10 km) radius.
March 12, 2011	05:44	The government ordered 6-mile (10-km) radius evacuation.
	18:25	The government ordered 12-mile (20-km) evacuation.
March 15, 2011	11:01	The government ordered to keep staying inside buildings in the area of 12–19 miles (20–30 km) from the plant.
March 25, 2011		The government requested voluntary evacuation in the area of 12–19 miles (20–30 km).
April 21, 2011		The government set the 12-mile (20 km) radius no-go area.

Source UNSCEAR (2014)

and classifying them according to the three categories defined by the type of team sent out to inspect the event, which is in turn determined by the level or degree of severity of the event. There are three such teams, namely the augmented inspection team (AIT), the incident inspection team (IIT), and the special inspection team (SIT). The AIT evaluates events that pose a 10-fold increase in risk, whereas the SIT investigates events that pose a 1,000-fold increase in risk level (Lochbaum 2010, 2011, 2012). In 2010, there were a total of 14 near-miss events (13 SIT and 1 AIT); in 2011, 15 (14 SIT and 1 AIT); and, in 2012, 14 (11 SIT and 3 AIT), figures that hint at the regularity with which such plants pose a risk to their surrounding areas.

1.3.2 Contamination Risks

Another risk associated with NPPs is that of contamination of the sort caused by the unmonitored and unplanned release of liquids. Commercial NPPs release radioactive materials into the environment, in either liquid or gaseous or both, on a routine basis. There have been incidents of unplanned and unmonitored leaks of liquids that occurred at Braidwood, Indian Point, Byron, and Dresden NPPs. The Liquid Radioactive Release Lessons Learned Task Force (LLTF) has been tasked with identifying the causes of such leaks (Richards et al. 2011). Their findings included the following: (1) The construction of plant components did not meet existing safety standards; (2) the components that caused or contributed to said leaks were not required to be monitored, nor were they subject to routine maintenance activities according to the NRC regulations; (3) some components associated with subterranean leaks or with spent-fuel pools were not readily accessible or physically visible; (4) liquid leaks can enter undetected into groundwater; (5) the contamination of groundwater may go undetected because it is not required to be monitored according to the existing NRC regulations; and (6) the contaminated groundwater could flow off-site undetected. The communities that host NPPs therefore are at perpetual risk due to the very presence of the NPPs.

Another potential risk associated with NPPs surrounds their spent fuel—the highly radioactive used fuel rods that are removed from a nuclear reactor. Nuclear fuel rods are made of fissionable materials and retain their inherent destructive potential long after their initial use. Approximately 74% of all spent fuels are stored at the reactor site, most often in spent-fuel pools, while the rest are stored in dry storage casks distributed across 33 states (US GAO 2012). The US Government Accountability Office (GAO) projects an annual 2,000-metric-ton increase in the amount of spent fuel stored on site, with the current 70,000-metric-ton spent-fuel stockpile growing to 140,000 metric tons within a decade. New storage facilities are expected to be ready to begin accepting spent fuel in 15–40 years. The potential risks associated with such facilities include (1) the release of radiation, which could have severe negative impacts on human health and (2) a self-sustaining fire if and when the water is drained and the fuel rods are exposed to air. The amount of hazardous materials stored in each state varies from a maximum of approximately 2,000 metric tons in Illinois to a minimum of less than 100 metric tons at the decommissioned St. Vrain NPP in Colorado (US NRC 2013a). Above all, the threats that are inherent to the 70,000-metric-ton spent-fuel stockpile cannot and must not be underestimated. To put into perspective the dire consequences of doing so, we must bear in mind that only approximately 400 metric tons of spent fuel were stored in the fuel pool at the Fukushima NPP, materials that are equivalent in their destructive potential to 14,000 Hiroshima bombs (The Asahi Shimbun 2013).

The most hazardous human-made materials do not sit forever in the spent-fuel pools or stay safely in the dry-storage casks housed at the reactor sites. When they are transported, they pose great risk not only to the host communities but also to the communities at large along the transit route. The purpose of transporting spent fuels is to relocate them to storage space shared by multiple reactor sites operated by the same owner (US NRC 2013d). According to Garrick (2003), between 1964 and 1997, a total of 3,025 shipments moved 829 metric tons of heavy metal (MTHM) by road and an additional 1,445 MTHM by rail. The transportation of such materials is closely supervised by the NRC and the US Department of Transportation (US NRC 2013d). Although regulated and supervised, the transportation—whether by road or by rail—has not been accident-free. Between 1971 and 1995, there were four accidents on the highways and four on the rail lines; one highway accident resulted in the death of a driver and detectable emissions of radiation (Garrick 2003; US NRC 2013g). All of the approximately 70,000 metric tons of spent fuels stored at reactor sites are in waiting to be transported to high-level repositories, when such facilities are open and available to accept said materials (US NRC 2013d). With the Obama administration's 2009 decision to withdraw funding for the Yucca Mountain High-Level Nuclear Waste Project, there is at present no permanent storage facility under development (Wald 2009).

Another risk that NPPs pose is inherent to their design and serves to highlight why it is critical that the design of a reactor that contains fissionable fuel rods during its normal operation be centered on safety. The Mark 1 containment vessel, first produced by GE in the 1960s and still in use at Fukushima NPP at the time of the accident there, contributed to the catastrophic events of 2011 (Zeller 2011). As early as 1975, it was known that there had not been sufficient testing of the containment design and that any flaws that might persist could compromise the safety of the plant and its surroundings. However, the warning did not lead to a halt in the operating of the design, given that by that time, it had been widely accepted throughout the nuclear power industry. However, Harold Denton, a retired NRC official, pointed out that the probability of a Mark 1-type reactor bursting if the fuel rods were to enter into meltdown was approximately 90% (Denton 1987). In the USA, there remain 23 Mark 1 reactors still in operation, located at 16 NPPs, including Oyster Creek, New Jersey; Dresden, Illinois; and Monticello, Minnesota. The host communities of NPPs that still operate Mark 1 reactors are at greater risk of exposure to a Fukushima-like disaster than sites with safer reactor designs. Nuclear power plants are also vulnerable to seismic events of the type that played a role in triggering the catastrophe at Fukushima. The NPPs sited along the US eastern seaboard, in particular, do not include in their designs any kind of measures meant to shield against the aftereffects of a strong earthquake (Koch 2011).

1.3.3 Nuclear Radiation Risks

Nuclear power plants, even under normal operating conditions, release limited amounts of radioactive contaminants into the atmosphere. In the event of a nuclear disaster, the quantities of radioactive materials can increase by orders of magnitude. Ionizing radiation can quickly travel more than one hundred kilometers (62 miles), carried on the wind in the form of a radioactive plume to outlying areas (Cyranoski and Brumfiel 2011). In general, the routine emissions of ionizing radiation originating from NPPs are known to be associated with elevated incident rates of a number of types of cancer, permanent damage to human vital organs, and death (Astakhova et al. 1998; Canu et al. 2008; Cardis et al. 2005). Populations living near an NPP, as well as those that may prove to be in the path of a radioactive effluent plume in the case of an accident, are vulnerable to exposure of high levels of radiation.

The communities that host such facilities shoulder the potential risks associated with nuclear power from the time the plant begins operation until the time it is decommissioned—for a typical reactor design, the use life is between 40 and 60 years (Green Peace 2010). After a plant is shut down, it takes under normal conditions, 50 years for the reactor to cool; this is followed by a period of decommissioning that normally takes another 40–60 years (Green Peace 2010).

The NRC has advanced three strategies for dealing with end of use at of NPPs: (1) DECON or decommissioning strategies; (2) SAFSTOR or deferred dismantling; and (3) ENTOMB or the permanent encasing on site of radioactive contaminants (US NRC 2013e). For example, Hallam Nuclear Power Plant, Nebraska, a NPP with a short use life, was decommissioned in 1971. All potential containments were at that time buried underground. Since then, the Department of Energy, working in conjunction with the Nebraska Department of Health, has monitored the entombment site for possible groundwater contamination and radiation and will continue to do so for a period of 119 years, until 2090 (Nebraska Energy Quarterly 1997).

1.3.4 Terrorist Attack Risks

Spent fuel, among the most hazardous material created by humans, is always vulnerable to terrorist attack (Holt and Andrews 2007). The terrorist attack on the World Trade Center in New York and the Pentagon building in Washington, D.C., on September 11, 2011—the so-called 9/11 attack—serves as reminder of the dire consequences and very real threat of a potential attack on NPPs. Afterward, evidence came to light that revealed Al Qaeda had considered targeting an NPP in their initial plan of attack (Holt and Andrews 2007).

According to the National Academy of Sciences (NAS), a successful attack on a spent-fuel storage facility would be difficult, but it is possible. In the event of such an attack, the spent fuel could become the source of a self-sustaining zirconium cladding fire and would release a massive amount of radioactive materials. Communities that host NPPs therefore live with the unpredictable risk of exposure to a massive release of radiation.

1.3.5 Inevitable Risks

Despite their years of operation, NPPs have yet to prove themselves the peaceful resource for the benefit of humankind that President Eisenhower first envisioned. In fact, NPPs use a technology that seeks to harness nuclear fissionable materials to produce heat, which is used to boil water, transforming it into steam that is then used to power electric generators. These nuclear fissionable materials were initially used in the manufacturing of nuclear weapons. Their powerful destructive potential is beyond imagination, as was revealed by the Trinity test—the first detonation of an atomic bomb, which occurred at 5:29:45 a.m., July 16, 1945, in the desert of New Mexico (Masco 2006). The destructive potential of nuclear fission can escalate beyond an NPP's ability to control it, for instance when a plant experiences an accident that damages its nuclear reactor, where the fuel rods made of fissionable materials are housed.

Above all, the potential risks associated with NPPs can be explained by Perrow's "normal accidents theory." As noted previously, NPPs are complex and tightly coupled systems, and complexity produces unknown risks; according to the Perrow's theory:

> Nothing is perfect, neither designs, equipment, procedures, operators, supplies, or the environment. Because we know this, we load our complex systems with safety devices in the form of buffers, redundancies, circuit breakers, alarms, bells, and whistles. Small failures go on continuously in the system since nothing is perfect, but the safety devices and the cunning of designers, and the wit and experience of the operating personnel, cope with them. Occasionally, however, two or more failures, none of them devastating in themselves in isolation, come together in unexpected ways and defeat the safety devices—the definition of a "normal accident" or system accident. If the system is also tightly coupled, these failures can cascade faster than any safety device or operator can cope with them… [I]f the accident brings down a significant part of the system, and the system has catastrophic potential, we will have a catastrophe. (Perrow 1999, pp. 356–357)

According to the Perrow's analysis, the risks associated with NPPs therefore should be seen as intrinsic to the system, normal, and as something that cannot be avoided.

References

Astakhova LN, Anspaugh LR, Beebe GW, Bouville A, Drozdovitch VV, Garber V et al (1998) Chernobyl-related thyroid cancer in children of Belarus: a case-control study. Rad Res 150 (3):349–356. Retrieved from WOS:000075670600010

Canu IG, Ellis ED, Tirmarche M (2008) Cancer risk in nuclear workers occupationally exposed to uranium—emphasis on internal exposure. Health Phys 94(1):1–17. Retrieved from WOS:000251786800001

Cardis E, Kesminiene A, Ivanov V, Malakhova I, Shibata Y, Khrouch V et al (2005) Risk of thyroid cancer after exposure to I-131 in childhood. J Natl Cancer Inst 97(10):724–732. Retrieved from WOS:000229856500011

Cooper M (2013) Nuclear aging: not so graceful. Retrieved from Bulletin of the Atomic Scientists http://thebulletin.org/nuclear-aging-not-so-graceful

Cosmic Convergence (2013) Fukushima: a nuclear catastrophe of epic proportions. Retrieved from the mind unleashed http://www.themindunleashed.org/2013/09/fukushima-nuclear-catastrophe-of-epic.html

Crooks E (2016) Uneconomic US nuclear plants at risk of being shut down. Retrieved from Energy http://www.ft.com/cms/s/0/da2a6bc6-98fa-11e3-a32f-00144feab7de.html#axzz430cDscgc

Cyranoski D, Brumfiel G (2011) Fukushima impact is still hazy. Nature 477(7363):139–140. Retrieved from WOS:000294603900005

Dotson S (2014) Lessons learned from Kewaunee's closing. Retrieved from Power Engineering http://www.power-eng.com/articles/npi/print/volume-7/issue-3/nucleus/lessons-learned-from-kewaunee-s-closing.html

Eisenhower DD (1953) Atoms for peace. Retrieved from Voices of Democracy http://voicesofdemocracy.umd.edu/eisenhower-atoms-for-peace-speech-text/

Ferguson CD (2011) Nuclear energy what everyone needs to know. Retrieved from Oxford University Press

Garrick J (2003) The current status, safety, and transportation of spent nuclear fuel. 33(3):11–18. Retrieved from National Academy of Sciences

Green Peace (2010) Nuclear power: decommissioning risks. Retrieved from Green Peace USA http://www.greenpeace.org/usa/en/media-center/reports/Nuclear-Power-Decommissioning-Risks/

Hatch MC, Wallenstein S, Beyea J, Nieves JW, Susser M (1991) Cancer rates after the three-Mile-Island nuclear accident and proximity of residence to the plant. Am J Public Health 81(6):719–724. Retrieved from WOS:A1991FV60700009

Holt M, Andrews A (2007) Nuclear power plants: vulnerability to terrorist attack. CSR Report for Congress

Koch W (2011) Earthquake readiness of U.S. nuclear power plants is unclear. USA Today

Levin RJ (2008) Incidence of thyroid cancer in residents surrounding the three mile island nuclear facility. Laryngoscope 118(4):618–628. Retrieved from ISI:000260662000009

Levin RJ, De Simone NF, Slotkin JF, Henson BL (2013) Incidence of thyroid cancer surrounding three mile island nuclear facility: the 30-year follow-up. Laryngoscope 123(8):2064–2071. Retrieved from WOS:000322004400041

Lochbaum D (2010) The NRC and Nuclear power plant safety in 2012: tolerating the intolerable. Union of Concerned Scientists: Citizens and Scientists for Environmental Solutions. Retrieved from: http://www.ucsusa.org/assets/documents/nuclear_power/NRC-nuclear-safety-2010-report.pdf

Lochbaum D (2011) The NRC and nuclear power plant safety in 2012: tolerating the intolerable. Union of Concerned Scientists: Citizens and Scientists for Environmental Solutions. Retrieved from http://www.ucsusa.org/assets/documents/nuclear_power/NRC-nuclear-safety-2011-report.pdf

Lochbaum D (2012) The NRC and nuclear power plant safety in 2012: tolerating the intolerable. Union of Concerned Scientists: Citizens and Scientists for Environmental Solutions. Retrieved from http://www.ucsusa.org/assets/documents/nuclear_power/NRC-nuclear-safety-2012-report.pdf

Masco J (2006) The nuclear borderlands the Manhattan Project in post-Cold War New Mexico. Princeton University Press, Princeton, N.J.

MIT (2003) Retrieved from Massachusetts Institute of Technology http://thebulletin.org/nuclear-aging-not-so-graceful

Morris-Suzuki T, Boilley D, McNeill D, Gundersen A (2012) Lessons from Fukushima. Greenpeace.org

Nebraska Energy Quarterly (1997) State's First Nuclear Plant Buried Near Lincoln. Retrieved from Nebraska Energy Quarterly Newsletter

Richards S, Frye T, Shepherd J, Nicholson T, Kuzo G, Shoop U et al (2011) Liquid radioactive release lessons learned task force final report. Retrieved from United States Nuclear Regulatory Commission www.nrc.org

Smith A (2013) Fukushima evacuation has killed more than earthquake and tsunami, survey says. Retrieved from NBC News http://www.nbcnews.com/news/other/fukushima-evacuation-has-killed-more-earthquake-tsunami-survey-says-f8C11120007

The Asahi Shimbun (2013) The long road ahead in ending Fukushima water problem. Retrieved from The Asahi Shimbun www.ajw.asahi.com

The Chernobyl Gallery (2016) The Sarcophagus. Retrieved from The Chernobyl Gallery http://chernobylgallery.com/chernobyl-disaster/sarcophagus/

Three Mile Island Alert (2013) Incident Chronology at TMI from NRC: 1979–2012. Retrieved from Three Mile Island Alert http://www.tmia.com/node/1586

UNSCEAR (2008) Maps of radionuclide deposition. Retrieved from United Nations Scientific Committee on the Effects of Atomic Radiation http://www.unscear.org/unscear/en/chernobylmaps.html

UNSCEAR (2014) Sources, effects, and risks of ironizing radiation: UNSECAR 2013 Report. New York: United Nations Scientific Committee on the Effects of Atomic Radiation. Retrieved from: http://www.unscear.org/docs/reports/2013/14-06336_Report_2013_Annex_A_Ebook_website.pdf

US DOE (1995) The History of Nuclear Energy (DOE/NE 0088). Washington D.C.: US Department of Energy, Office of Nuclear Energy, Science, and Technology. Retrieved from: http://www.energy.gov/sites/prod/files/The%20History%20of%20Nuclear%20Energy_0.pdf

US EIA (2015) Nuclear explained. Retrieved from US Energy Information Administration https://www.eia.gov/energyexplained/index.cfm?page=nuclear_home

US EIA (2016) How many nuclear power plants are in the United States, and where are they located. Retrieved from US Energy Information Administration https://www.eia.gov/tools/faqs/faq.cfm?id=207&t=3

US EPA (2016) Summary of the National Environmental Policy Act. Retrieved from: https://www.epa.gov/laws-regulations/summary-national-environmental-policy-act

US GAO (2012) Spent nuclear fuel: accumulating quantities at commercial reactors present storage and other challenges (GAO-12-797). United States Government Accountable Office (US GAO)

US NRC (2004) Nuclear Power Plant Licensing Process. Retrieved from US Nuclear Regulatory Commission http://www.nrc.gov/reading-rm/doc-collections/nuregs/brochures/br0298/br0298r2.pdf

US NRC (2010) 2009–2010 Information Digest US Nuclear Regulatory Commission. Retrieved from: http://www.nrc.gov/reading-rm/doc-collections/nuregs/staff/sr1350/v21/sr1350v21.pdf

US NRC (2011) Nuclear Regulatory Commission. Retrieved from US Nuclear Regulatory Commission www.nrc.gov

US NRC (2013a) 2012–2013 Information Digest (NUREG-1350, Volume 24). US Nuclear Regulatory Commission. Retrieved from http://www.nrc.gov/reading-rm/doc-collections/nuregs/staff/sr1350/v24/sr1350v24.pdf

US NRC (2013b) Backgrounder on chernobyl nuclear power plant accident. Retrieved from US Nuclear Regulatory Commission http://www.nrc.gov/reading-rm/doc-collections/fact-sheets/chernobyl-bg.html

US NRC (2013c) Backgrounder on the three mile island accident. Retrieved from US Nuclear Regulatory Commission http://www.nrc.gov/reading-rm/doc-collections/fact-sheets/3mile-isle.html

US NRC (2013d) Backgrounder on transportation of spent fuel and radioactive materials. Retrieved from US Nuclear Regulatory Commission http://www.nrc.gov/reading-rm/doc-collections/fact-sheets/transport-spenfuel-radiomats.pdf

US NRC (2013e) Decommissioning nuclear power plants. Retrieved from US Nuclear Regulatory Commission http://www.nrc.gov/reading-rm/doc-collections/fact-sheets/decommissioning.html

US NRC (2013f) Fact Sheet on Summary of Japan Events in March 2011 and NRC Response. Retrieved from US Nuclear Regulatory Commission http://www.nrc.gov/reading-rm/doc-collections/fact-sheets/fs-japan-events.html

US NRC (2013g) Spent Fuel Transportation: Risk Assessment (NUREG-2125). Office of Nuclear Material Safety and Safeguards, US Nuclear Regulatory Commission. Retrieved from: http://www.nrc.gov/reading-rm/doc-collections/fact-sheets/transport-spenfuel-radiomats.pdf

US NRC (2015a) 2014–2015 Information Digest. US Nuclear Regulatory Commission. Retrieved from: http://www.nrc.gov/reading-rm/doc-collections/nuregs/staff/sr1350/

US NRC (2015b) SONGS steam generator tube degradation. U.S. Nuclear Regulatory Commission. Retrieved from: http://www.nrc.gov/reactors/operating/ops-experience/tube-degradation.html

US NRC (2016a) 2015–2016 Information Digest. US Nuclear Regulatory Commission. Retrieved from: http://www.nrc.gov/reading-rm/doc-collections/nuregs/staff/sr1350/v26/sr1350v26.pdf

US NRC (2016b) Glossary. U.S. Nuclear Regulatory Commission. Retrieved from: http://www.nrc.gov/reading-rm/basic-ref/glossary.html

US NRC (2016c) History. Retrieved from US Nuclear Regulatory Commission http://www.nrc.gov/about-nrc/emerg-preparedness/history.html

US NRC (2016d) Licensing. U.S. Nuclear Regulatory Commission. Retrieved from: http://www.nrc.gov/about-nrc/regulatory/licensing.html

Wald ML (2009) Future dim for nuclear waste repository. Retrieved from The New York Times http://www.nytimes.com/2009/03/06/science/earth/06yucca.html?ref=yuccamountain&pagewanted=print

Wendle J (2016) Chernobyl's radioactive ruins get a new tomb. Retrieved from National Geographic http://news.nationalgeographic.com/2016/04/042516-chernobyl-new-safe-confinement-contains-radiation/

World Nuclear Association (2009) Sequence of events: chernobyl nuclear power accident. Retrieved from World Nuclear Association http://www.world-nuclear.org/information-library/safety-and-security/safety-of-plants/appendices/chernobyl-accident-appendix-1-sequence-of-events.aspx

World Nuclear Association (2012) Fukushima: background on reactors. Retrieved from World Nuclear Association http://www.world-nuclear.org/information-library/safety-and-security/safety-of-plants/appendices/fukushima-reactor-background.aspx

World Nuclear Association (2016a) Chernobyl Accident 1986. Retrieved from World Nuclear Association http://www.world-nuclear.org/information-library/safety-and-security/safety-of-plants/chernobyl-accident.aspx

World Nuclear Association (2016b) Fukushima accident. Retrieved from World Nuclear Association http://www.world-nuclear.org/information-library/safety-and-security/safety-of-plants/fukushima-accident.aspx

Zeller T (2011) Experts had long criticized potential weakness in design of stricken reactor. Retrieved from New York Times http://www.nytimes.com/2011/03/16/world/asia/16contain.html?_r=0&pagewanted=print

Chapter 2
Communities Hosting US Commercial Nuclear Power Plants

2.1 Setting Boundaries in Host Communities

Communities located around a nuclear power plant shoulder inevitable risks associated with the nuclear power plants. These inevitable risks include nuclear contamination, radiation, and nuclear core-meltdown accidents. In case of the worst accident scenario, an event with a core-meltdown accident at a nuclear power plant, communities near the site could face the risk of exposure to highly toxic radioactive plume. The areas that are prone to such radiation risks depend on how far the plume could travel and the weather conditions in that particular day. The NRC has designated two types of emergency planning zones (EPZs): a 10-mile radius zone and a 50-mile radius from a nuclear power plant (US NRC 2016). The first 10-mile radius zone around the nuclear power plant is called the plume exposure pathway, whereas the latter one is named the ingestion exposure pathway. The difference between the two zones is the emergency management plan for each. The plume exposure zone (PEZ) requires a mandatory and detailed emergency management plan, whereas the ingestion exposure zone (IEZ) is not required a mandatory emergency management plan. However, the two zones share a common goal of avoiding or reducing potential radiation. The PEZ with predetermined protective action plans is designed to avoid or reduce radiation by sheltering, evacuation, and the use of potassium iodide, whereas the IEZ is designed to avoid or reduce potential ingestion.

To define the host community of a nuclear power plant, I consider the 50-mile distance based on the NRC's predetermined two emergency zones: a 10-mile radius and 50-mile radius. Therefore, in this chapter, communities hosting nuclear power plants are defined as those communities living around a 50-mile radius from a nuclear power plant, as designated in both exposure pathway and ingestion pathway by the NRC (US NRC 2016). However, a group of more than 37 environmental and civic organizations argued that the 50-mile radius is not a sufficient distance because the radioactive plume could travel longer than the specified distance, as evident in the events of Chernobyl and Fukushima (NIRS 2013; Kyne 2014). In acknowledging

D. Kyne, *Nuclear Power Plant Emergencies in the USA*,
DOI 10.1007/978-3-319-50343-1_2

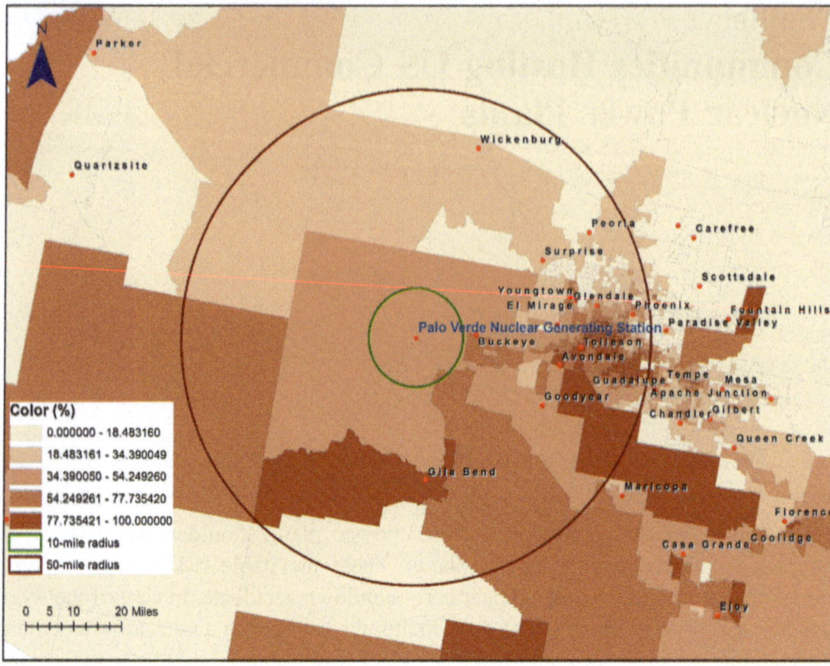

Fig. 2.1 Communities hosting the Palo Verde Nuclear Generating Station (PVNGS) within a 50-mile radius from the power plant

the controversial debate on the definition of impacted areas, this study uses the most legal definition of an impacted area, defined as an area within a 50-mile radius of an NPP. For example, a community living around a 50-mile radius from the Palo Verde Nuclear Generating Station (PVNGS) is depicted in Fig. 2.1.

2.2 Invisible Risks and Unknown Consequences

This session takes a closer look at the communities living around the nuclear power plants, to enhance understanding on environmental justice issues and current knowledge of consequences from constant exposure to low-level radiation.

2.2.1 Nuclear Power Plant Siting and Environmental Justice

Communities hosting the nuclear power plants are in fact embracing all different types of potential risks. There are certain types of risks that simply cannot be avoided, as discussed earlier in Chap. 1. The potential and inevitable risks include: core meltdown, contamination, nuclear radiation, and terrorist attack risks. NPPs utilize nuclear fuel rods containing fissionable materials and their destructive potentials are unpredictable in nature; their profound impacts on environments and humankind cannot be underestimated (Kyne 2014). Due to the potential destructive nature and unpredictable long-term impacts on environments and humankind, it is necessary for the Federal Government authorities to regulate all NPPs in the U.S. (Kyne 2014). Responsible regulatory authorities must monitor every step from selecting a plant site, the granting of construction licenses, and the warranting of operating a plant. Once an NPP is built within a community, that particular population in the area becomes the host community and begins to embrace the above-mentioned potential risks (Kyne 2014).

Given the fact that such nuclear power facilities have the potential to cause powerful destruction, governmental intervention plays a vital role in protecting human life and the environment. The Environmental Protection Agency (EPA) was established under the National Environmental Policy Act (NEPA) in 1970, with a mission to incorporate environmental protection policy into all federally sponsored projects. The EPA established goal is "to provide an environment where all people enjoy the same degree of protection from environmental and health hazards and equal access to the decision-making process to maintain a healthy environment in which to live, learn, and work (US EPA 2013)." To achieve this goal, the EPA promotes one of the key elements which is "public involvement." It is vital for communities that are likely to host a nuclear power plant to actively participate and engage in every step of environmental decision-making process, from site selection to issuing operating licenses.

Previously, when issuing the construction and operation licenses to all previous U.S. commercial nuclear power plants, environmental justice concepts were not incorporated, and the AEC has issued 126 NPP operating licenses without taking into account any environmental impacts or guidelines (US NRC 2004; Kyne 2014). During the AEC administration era, the agency struggled dual conflicting roles—on one hand, it was required to regulate the nuclear power industry; on the other hand, it was required to promote the industry growth. After receiving criticism for its dual roles and ineffectiveness, the AEC was dissolved in 1975. The NRC became its successor under the Energy Reorganization Act of 1974 (US NRC 2011; Kyne 2014). Since then, the NRC became the agency which has authority and responsibility in administering the licensing process (US NRC 2004; Kyne 2014). The licensing process in fact encourages public participation through adjudicatory, or courtroom-style hearings. In public meetings, proposed NPP projects are presented and their quality of design conform within the current existing environmental laws. Participants have opportunities to make comments on any design flaws, raise any environmental concerns and engage in debates on potential impacts (Mariottee 2006; Kyne 2014). There still remains the unanswered question with regard to public participation in the licensing process: Does the licensing process encourage public participation and if so, to what extend does the public participate; and to what extend the public concerns are taken into serious consideration by the NRC and other concerned authorities? Systematic studies on evaluation of public participation in nuclear power plant licensing processes are rarely found.

One notion is that there was a transformation taking place in the licensing process after 1992. Previously, all prospective owners of NPPs were required to seek the construction license first and then the operation license. The two steps were combined into one-step process in which the prospective owners or operators of NPPs could obtain both licenses. The one-step process was well received as it is shorter and provides encouragement to the prospective owners and operators to invest in NPPs. On the other hand, it also received criticism for placing increased constraints on public participation. The main barriers to public participation was observed as the creation of public contentions and its legitimate requirements to the ASLB. In public hearings, any public contentions must be decided by the three-person panel of NRC employees which consists of two technical experts and one attorney. Any citizen who would like to content an objection must bear substantial costs which include expert and attorney fees (an estimated cost of $100,000 to $500,000) who normally testify in support of the objecting party (Mariottee 2006; Kyne 2014). The barriers to public participation could undermine promotion of procedural equity which is essential for effective environmental decision-making processes and outcomes; thus, in turn, it could negatively affect the people living near NPPs (Kyne 2014).

In the history of US commercial nuclear power plant licensing, NPPs have not been well received by communities. There were many objections and resistance by the public who feared the unpredictable risks associated with nuclear power production. When nuclear power energy was introduced, there was a tension between two groups of scientists—those supporting the advantages of nuclear power and those who foresaw the risks associated with fuel rods, spent fuels, and radiation. During 1970s, the sentiments against nuclear power gained momentum and received significant public attention. Collective public action placed a haul or a delay of a number of the proposed NPPs at the time (Nuclear Heritage 2013; Kyne 2014). For example, there were more than 70,000 people who protested against nuclear power to Washington, DC, in May of 1979. For nuclear power energy, the protesters were not fully granted their constitutional rights of freedom of speech, but there were more than 1,414 people who protested against the construction of Seabrook NPP in New Hampshire, where they were arrested on May 2, 1997 (Nuclear Heritage 2013; Kyne 2014).

2.2.2 *Embracing the Low-Level Radiation*

NPPs regularly or from time to time release radioactive materials in gaseous or liquid forms from their normal operations, and from temporary shutdowns and abnormal situations such as a fuel-rod meltdown (Ottaviani and Wehe 1989).

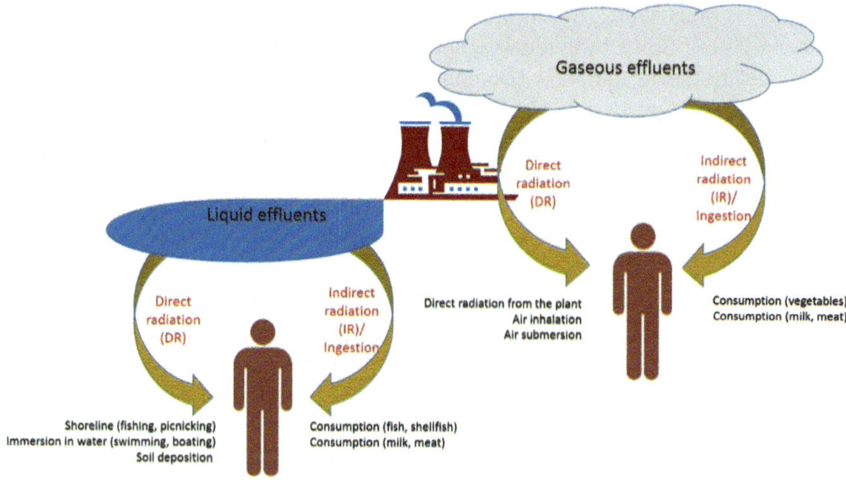

Fig. 2.2 Pathways for exposures to radiation from effluent releases from nuclear power plants. *Source* Adapted from (National Research Council (USA) & Committee on the Analysis of Cancer Risks in Populations near Nuclear Facilities 2012)

Individuals living in the host communities are involuntarily exposed to constant low-level radiation on a regular basis. In addition, these individuals involuntarily accept negative consequences from constantly being exposed to low-level radiation. Figure 2.2 depicts how an NPP releases gaseous and liquid radioactive effluents during normal operation.

According to the National Research Council (USA) and Committee on the Analysis of Cancer Risks in Populations near Nuclear Facilities (2012), there are mainly two radiation exposure pathways: external radiation and internal radiation. External radiation pathways include exposure to radionuclides that emit high-level radiation that could penetrate the human body. Possible sources are directly from a nuclear power facility, from ionizing radioactive particles present in the air, radioactive materials, and contaminated water body from radioactive liquid effluents from the nuclear power plant. Internal radiation pathway includes individuals receiving radiation thorough intake of radioactive materials into the body. For example, individuals could inhale air containing radioactive particles, or by consuming foods that were ingested with radiative materials.

2.2.3 Consequence of Constant Exposure to Low-Level Radiation

Many scientists have attempted to prove the casual relationships between low-level radiation and impacts on health. They have specified some negative health consequences that are related to permanent cellular damage and chronic illnesses, including various types of cancer, such as leukemia, thyroid, and other cancers (Ramana 2009). Some individuals could even face the worst consequence of premature death due to radiation (Denton 1987). Many researchers face great challenges in attempting to establish a correlation between the exposure to low-level radiation and the incidence of certain kinds of cancer (e.g., Bouges et al. 1999; Canu et al. 2008; Cardis et al. 1995, 2005; Sexton and Adgate 1999). The primary reason for this challenge is lack of data that could be used for investigations related to the attempted establishment in their studies. Some critical data that are not readily available for studies are data collected for the effects of prolonged exposure to low-level radiation, varying radiation doses, and health conditions of individuals prior to moving to the host communities (Kyne 2014).

In the existing studies, there was one study with a focus on cancer incident rates in the host communities before and after the NPP closure, after decommissioning (Mangano and Sherman 2013; US NRC 2013a). The two researchers focused their subject of the study on the Rancho Seco NPP, in Sacramento County, California, where they evaluated the population living in this particular host community. They collected two types of cancer incident rates—one consists of incident rates data collected during the last 2 years of the plant's operation in 1988 and 1989. The other included incident rates data of the communities two decades after decommissioning period. The study's findings show that cancer incident rates decline in 18 out of 31 categories. The study concluded that people in the communities who were exposed to prolonged low-level radiation were more likely to develop illnesses related to cancer (Mangano and Sherman 2013; Kyne 2014). The scientific study of the effects of exposure to radiation began with the injection of plutonium into 18 men, women, and children—performed in a U.S. hospital ward in 1945, and without informing the unwitting subjects of the potential risks and dangers—and has continued with present-day studies that have attempted to better understand the human body's reaction to various radioisotopes (Welsome 1999).

2.2.4 Incomplete Knowledge of Low-Level Radiation Consequences

It is surprising that nuclear power plants have been in existence and in operation for more than six decades, but the NRC does not have a solid and complete knowledge on negative health consequences for individuals exposed to low-level radiation for a long period of time. The National Research Council (USA) and Committee on the Analysis of Cancer Risks in Populations near Nuclear Facilities (2012) conducted a study design that would allow for the accurate assessing of cancer incident rates among populations living near NPPs. Their findings include that (1) many epidemiologic studies face challenges because of (a) lack of availability and quality of data on cancer mortality and incidence at geographic levels smaller than a county, (b) lack of quality data on nuclear facility effluent releases, (c) lack of population mobility risks factors, and (d) low expected statistical power; (2) different study designs such as risk projection models, ecologic studies, as cohort studies have strengths and limitations; and (3) effluent release, direct exposure, and meteorology data are needed to better understand what effect radiation has, depending on distance and direction from a nuclear power plant. The committee recommends that the NRC should carry out an epidemiologic study of cancer risks on two areas: (1) an ecologic study of multiple cancer types and (2) case–control study of cancers in children born near the nuclear facility.

In their second phase (National Research Council (USA) and Committee on the Analysis of Cancer Risks in Populations near Nuclear Facilities 2014), seven nuclear facilities were selected for the pilot study of cancer risks:

1. Dresden Nuclear Power Station, Morris, Illinois,
2. Millstone Power Station, Waterford, Connecticut,
3. Oyster Creek Nuclear Generating Station, Forked River, New Jersey,
4. Haddam Neck Plant, Haddam Neck, Connecticut,
5. Big Rock Point Nuclear Plant, Charlevoix, Michigan,
6. San Onofre Nuclear Generating Station, San Clemente, California, and
7. Nuclear Fuel Services, Erwin, Tennessee.

As recommended in the Phase I study, the committee will carry out two studies: a case–control and an ecologic study in Phase II (National Research Council (USA) and Committee on the Analysis of Cancer Risks in Populations near Nuclear Facilities 2014). The first study examines cancer types of all ages living within 30-mile radius from the nuclear power plant.

(1) "Do cancer incidence and mortality vary by proximity to nuclear facilities?
(2) Does cancer incidence or mortality reflect patterns of radiation exposure associated with the nuclear facility? (National Research Council (USA) and Committee on the Analysis of Cancer Risks in Populations near Nuclear Facilities 2014)."

In the latter study, the subjects of the study are children younger than 15 years of age and their mother lived in areas within a 30-mile radius. The study attempts to hypothesize that children born by mothers living closer to the nuclear facilities are more likely to develop cancer than their counterparts who live further away, from the 30-mile radius. The study's findings are expected to provide answers to the following two research questions.

(1) "Is a mother's residential proximity to a nuclear facility at time of delivery associated with cancer in her children?
(2) Is estimated radiation exposure during pregnancy, early infancy, or childhood associated with childhood cancer occurrence? (National Research Council (USA) and Committee on the Analysis of Cancer Risks in Populations near

Nuclear Facilities 2014)" The current estimated time frame to complete this pilot study was 16 months, another 30 months to assess feasibility of nationwide study, and another 30 months to estimate the risk in a nationwide study; a total of 66 months is projected.

Above all, the agency who is responsible and accountable for regulating the nuclear power industry does not have a compete knowledge on constant exposure to low-level radiation. Without a complete knowledge, the authorities seem being blindfolded and walking on the regulation pathway without any direction. As a result, this has placed individuals living in the host communities at greater risk.

2.2.5 *Nuclear Power Facilities Living Longer Than Human Average Longevity*

The longevity of a commercial NPP is about 120 years, which normally counts the day operation began until the day its decommission is successfully completed, and includes 60 years from inception to the decommission; 50 years to cool down the reactor after its permanent shutdown; and 40 to 60 years to decommission the plant (Green Peace 2010; Kyne 2014). In general, the existence of nuclear power plants in host communities could be more than 120 years (Kyne 2014). Activities to remove an NPP after it has been shut down requires regulated monitoring because of the removal of several thousand tons of radioactive materials, such as spent fuel rods. Thus, NRC has adopted three advanced strategies to decommission an NPP:

(1) DECON, or decommissioning strategies;
(2) SAFSTOR, or deferred dismantling; and,
(3) ENTOMB, or the permanent encasing on site of radioactive contaminants (US NRC 2013a; Kyne 2014).

Among the three strategies, entombing includes permanently encasing the radioactive materials. In some cases, it requires several years to monitor both underground water and soil contamination. Hallam Nuclear Power Plant in Nebraska operated for a short time period and later it was decommissioned in 1971. It was entombed and the radioactive contaminated materials were buried underground. Due to the risks for underground water contamination, the Nebraska Department of Health has been monitoring possible underground water contamination and radiation. The monitoring will last for a period of 119 years, until 2090 (Nebraska Energy Quarterly 1997; Kyne 2014).

It is evident that after an NPP is introduced to the host communities, the individuals living in the area have to bear the risks of exposure to low-level radiation, and face health risks and long-term negative impacts on their environment (Kyne 2014).

2.3 Data and Research Methods

2.3.1 Study Questions

The primary goal of this chapter is to understand the communities hosting nuclear power plants. The understanding is necessary for effectively managing nuclear emergency and response. To achieve the primary goal, the following research questions are determined as follows:

(1) What are demographic characteristics of the individuals living in the communities hosting nuclear power plants in 1990, 2000, and 2010?
(2) Do the demographic characteristics vary with proximity to the nuclear power plants?
(3) Do the demographic characteristics vary from urban to non-urban areas?
(4) Do the demographic characteristics vary at individual nuclear power plants?
(5) Did the demographic characteristics change between 1900 and 2000, and between 2000 and 2010?

2.3.2 Study Variables

This chapter attempts to investigate the demographic characteristics of populations living in either of the two study areas described above. The demographic variables selected for this study purpose include both racial/ethnic and socioeconomic variables.

First, the racial/ethnic variables included the categories White, Black, Asian, Hispanic, Native American, Other, and Color. They are defined as follows:

(1) Percent White as percent of all non-Hispanic Whites;
(2) Percent Black is as percent of non-Hispanic Blacks or African Americans;
(3) Percent Asian as percent of Asian, and Native Hawaiian or Other Pacific Islander;
(4) Percent Native American as percent of American Indian or Alaska Native;
(5) Percent Other as percent of Some Other Race whose are not included in the White, Black, or African American, American Indian or Alaska Native, Asian, and Native Hawaiian or Other Pacific Islander race categories;
(6) Percent Color as percent of all other races except non-Hispanic Whites; and
(7) Percent Hispanic as percent of people who are Hispanic origin.

Second, the socioeconomic variables selected for inclusion were as follows:

(1) renter-occupied housing,
(2) unemployment rate,
(3) percent living in poverty,

(4) mean household income, and
(5) percent composition at the census-tract level.

The construction of the study variables is depicted in Appendix A (Table A.1).

In addition to the above-mentioned demographic variables, this study also includes distance from the NPP in order to analyze proximity-based disparities in any of the given study variables. The distance from the NPP was measured in Euclidean distance, in miles, between the census-tract center point and the NPP.

2.3.3 *Study Data*

To answer the research questions above, multiple datasets were obtained. First, for demographic characteristics, the US Census Bureau and GeoLytics datasets are acquired:

(1) US Census 1990, long form, normalized to 2010 census-tract boundaries (GeoLytics 2012b);
(2) US Census 2000, long form, normalized to 2010 census-tract boundaries (GeoLytics 2012a); and
(3) US Census 2010/American Community Survey, 5-year estimate (GeoLytics 2012c).

The 1990 and 2000 data containing census-tract level surveys are normalized to those of 2010 census boundaries. This normalization allows for the comparison of study variables across the study years. The geographical unit of analysis in this study is the census tract. The selected datasets for 1990, 2000, and 2010 contain racial and ethnic subgroups and socioeconomic data.

In order to conduct spatial analyses including tracking the changes in study variables across the study years, this study used the TIGER/Line shapefiles and the 2010 US Census Summary File 1 Demographic Profile (DP1) for the USA and Puerto Rico from the US Census Bureau's Web site. (The US Census Bureau 2010b). The selected study variables were merged with the selected shapefiles in the ArcGIS 10.2 program.

Geographical locations of the 99 nuclear reactors which are currently in operation were created in the shapefiles to conduct spatial analysis. Information was obtained from the NRC's Web site (NRC 2012) in order to create information on the following: geographical locations of the reactors, data on reactor type and containment type, design type, docket number, licensee, operating license issue date, commercial operation start date, renewed operation license, issue date, and operating license expiration date. However, since the September 11, 2011 event, the NRC did not provide latitude and longitude information for safety reason. To obtain latitude and longitude information for individual NPPs, individual planta was located in the Google Earth program and the XY coordinates were obtained manually.

2.3.4 Study Methods

This study covers populations living within a 50-mile radius of an NPP and considers this distance to define a host community. Those living at a distance greater than a 50-mile radius from an NPP are thus not considered a host community, for the purposes of this study. The selection of a 50-mile distance reflects the need to consider two emergency zones, namely the plume exposure pathway zone, which includes a radius of approximately 10 miles from the reactor site, and the ingestion exposure pathway zone, which has a radius of approximately 50 miles from the reactor site as defined by the NRC (US NRC 2013b).

However, the NRC's use of the 50-mile distance is considered to be controversial, given that more than 37 environmental and civic organizations argued that the 50-mile distance is not adequate to allow for full emergency preparedness, in reference to the Chernobyl and Fukushima Daiichi accidents (NIRS 2013). The selection of a 50-mile distance reflects the legal definition and boundaries of emergency zones. To determine the host communities surrounding an NPP, the census tracts that are within a 50-mile distance from the power plant are included as host communities (Fig. 2.3).

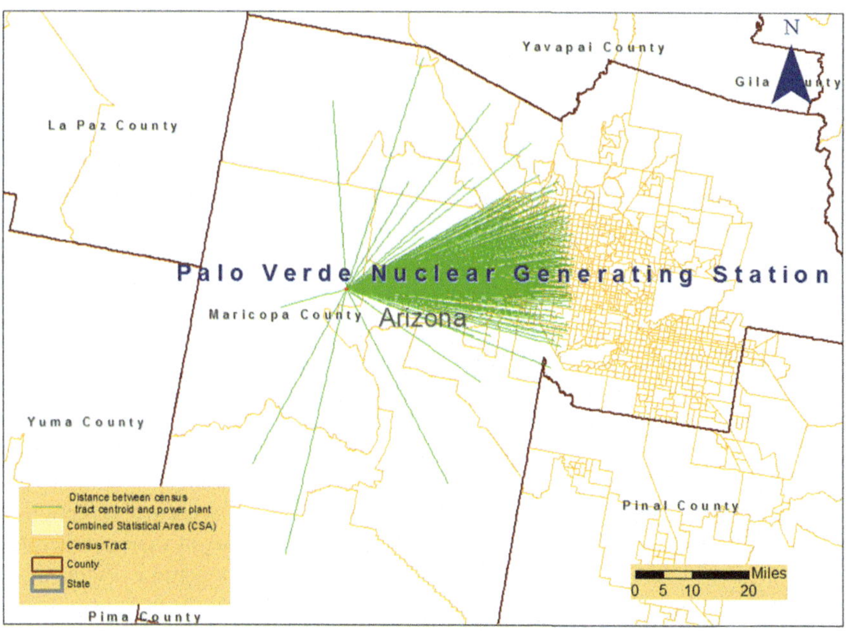

Fig. 2.3 Distance between census-tract centroid and Palo Verde Nuclear Generating Station within 50-mile distance; *Source* (Kyne 2014)

2.4 Overall Demographic Composition by Distance

This section explains findings of demographic characteristics in communities hosting 61 nuclear power plants. As stated, the overall population is the sum of population at all census tracts living around the nuclear power plants within a 50-mile radius. In some locations, when the two power plants are located close to each other, or a distance less than 50 miles from one another, some census tracts could be included in both host communities for each of the power plants. In such cases, the census tracts are counted only once. In other words, the total population living in the communities hosting the 61 power plants include only those individuals living in unique census tracts. To understand the disparities, the study measures the socio-demographic characteristics of the host communities by the six distance categories, namely 0–10 miles from the nearest NPP, 11–20 miles, 21–30 miles, 31–40 miles, 41–50 miles, 0–50 miles, and those more than 50 miles from the nearest NPP (outside), included in the study periods, 1990, 2000, and 2010. Population living in those communities outside of a 50-mile radius within the state are also described in the table (outside areas) to compare the demographic characteristics of population between host communities and non-host communities.

There are 87.47 million in 2010, 82.93 million in 2000, and 75.56 million individuals living in areas within a 50-mile radius from the 61 nuclear power plants (Tables 2.1, 2.2, and 2.3). Looking at each racial and ethnic group, the total population consists of 71% White, 17% Black, 4.4% Asian, 6.9% Other, 13.18% Hispanic, and 36.32% Color. When comparing with the outside areas in 2010, the 71% White were in host communities, whereas there are 75% White in outside areas. This implies that the present of White individual is higher in outside areas than their counterparts in the host communities. In other words, more White people prefer to live in non-host communities. In contrast, more Black people tend to live in host communities. There are 17.30% Black in the host communities, whereas only 10.55% of their counterparts present in outside areas. Unlike Black, fewer Hispanics are found in the host communities. There are 13.18% Hispanic living in the host communities, whereas 16% of their counterparts live in outside areas. Findings are similar in other racial groups including Asian, Native Americans, and the "Other" category. "People of Color" is a category that includes the total population excluding non-Hispanic Whites (see Appendix A, Table A.1) and was observed as a more prevalent racial group among communities within a 50-mile radius of an NPP (36%) than their counterparts living among communities in outside areas (35%) (Table 2.1). Findings indicate that the presence of People of Color in the outside areas was slightly lower than the presence of their counterparts in the host communities. The following observations on populations living in the host communities in comparison with the communities living in the outside areas could be made from the study findings: They (1) earn higher than the average household income; (2) are more likely to live in renter-occupied housing units;

Table 2.1 Demographic composition according to area, sorted by distance from any one of the 61 US-based commercial NPPs, based on 2010 American community survey data

	0–10	11–20	21–30	31–40	41–50	0–50	#Outside
Tract	881	3,123	5,902	7,267	4,452	21,625	51,432
Tract area (sq. mile)	17,411	53,061	77,062	89,543	49,638	286,714	3,510,028
Total population	3,785,622	12,924,633	24,002,206	29,428,636	17,329,084	87,470,176	216,495,088
White	3,149,990	10,331,676	17,654,848	20,074,272	10,964,530	62,175,316	162,720,384
Black	403,053	1,574,326	4,038,444	5,493,675	3,623,037	15,132,535	22,846,216
Asian	76,117	347,859	782,670	1,464,925	1,174,769	3,846,340	10,830,826
Native American	9,872	38,608	69,544	110,421	59,304	287,749	2,192,716
Others	146,590	632,164	1,456,700	2,285,341	1,507,444	6,028,239	17,904,950
Hispanic	309,223	1,287,201	2,964,844	4,326,143	2,640,485	11,527,896	36,199,636
Color	844,011	3,417,169	8,130,174	11,683,426	7,698,411	31,773,192	75,619,312
Female	1,922,927	6,609,058	12,335,718	15,167,198	8,940,031	44,974,931	109,591,612
Old (65+ years)	419,393	1,372,779	2,302,602	2,648,012	1,354,474	8,085,141	20,575,376
Kid (<5 years)	25,838	100,919	260,071	346,928	234,982	968,684	1,533,545
White (%)	83.21	79.94	73.56	68.21	63.27	71.08	75.16
Black (%)	10.65	12.18	16.83	18.67	20.91	17.30	10.55
Asian (%)	2.01	2.69	3.26	4.98	6.78	4.40	5.00
Native American (%)	0.26	0.30	0.29	0.38	0.34	0.33	1.01
Others (%)	3.87	4.89	6.07	7.77	8.70	6.89	8.27
Hispanic (%)	8.17	9.96	12.35	14.70	15.24	13.18	16.72
Color (%)	22.30	26.44	33.87	39.70	44.42	36.32	34.93
Female (%)	50.80	51.14	51.39	51.54	51.59	51.42	50.62
Old (65+ years) (%)	13.31	13.29	13.04	13.19	12.35	13.00	12.64
Kid (<5 years) (%)	6.41	6.41	6.44	6.32	6.49	6.40	6.71

(continued)

Table 2.1 (continued)

	0–10	11–20	21–30	31–40	41–50	0–50	#Outside
Native-born (%)	92.57	90.93	88.08	84.89	79.61	85.94	87.82
Renter housing units (%)	21.72	25.60	28.71	33.24	34.81	30.66	28.80
College degree or higher (%)	27.96	29.86	29.46	30.46	30.16	29.93	27.07
Unemployment (%)	7.19	7.33	8.02	8.43	8.21	8.06	7.87
Poverty (%)	10.10	11.39	12.98	13.72	13.53	12.98	14.16
Mean household income ($)	74,831	75,869	73,736	75,122	76,236	75,057	69,179

Table 2.2 Demographic composition according to area, sorted by distance from any one of the 61 US-based commercial NPPs, based on 2000 American community survey data normalized to 2010 US census boundaries

	0–10	11–20	21–30	31–40	41–50	0–50	#Outside
Tract	881	3,123	5,902	7,267	4,452	21,625	51,432
Tract area (sq. mile)	17,411	53,061	77,062	89,543	49,638	286,714	3,510,028
Total population	3,320,806	11,798,208	22,752,148	28,365,748	16,693,309	82,930,216	198,491,696
White	2,845,596	9,728,702	17,187,364	19,820,268	10,866,363	60,448,296	150,905,424
Black	317,882	1,344,595	3,741,738	5,233,932	3,497,843	14,135,990	20,225,750
Asian	42,950	218,981	533,037	1,084,885	840,199	2,720,052	7,830,550
Native American	10,232	40,587	75,776	118,638	61,272	306,505	2,141,484
Others	104,272	465,182	1,214,213	2,108,042	1,427,585	5,319,294	17,388,556
Hispanic	165,809	803,233	2,132,320	3,296,326	2,085,288	8,482,976	26,755,504
Color	571,357	2,540,502	6,766,977	10,081,858	6,765,774	26,726,468	60,181,296
Female	1,690,846	6,038,254	11,738,862	14,667,990	8,638,993	42,774,942	100,730,780
Old (65+ years)	365,327	1,255,586	2,227,851	2,563,851	1,323,869	7,727,279	18,533,455
Kid (<5 years)	25,705	106,589	301,085	415,214	286,593	1,133,983	1,668,418
White (%)	85.69	82.46	75.54	69.87	65.09	72.89	76.03
Black (%)	9.57	11.40	16.45	18.45	20.95	17.05	10.19
Asian (%)	1.29	1.86	2.34	3.82	5.03	3.28	3.95
Native American (%)	0.31	0.34	0.33	0.42	0.37	0.37	1.08
Others (%)	3.14	3.94	5.34	7.43	8.55	6.41	8.76
Hispanic (%)	4.99	6.81	9.37	11.62	12.49	10.23	13.48
Color (%)	17.21	21.53	29.74	35.54	40.53	32.23	30.32
Female (%)	50.92	51.18	51.59	51.71	51.75	51.58	50.75
Old (65+ years) (%)	12.84	12.91	12.96	12.94	12.18	12.78	12.28
Kid (<5 years) (%)	8.09	7.93	8.05	7.93	8.19	8.02	8.25

(continued)

Table 2.2 (continued)

	0–10	11–20	21–30	31–40	41–50	0–50	#Outside
Native-born (%)	94.71	92.87	89.78	86.75	81.53	87.72	89.46
Renter housing units (%)	22.79	26.59	30.42	35.42	37.35	32.64	30.00
College degree or higher (%)	23.76	26.02	25.59	26.41	25.77	25.89	23.77
Unemployment (%)	4.76	4.85	5.71	6.02	6.40	5.79	5.77
Poverty (%)	8.90	9.43	11.59	12.38	12.95	11.72	12.65
Mean household income ($)	73,480	75,633	73,474	74,511	75,395	74,520	68,904

Table 2.3 Demographic composition according to area, sorted by distance from any one of the 61 US-based commercial NPPs, based on 1990 American community survey data normalized to 2010 US census boundaries

	0–10	11–20	21–30	31–40	41–50	0–50	#Outside
Tract	881	3,123	5,902	7,267	4,452	21,625	51,432
Tract area (sq. mile)	17,411	53,061	77,062	89,543	49,638	286,714	3,510,028
Total population	2,872,653	10,412,267	20,904,464	26,161,414	15,209,935	75,560,736	173,148,624
White	2,547,599	9,011,480	16,755,732	19,829,114	10,838,429	58,982,352	140,844,256
Black	261,483	1,113,991	3,288,618	4,674,866	3,189,857	12,528,815	17,401,612
Asian	25,874	131,508	337,458	668,567	514,220	1,677,627	5,549,255
Native American	9,013	30,139	58,238	95,796	47,295	240,481	1,774,563
Others	28,784	124,947	464,452	893,046	620,151	2,131,380	7,578,717
Hispanic	89,271	395,405	1,412,975	2,101,466	1,351,832	5,350,949	16,549,152
Color	381,220	1,659,713	5,037,307	7,364,445	4,983,126	19,425,812	40,859,180
Female	1,460,436	5,366,463	10,849,682	13,599,014	7,906,815	39,182,411	88,354,808
Old (65+ years)	321,785	1,144,949	2,194,278	2,595,350	1,357,899	7,613,219	17,441,976
Kid (<5 years)	22,867	96,001	284,830	398,670	272,221	1,075,803	1,554,033
White (%)	88.68	86.55	80.15	75.80	71.26	78.06	81.34
Black (%)	9.10	10.70	15.73	17.87	20.97	16.58	10.05
Asian (%)	0.90	1.26	1.61	2.56	3.38	2.22	3.20
Native American (%)	0.31	0.29	0.28	0.37	0.31	0.32	1.02
Others (%)	1.00	1.20	2.22	3.41	4.08	2.82	4.38
Hispanic (%)	3.11	3.80	6.76	8.03	8.89	7.08	9.56
Color (%)	13.27	15.94	24.10	28.15	32.76	25.71	23.60
Female (%)	50.84	51.54	51.90	51.98	51.98	51.86	51.03
Old (65+ years) (%)	12.63	12.71	13.10	13.09	12.53	12.91	12.38
Kid (<5 years) (%)	8.75	8.62	8.66	8.53	8.53	8.59	8.93

(continued)

Table 2.3 (continued)

	0–10	11–20	21–30	31–40	41–50	0–50	#Outside
Native-born (%)	96.32	95.18	92.06	90.39	86.66	90.99	92.52
Renter housing units (%)	24.68	28.03	31.89	36.97	38.74	34.20	31.33
College degree or higher (%)	18.52	21.32	21.01	22.12	21.24	21.39	19.87
Unemployment (%)	5.55	5.20	6.20	6.60	6.60	6.26	6.34
Poverty (%)	9.60	9.70	12.02	12.62	12.29	11.87	13.66
Mean household income ($)	64,953	68,022	66,607	67,859	69,318	67,717	61,101

(3) are more likely to hold college degree or higher; (4) are less likely to be employed; and (5) are less likely to be native-born citizens. However, findings do not show any significant differences between the two areas in age or gender study variables. Similar demographic patterns are observed in the years 2000 (Table 2.2) and 1990 (Table 2.3).

Looking into proximity-based disparities in demographic characteristics, the findings indicate that there is a negative association between the percent White and the distance. In the 2010 data, the percent of White population was observed as 83.21, 79.94, 73.56, 68.21, and 63.27%, respectively, in the 0–10 miles, 11–20 miles, 21–30 miles, 31–40 miles, and 41–50 mile categories, respectively (Table 2.1). In contrast, the positive association is observed between the percent in other racial groups, namely Black, Asian, Hispanic, and Native American, and their distance from the NPP. Similarly, the positive association is observed between the percent of socio-demographic study variables, namely poverty, unemployment rate, education attainment, and household income (Table 2.1, Fig. 2.4). Similar patterns in the study variables are observed in the US Census data for the years 2000 (Table 2.2) and 1990 (Table 2.3).

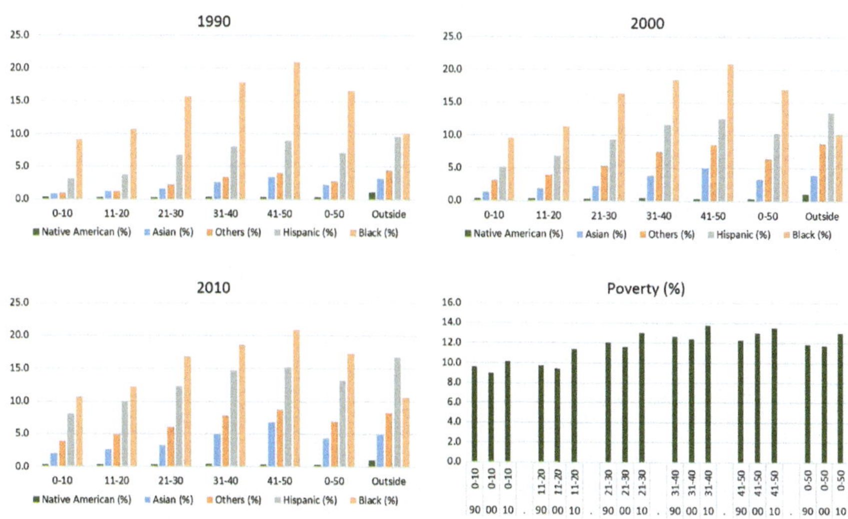

Fig. 2.4 Demographic trends (race/ethnicity) in the areas surrounding any one of the 61 US-based commercial NPPs, as sorted by distance

2.5 Hosting Communities in Urban and Non-urban Areas

To define the boundaries of metropolitan areas, this study used the Combined Statistical Areas (CSAs) 2010 TIGER Shape file (The US Census Bureau 2010b). The Shape file consists of metropolitan statistical areas which are created based on urbanized areas (50,000 people or more) and micropolitan statistical areas, which are based on urban clusters (between 10,000 and fewer than 50,000 people) (The US Census Bureau 2010a).

In 2010, out of the 87 million people living within a 50-mile radius of an NPP, 77 million lived in a metropolitan area (Table 2.4). Within that same group, those living within a 50-mile radius of an NPP, there was a higher percent Black, Asian, and Hispanic in metropolitan areas as compared to non-metropolitan areas. A higher percent of Color people, 37%, lived in metropolitan areas than that lived in non-metropolitan areas, 31%. There was a pattern in the socioeconomic data, such that metropolitan areas included more people with a college degree or who had completed some other form of higher education, a greater percentage of unemployed individuals, and fewer people living in poverty. Further, the mean household income was higher than as compared to demographically similar households in non-metropolitan areas. Similar patterns were observed in the US Census data for the years 2000 (Table 2.5) and 1990 (Table 2.6), respectively.

Over the past three decades, in communities sited within a 50-mile radius of an NPP, there has been a sharp decrease in the percent White among the population, with a compensating dramatic increase in the percent Hispanic and percent Color (Fig. 2.5), in both metropolitan and non-metropolitan areas. This same pattern is observed in non-metropolitan areas when considering the demographic data for the outlying areas.

2.6 Hosting Communities at Individual Nuclear Power Plants

In this section, the demographic characteristics of the communities surrounding individual NPPs in areas within a 10-mile radius and within a 50-mile radius are discussed. Second, the percent change in total population of the communities in proximity to an individual NPP during the periods 1990–2000 and 2000–2010 is listed.

Analysis and discussion of the socio-demographic composition of the host communities consider those populations living in communities within a 10-mile radius and a 50-mile radius of a NPP, respectively. First, each NPP was ranked according to the total population living within a 10-mile radius according to 2010 US Census data. The total such population varies from just 2,654, near South Texas Project, Texas, to as much as 257,462, at Limerick Generating Station, Pennsylvania (Table 2.7)—approximately 0.06% of the total US population lives within a 10-mile radius of the Limerick Generating Station. The ten NPPs with the largest population living in such close proximity as of 2010 were as follows:

Table 2.4 Demographic composition in metropolitan areas within a 50-mile radius of any one of the 61 US commercial NPPs, based on 2010 American community survey data

	Urban	Non-urban	Urban outside	Non-urban outside
Tract	19,222	2,403	35,148	16,284
Tract area (sq. mile)	173,650	113,064	764,061	2,745,968
Total population	77,488,104	9,982,079	152,552,224	63,942,864
White	54,565,688	7,609,628	111,465,152	51,255,228
Black	13,487,840	1,644,695	17,367,044	5,479,172
Asian	3,663,882	182,458	8,893,478	1,937,348
Native American	230,165	57,584	1,038,965	1,153,751
Others	5,540,525	487,714	13,787,588	4,117,362
Hispanic	10,477,473	1,050,423	27,627,094	8,572,543
Color	28,723,434	3,049,757	57,342,840	18,276,470
White (%)	70.42	76.23	73.07	80.16
Black (%)	17.41	16.48	11.38	8.57
Asian (%)	4.73	1.83	5.83	3.03
Native American (%)	0.30	0.58	0.68	1.80
Others (%)	7.15	4.89	9.04	6.44
Hispanic (%)	13.52	10.52	18.11	13.41
Color (%)	37.07	30.55	37.59	28.58
Female (%)	51.49	50.83	50.79	50.22
Old (65+ years) (%)	12.88	13.96	11.92	14.38
Kid (<5 years) (%)	6.37	6.65	6.80	6.49
Native-born (%)	84.99	93.34	85.80	92.62
Renter housing units (%)	31.29	25.94	30.40	25.35
College degree or higher (%)	30.93	22.02	29.28	21.86
Unemployment (%)	8.12	7.49	7.94	7.69
Poverty (%)	12.71	15.09	13.51	15.71
Mean household income ($)	76,850	61,118	73,392	59,404

Table 2.5 Demographic composition in metropolitan areas within a 50-mile radius of any one of the 61 US commercial NPPs, based on 2000 American community survey data

	Urban	Non-urban	Urban outside	Non-urban outside
Tract	19,222	2,403	35,148	16,284
Tract area (sq. mile)	173,650	113,064	764,061	2,745,968
Total population	73,925,584	9,004,636	138,833,968	59,657,720
White	53,513,884	6,934,410	103,039,168	47,866,268
Black	12,672,128	1,463,862	15,169,283	5,056,467
Asian	2,601,777	118,275	6,327,916	1,502,634
Native American	257,619	48,886	1,033,967	1,107,517
Others	4,880,127	439,167	13,263,712	4,124,844
Hispanic	7,868,239	614,737	20,485,020	6,270,484
Color	24,394,302	2,332,166	45,308,664	14,872,632
White (%)	72.39	77.01	74.22	80.23
Black (%)	17.14	16.26	10.93	8.48
Asian (%)	3.52	1.31	4.56	2.52
Native American (%)	0.35	0.54	0.74	1.86
Others (%)	6.60	4.88	9.55	6.91
Hispanic (%)	10.64	6.83	14.76	10.51
Color (%)	33.00	25.90	32.64	24.93
Female (%)	51.64	51.07	50.87	50.47
Old (65+ years) (%)	12.64	13.93	11.53	14.03
Kid (<5 years) (%)	8.04	7.89	8.43	7.83
Native-born (%)	86.80	95.32	87.55	93.91
Renter housing units (%)	33.52	25.66	32.07	25.64
College degree or higher (%)	26.69	19.29	25.87	18.92
Unemployment (%)	5.81	5.55	5.63	6.11
Poverty (%)	11.51	13.51	11.90	14.41
Mean household income ($)	76,287	60,094	73,621	58,173

Table 2.6 Demographic composition in metropolitan areas within a 50-mile radius of any one of the 61 US commercial NPPs, based on 1990 American community survey data

	Urban	Non-urban	Urban outside	Non-urban outside
Tract	19,222	2,403	35,148	16,284
Tract area (sq. mile)	173,650	113,064	764,061	2,745,968
Total population	67,602,272	7,958,464	120,016,720	53,131,904
White	52,568,548	6,413,805	96,121,016	44,723,232
Black	11,264,808	1,264,007	12,908,319	4,493,294
Asian	1,598,934	78,693	4,243,576	1,305,679
Native American	201,596	38,885	842,528	932,035
Others	1,968,337	163,043	5,901,096	1,677,621
Hispanic	5,051,194	299,755	12,640,686	3,908,466
Color	17,752,876	1,672,936	30,344,224	10,514,956
White (%)	77.76	80.59	80.09	84.17
Black (%)	16.66	15.88	10.76	8.46
Asian (%)	2.37	0.99	3.54	2.46
Native American (%)	0.30	0.49	0.70	1.75
Others (%)	2.91	2.05	4.92	3.16
Hispanic (%)	7.47	3.77	10.53	7.36
Color (%)	26.26	21.02	25.28	19.79
Female (%)	51.89	51.55	51.10	50.87
Old (65+ years) (%)	12.79	13.94	11.62	14.12
Kid (<5 years) (%)	8.57	8.69	9.02	8.74
Native-born (%)	90.24	97.37	91.02	95.91
Renter housing units (%)	35.05	27.20	33.54	26.73
College degree or higher (%)	22.01	15.96	21.74	15.63
Unemployment (%)	6.29	5.94	6.10	6.93
Poverty (%)	11.51	15.00	12.45	16.42
Mean household income ($)	69,394	53,429	65,690	50,809

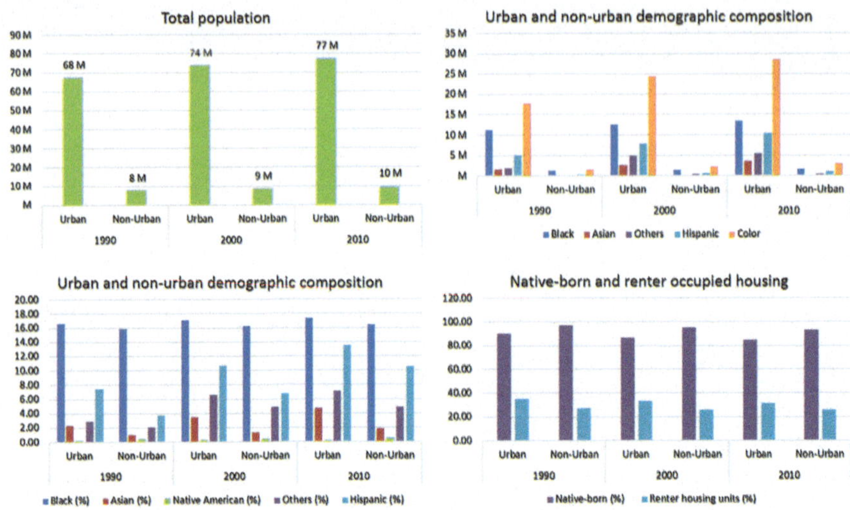

Fig. 2.5 Demographic trends (race/ethnicity) in the areas surrounding any one of the 61 US-based commercial NPPs, as sorted by urban and non-urban areas

(1) Limerick Generating Station, Pennsylvania, 257,462; (2) FitzPatrick Nuclear Power Plant, New York, 253,977; (3) St. Lucie Plant, Florida, 209,961; (4) Catawba Nuclear Station, South Carolina, 200,869; (5) Three Mile Island Nuclear Station, Pennsylvania, 191,325; (6) McGuire Nuclear Station, North Carolina, 188,937; (7) Turkey Point Nuclear Generating, Florida, 147,705; (8) Surry Nuclear Power Station, Virginia, 133,856; (9) Oyster Creek Nuclear Generating Station, New Jersey, 128,893; and (10) Seabrook Station, New Hampshire, 120,876. The three NPPs with the smallest total populations living within a 10-mile radius as of 2010 were as follows: (1) South Texas Project, Texas, 2,654; (2) Palo Verde Nuclear Generating Station, Arizona, 3,090; and (3) Columbia Generating Station, Washington, 6,007 (Table 2.7).

In 2000, among the ten NPPs with the largest such populations as of 2010, Beaver Valley Power Station, Pennsylvania, and Millstone Power Station, Connecticut, replaced Turkey Point Nuclear Generating, Florida, and Oyster Creek Nuclear Generating Station, New Jersey (Table 2.7); the three NPPs with the smallest such populations were the same in 2000 (Table 2.7). In 1990, among the top ten such NPPs as of 2000, Oyster Creek Nuclear Generating Station, New Jersey, replaced Millstone Power Station, Connecticut, while the bottom-three NPPs in terms of such population again were unchanged (Table 2.7).

The total population living within a 50-mile radius of an NPP in 1990, 2000, and 2010 are also depicted in Table 2.8. The total population of communities sited in such areas ranged between 111,113, and 15 million (Table 2.8); approximately 5% of the total US population as of 2010 lived within a 50-mile radius of Indian Point Nuclear Generating Station. The ten NPPs with the largest total population living within a 50-mile radius were as follows: (1) Indian Point Nuclear Generating,

Table 2.7 Total population living in census tracts within a 10-mile radius of a NPP as of 2010, as measured from census-tract centroid point to NPP

Index	Plant name	State	No. of tract	Tract area	Total population (0–10 mile)		2010	Rank
			(10-mile)	(sq. mile)	1990	2000		2010
1	Browns Ferry	Alabama	10	302	36,259	38,640	40,292	31
2	Farley	Alabama	5	358	15,104	15,882	17,021	50
3	Palo Verde	Arizona	1	120	1,365	2,112	3,090	60
4	Arkansas Nuclear	Arkansas	8	196	32,038	37,734	43,212	29
5	Diablo Canyon	California	7	358	39,157	40,444	27,776	40
6	Millstone	Connecticut	32	266	118,811	115,203	118,025	11
7	Saint Lucie	Florida	35	234	132,150	157,349	206,961	3
8	Turkey Point	Florida	26	208	89,348	99,529	147,705	7
9	Hatch	Georgia	3	333	9,459	10,524	10,422	55
10	Vogtle	Georgia	2	325	6,629	7,929	8,333	58
11	Braidwood	Illinois	10	327	28,159	32,863	35,921	32
12	Byron	Illinois	5	300	19,736	22,901	23,750	43
13	Clinton	Illinois	4	254	13,021	13,049	12,773	51
14	Dresden	Illinois	17	251	44,647	54,839	73,784	18
15	La Salle	Illinois	3	188	10,831	11,165	12,195	52
16	Quad Cities	Illinois	7	272	33,742	32,764	32,223	35
17	Duane Arnold	Iowa	22	291	77,265	90,277	95,814	14
18	Wolf Creek	Kansas	3	654	8,404	8,865	8,587	57
19	River Bend	Louisiana	5	239	16,273	18,311	18,786	49
20	Waterford	Louisiana	18	285	61,978	67,869	73,071	19
21	Calvert Cliffs	Maryland	10	258	33,438	48,206	48,285	26
22	Pilgrim	Massachusetts	15	453	59,169	66,086	69,440	20
23	D.C. Cook	Michigan	16	183	51,729	53,837	53,507	25

(continued)

Table 2.7 (continued)

Index	Plant name	State	No. of tract	Tract area	Total population (0–10 mile)		2010	Rank
			(10-mile)	(sq. mile)	1990	2000		2010
24	FermiFermi	Michigan	24	268	78,888	82,873	85,991	15
25	Palisades	Michigan	8	183	34,133	34,594	33,432	34
26	Monticello	Minnesota	8	308	28,856	42,437	56,001	23
27	Prairie Island	Minnesota	5	226	22,549	24,295	25,260	41
28	Grand Gulf	Mississippi	4	662	13,756	14,443	12,168	53
29	Callaway	Missouri	2	237	8,124	10,726	11,642	54
30	Cooper	Nebraska	3	670	11,078	10,205	9,770	56
31	Fort Calhoun	Nebraska	5	173	15,687	17,820	18,933	48
32	Seabrook	New Hampshire	30	270	101,342	117,140	120,876	10
33	Hope Creek	New Jersey	15	332	34,738	40,224	47,275	27
34	Oyster Creek	New Jersey	29	287	91,665	114,293	128,893	9
35	Salem	New Jersey	15	332	34,738	40,224	47,275	28
36	FitzPatrick	New York	55	292	229,882	242,165	253,977	2
37	Ginna	New York	9	374	30,260	29,878	29,580	36
38	Indian Point	New York	9	374	30,260	29,878	29,580	37
39	Nine Mile Point	New York	13	139	49,871	59,508	65,466	21
40	Brunswick	North Carolina	11	160	13,649	18,639	28,812	39
41	McGuire	North Carolina	47	322	61,535	117,264	188,937	6
42	Shearon Harris	North Carolina	19	272	24,724	54,745	84,598	16
43	Davis-Besse	Ohio	6	416	22,999	23,501	23,312	45
44	Perry	Ohio	22	141	68,920	75,998	80,167	17
45	Beaver Valley	Pennsylvania	33	279	119,111	115,882	108,656	12
46	Limerick	Pennsylvania	62	350	184,712	222,564	257,462	1

(continued)

Table 2.7 (continued)

Index	Plant name	State	No. of tract	Tract area	Total population (0–10 mile)		2010	Rank
			(10-mile)	(sq. mile)	1990	2000		2010
47	Peach Bottom	Pennsylvania	10	317	35,760	40,396	43,032	30
48	Susquehanna	Pennsylvania	14	287	51,798	53,318	55,117	24
49	Three Mile Island	Pennsylvania	41	297	156,537	177,279	191,325	5
50	Catawba	South Carolina	46	309	101,221	137,392	200,869	4
51	Oconee	South Carolina	8	309	30,342	32,295	33,936	33
52	Robinson	South Carolina	15	281	51,644	58,346	64,579	22
53	Summer	South Carolina	5	457	14,256	18,174	22,287	46
54	Sequoyah	Tennessee	20	312	76,278	90,338	104,716	13
55	Watts Bar	Tennessee	5	417	17,803	22,704	24,916	42
56	Comanche Peak	Texas	6	356	16,383	23,844	28,892	38
57	South Texas	Texas	1	25	3,233	2,848	2,654	61
58	North Anna	Virginia	6	363	14,747	19,529	23,603	44
59	Surry	Virginia	29	356	108,530	121,515	133,856	8
60	Columbia	Washington	1	209	3,530	4,364	6,007	59
61	Point Beach	Wisconsin	5	192	20,905	20,819	19,752	47
	All 61 plants in operation (0–10)		881	17,411	2,872,653	3,320,806	3,785,622	
	All 61 plants in operation (0–50)		21625	286,714	75,560,733	82,930,217	87,470,180	

New York, 15,118,181; (2) Limerick Generating Station, Pennsylvania, 6,559,209; (3) Dresden Nuclear Power Station, Illinois, 4,700,949; (4) Hope Creek Generating Station, New Jersey, 4,372,701; (5) Fermi, Michigan, 4,368,235; (6) Salem Nuclear Generating Station, New Jersey, 4,328,469; (7) Peach Bottom Atomic Power Station, Pennsylvania, 4,124,024; (8) Seabrook Station, New Hampshire, 3,767,215; (9) Turkey Point Nuclear Generating, Florida, 3,318,548; and (10) Pilgrim Plant, Massachusetts, 3,130,058.

The three NPPs with the smallest such populations were as follows: (1) Cooper Nuclear Station, Nebraska, 111,113; (2) Wolf Creek Generating Station, Kansas, 138,464; and (3) Arkansas Nuclear One, Arkansas, 194,963. In 2000, among the ten NPPs with the largest such populations as of 2010, Pilgrim Nuclear Power Station, Massachusetts, replaced Turkey Point Nuclear Generating, Florida (Table 2.8). In 1990, among the ten NPPs with the largest such populations as of 2000, Turkey Point Nuclear Generating, Florida, replaced Pilgrim Nuclear Power Station, Massachusetts (Table 2.8). The three NPPs with the smallest such populations were unchanged relative to 2010 in both 2000 (Table 2.8) and 1990 (Table 2.8).

2.7 Demographic Changes in 1990–2000 and 2000–2010

The percent change in population size within both a 10-mile radius and a 50-mile radius during the periods 1990–2000 and 2000–2010 was tracked. Among communities within a 10-mile radius of an NPP from 2000 to 2010, there was an increase in total population at 44 NPPs, while at the remaining 15 NPPs, the total such population declined. The percent change ranged from a 61% increase, in the area surrounding McGuire Nuclear Station, North Carolina (Fig. 2.6), to a 31% decrease among the communities surrounding Diablo Canyon Nuclear Power Plant, California. Large positive percent changes during the period 2000–2010 also occurred in the areas immediately surrounding Brunswick Steam Electric Plant, North Carolina, 55%, and Shearon Harris Nuclear Power Plant, North Carolina, 55%; a large negative percent change also occurred in the area surrounding Grand Gulf Nuclear Station, Mississippi (−16%). From 1990 to 2000, the largest percent change in total population living within a 10-mile radius of a NPP occurred at Shearon Harris Nuclear Power Plant, North Carolina (121%); the communities surrounding McGuire Nuclear Station, North Carolina, also experienced dramatic population growth (91%).

Fifty-six NPPs saw a positive percent change, or an increase in the population living in communities within a 50-mile radius; the remaining 9 NPPs saw a decrease in such populations from 2000 to 2010 (Fig. 2.7). The percent change range varied between 51 and −7%. The largest percent change occurred among the communities within a 50-mile radius of Palo Verde Nuclear Generation Station, Arizona (51%); the same NPP also saw the largest percent change in the growth of its same such population during the period 1990–2000 (53%). LaSalle County Station, Illinois (28%), and Brunswick Steam Electric Plant, North Carolina (27%), also saw significant growth in the population living within a 50-mile radius of each respective

Table 2.8 Total population living in census tracts within a 50-mile radius of a NPP as of 2010, as measured from census-tract centroid point to NPP

Index	Plant name	State	Tract	Tract area	Total population (50 mile)		2010	Rank
			(50-mile)	(sq. mile)	1990	2000		2010
1	Browns Ferry	Alabama	207	6,353	716,311	812,964	885,374	38
2	Farley	Alabama	94	6,642	316,989	343,659	360,352	55
3	Palo Verde	Arizona	534	6,942	1,295,802	1,799,929	2,207,889	17
4	Arkansas Nuclear	Arkansas	45	6,046	154,290	182,869	194,963	59
5	Diablo Canyon	California	90	3,048	361,432	409,562	448,292	51
6	Millstone	Connecticut	395	5,103	1,522,087	1,589,650	1,667,925	24
7	Saint Lucie	Florida	266	3,296	732,319	930,833	1,151,024	30
8	Turkey Point	Florida	691	4,431	2,489,137	3,024,321	3,318,548	9
9	Hatch	Georgia	72	6,586	243,782	284,616	304,217	56
10	Vogtle	Georgia	139	6,248	517,981	587,317	628,353	46
11	Braidwood	Illinois	660	5,806	2,367,196	2,718,397	3,018,399	11
12	Byron	Illinois	199	5,627	706,621	773,547	821,645	40
13	Clinton	Illinois	154	5,851	548,176	577,409	608,176	47
14	Dresden	Illinois	1037	5,838	3,901,073	4,433,036	4,700,949	3
15	La Salle	Illinois	235	5,775	683,266	886,835	1,137,865	31
16	Quad Cities	Illinois	154	5,824	539,561	547,635	544,388	48
17	Duane Arnold	Iowa	129	5,795	462,956	510,461	543,720	49
18	Wolf Creek	Kansas	39	5,973	133,789	140,997	138,464	60
19	River Bend	Louisiana	161	6,869	656,582	737,814	814,822	41
20	Waterford	Louisiana	487	6,269	1,655,821	1,778,252	1,656,164	25
21	Calvert Cliffs	Maryland	438	6,082	1,460,569	1,605,592	1,719,445	22
22	Pilgrim	Massachusetts	747	4,394	2,907,647	3,078,452	3,130,058	10
23	D.C. Cook	Michigan	227	5,853	809,517	861,818	873,728	39

(continued)

Table 2.8 (continued)

Index	Plant name	State	Tract	Tract area	Total population (50 mile)		2010	Rank
			(50-mile)	(sq. mile)	1990	2000		2010
24	Fermi	Michigan	1312	4,552	4,489,966	4,542,979	4,368,235	5
25	Palisades	Michigan	226	6,785	794,744	882,586	899,600	36
26	Monticello	Minnesota	450	5,299	1,544,634	1,781,553	1,950,169	19
27	Prairie Island	Minnesota	450	5,643	1,465,322	1,686,829	1,800,095	20
28	Grand Gulf	Mississippi	56	6,076	214,616	218,887	210,536	58
29	Callaway	Missouri	81	5,752	307,345	357,726	392,917	53
30	Cooper	Nebraska	38	6,168	117,399	115,868	111,113	61
31	Fort Calhoun	Nebraska	269	5,587	731,120	813,760	890,963	37
32	Seabrook	New Hampshire	868	4,195	3,465,615	3,683,764	3,767,215	8
33	Hope Creek	New Jersey	1097	5,684	4,052,631	4,189,795	4,372,701	4
34	Oyster Creek	New Jersey	583	3,412	2,132,640	2,387,006	2,538,550	13
35	Salem	New Jersey	1088	5,715	4,011,786	4,147,881	4,328,469	6
36	FitzPatrick	New York	212	4,938	757,206	745,522	747,467	43
37	Ginna	New York	264	4,515	987,092	1,020,186	1,035,214	33
38	Indian Point	New York	3760	5,810	13,786,298	14,894,006	15,118,181	1
39	Nine Mile Point	New York	211	4,908	754,261	742,377	744,267	44
40	Brunswick	North Carolina	108	3,040	232,337	316,773	402,395	52
41	McGuire	North Carolina	613	6,366	1,636,412	2,072,499	2,517,850	14
42	Shearon Harris	North Carolina	486	6,217	1,393,352	1,829,822	2,222,038	16
43	Davis-Besse	Ohio	363	4,476	1,343,578	1,354,016	1,349,593	27
44	Perry	Ohio	531	4,905	1,766,037	1,788,572	1,699,459	23
45	Beaver Valley	Pennsylvania	808	5,647	2,824,537	2,757,925	2,655,595	12
46	Limerick	Pennsylvania	1609	5,759	5,986,390	6,282,721	6,559,209	2

(continued)

Table 2.8 (continued)

Index	Plant name	State	Tract (50-mile)	Tract area (sq. mile)	Total population (50 mile) 1990	2000	2010	Rank 2010
47	Peach Bottom	Pennsylvania	1008	5,794	3,569,625	3,845,576	4,124,024	7
48	Susquehanna	Pennsylvania	303	5,863	1,102,337	1,116,488	1,136,292	32
49	Three Mile Island	Pennsylvania	448	5,941	1,718,797	1,906,037	2,068,685	18
50	Catawba	South Carolina	556	6,403	1,437,561	1,819,689	2,235,497	15
51	Oconee	South Carolina	276	6,175	873,312	1,031,201	1,152,820	29
52	Robinson	South Carolina	152	6,363	568,041	616,110	649,720	45
53	Summer	South Carolina	251	6,436	749,396	865,550	971,019	34
54	Sequoyah	Tennessee	205	6,227	747,693	855,343	937,320	35
55	Watts Bar	Tennessee	178	6,455	616,648	731,267	812,206	42
56	Comanche Peak	Texas	268	6,086	881,604	1,050,191	1,220,100	28
57	South Texas	Texas	52	4,663	199,113	215,547	216,206	57
58	North Anna	Virginia	364	6,181	1,085,119	1,322,489	1,553,358	26
59	Surry	Virginia	450	5,599	1,523,998	1,645,196	1,741,096	21
60	Columbia	Washington	71	4,782	234,431	302,914	365,409	54
61	Point Beach	Wisconsin	114	5,150	418,292	461,723	485,589	50
	All 61 plants in operation (0–50)		21625	286,714	75,560,733	82,930,217	87,470,180	
	All 61 plants in operation outside		51432	3,510,028	173,148,625	198,491,689	216,495,092	

Index	Plant name	State	(10-mile)	(sq. mile)	1990-2000	2000-2010	(50-mile)	(sq. mile)	1990-2000	2000-2010
41	McGuire	North Carolina	47	322	90.56%	61.12%	613	6,366	26.65%	21.49%
40	Brunswick	North Carolina	11	160	36.56%	54.58%	108	3,040	36.34%	27.03%
42	Shearon Harris	North Carolina	19	272	121.42%	54.53%	486	6,217	31.33%	21.43%
8	Turkey Point	Florida	26	208	11.39%	48.40%	691	4,431	21.50%	9.73%
3	Palo Verde	Arizona	1	120	54.73%	46.31%	534	6,942	38.90%	22.67%
50	Catawba	South Carolina	46	309	35.73%	46.20%	556	6,403	26.58%	22.85%
60	Columbia	Washington	1	209	23.63%	37.65%	71	4,782	29.21%	20.63%
14	Dresden	Illinois	17	251	22.83%	34.55%	1037	5,838	13.64%	6.04%
26	Monticello	Minnesota	8	308	47.06%	31.96%	450	5,299	15.34%	9.46%
7	Saint Lucie	Florida	35	234	19.07%	31.53%	266	3,296	27.11%	23.66%
53	Summer	South Carolina	5	457	27.48%	22.63%	251	6,436	15.50%	12.19%
56	Comanche Peak	Texas	6	356	45.54%	21.17%	268	6,086	19.12%	16.18%
58	North Anna	Virginia	6	363	32.43%	20.86%	364	6,181	21.88%	17.46%
33	Hope Creek	New Jersey	15	332	15.79%	17.53%	1097	5,684	3.38%	4.37%
35	Salem	New Jersey	15	332	15.79%	17.53%	1088	5,715	3.39%	4.35%
54	Sequoyah	Tennessee	20	312	18.43%	15.92%	205	6,227	14.40%	9.58%
46	Limerick	Pennsylvania	62	350	20.49%	15.68%	1609	5,759	4.95%	4.40%
4	Arkansas Nuclear	Arkansas	8	196	17.78%	14.52%	45	6,046	18.52%	6.61%
34	Oyster Creek	New Jersey	29	287	24.69%	12.77%	583	3,412	11.93%	6.35%
52	Robinson	South Carolina	15	281	12.98%	10.68%	152	6,363	8.46%	5.46%
59	Surry	Virginia	29	356	11.96%	10.16%	450	5,599	7.95%	5.83%
39	Nine Mile Point	New York	13	139	19.32%	10.01%	211	4,908	-1.58%	0.25%
55	Watts Bar	Tennessee	5	417	27.53%	9.74%	178	6,455	18.59%	11.07%
11	Braidwood	Illinois	10	327	16.71%	9.31%	660	5,806	14.84%	11.04%
15	La Salle	Illinois	3	188	3.08%	9.23%	235	5,775	29.79%	28.31%
29	Callaway	Missouri	2	237	32.03%	8.54%	81	5,752	16.39%	9.84%
49	Three Mile Island	Pennsylvania	41	297	13.25%	7.92%	448	5,941	10.89%	8.53%
20	Waterford	Louisiana	18	285	9.50%	7.66%	487	6,269	7.39%	-6.87%
2	Farley	Alabama	5	358	5.15%	7.17%	94	6,642	8.41%	4.86%
47	Peach Bottom	Pennsylvania	10	317	12.96%	6.53%	1008	5,794	7.73%	7.24%
31	Fort Calhoun	Nebraska	5	173	13.60%	6.25%	269	5,587	11.30%	9.49%
17	Duane Arnold	Iowa	22	291	16.84%	6.13%	129	5,795	10.26%	6.52%
44	Perry	Ohio	22	141	10.27%	5.49%	531	4,905	1.28%	-4.98%
10	Vogtle	Georgia	2	325	19.61%	5.10%	139	6,248	13.39%	6.99%
51	Oconee	South Carolina	8	309	6.44%	5.08%	276	6,175	18.08%	11.79%
22	Pilgrim	Massachusetts	15	453	11.69%	5.08%	747	4,394	5.87%	1.68%
36	FitzPatrick	New York	55	292	5.34%	4.88%	212	4,938	-1.54%	0.26%
1	Browns Ferry	Alabama	10	302	6.57%	4.28%	207	6,353	13.49%	8.91%
27	Prairie Island	Minnesota	5	226	7.74%	3.97%	450	5,643	15.12%	6.71%
24	Fermi	Michigan	24	268	5.05%	3.76%	1312	4,552	1.18%	-3.85%
12	Byron	Illinois	5	300	16.04%	3.71%	199	5,627	9.47%	6.22%
48	Susquehanna	Pennsylvania	14	287	2.93%	3.37%	303	5,863	1.28%	1.77%
32	Seabrook	New Hampshire	30	270	15.59%	3.19%	868	4,195	6.29%	2.27%
19	River Bend	Louisiana	5	239	12.52%	2.59%	161	6,869	12.37%	10.44%
6	Millstone	Connecticut	32	266	-3.04%	2.45%	395	5,103	4.44%	4.92%
21	Calvert Cliffs	Maryland	10	258	44.17%	0.16%	438	6,082	9.93%	7.09%
23	D.C. Cook	Michigan	16	183	4.08%	-0.61%	227	5,853	6.46%	1.38%
43	Davis-Besse	Ohio	6	416	2.18%	-0.80%	363	4,476	0.78%	-0.33%
9	Hatch	Georgia	3	333	11.26%	-0.97%	72	6,586	16.75%	6.89%
37	Ginna	New York	9	374	-1.26%	-1.00%	264	4,515	3.35%	1.47%
38	Indian Point	New York	9	374	-1.26%	-1.00%	3760	5,810	8.03%	1.51%
16	Quad Cities	Illinois	7	272	-2.90%	-1.65%	154	5,824	1.50%	-0.59%
13	Clinton	Illinois	4	254	0.22%	-2.12%	154	5,851	5.33%	5.33%
18	Wolf Creek	Kansas	3	654	5.49%	-3.14%	39	5,973	5.39%	-1.80%
25	Palisades	Michigan	8	183	1.35%	-3.36%	226	6,785	11.05%	1.93%
30	Cooper	Nebraska	3	670	-7.88%	-4.26%	38	6,168	-1.30%	-4.10%
61	Point Beach	Wisconsin	5	192	-0.41%	-5.13%	114	5,150	10.38%	5.17%
45	Beaver Valley	Pennsylvania	33	279	-2.71%	-6.24%	808	5,647	-2.36%	-3.71%
57	South Texas	Texas	1	25	-11.91%	-6.81%	52	4,663	8.25%	0.31%
28	Grand Gulf	Mississippi	4	662	4.99%	-15.75%	56	6,076	1.99%	-3.82%
5	Diablo Canyon	California	7	358	3.29%	-31.32%	90	3,048	13.32%	9.46%
	All 61 plants in operation (0-50)		881	17,411	15.60%	14.00%	21625	286,714	9.75%	5.47%
	All 61 plants in operation-outside		21625	286,714	9.75%	5.47%	51432	3,510,028	14.64%	9.07%

Fig. 2.6 Percent change of the total population living in census tracts within a 10-mile radius and 50-mile radius, as measured from census-tract centroid point to the nearest NPP, during the periods 1990–2000 and 2000–2010, sorted according to percent change within a 10-mile radius

Index	Plant name	State	(10-mile)	(sq. mile)	1990-2000	2000-2010	(50-mile)	(sq. mile)	1990-2000	2000-2010
41	McGuire	North Carolina	47	322	90.56%	61.12%	613	6,366	26.65%	21.49%
40	Brunswick	North Carolina	11	160	36.56%	54.58%	108	3,040	36.34%	27.03%
42	Shearon Harris	North Carolina	19	272	121.42%	54.53%	486	6,217	31.33%	21.43%
8	Turkey Point	Florida	26	208	11.39%	48.40%	691	4,431	21.50%	9.73%
3	Palo Verde	Arizona	1	120	54.73%	46.31%	534	6,942	38.90%	22.67%
50	Catawba	South Carolina	46	309	35.73%	46.20%	556	6,403	26.58%	22.85%
60	Columbia	Washington	1	209	23.63%	37.65%	71	4,782	29.21%	20.63%
14	Dresden	Illinois	17	251	22.83%	34.55%	1037	5,838	13.64%	6.04%
26	Monticello	Minnesota	8	308	47.06%	31.96%	450	5,299	15.34%	9.46%
7	Saint Lucie	Florida	35	234	19.07%	31.53%	266	3,296	27.11%	23.66%
53	Summer	South Carolina	5	457	27.48%	22.63%	251	6,436	15.50%	12.19%
56	Comanche Peak	Texas	6	356	45.54%	21.17%	268	6,086	19.12%	16.18%
58	North Anna	Virginia	6	363	32.43%	20.86%	364	6,181	21.88%	17.46%
33	Hope Creek	New Jersey	15	332	15.79%	17.53%	1097	5,684	3.38%	4.37%
35	Salem	New Jersey	15	332	15.79%	17.53%	1088	5,715	3.39%	4.35%
54	Sequoyah	Tennessee	20	312	18.43%	15.92%	205	6,227	14.40%	9.58%
46	Limerick	Pennsylvania	62	350	20.49%	15.68%	1609	5,759	4.95%	4.40%
4	Arkansas Nuclear	Arkansas	8	196	17.78%	14.52%	45	6,046	18.52%	6.61%
34	Oyster Creek	New Jersey	29	287	24.69%	12.77%	583	3,412	11.93%	6.35%
52	Robinson	South Carolina	15	281	12.98%	10.68%	152	6,363	8.46%	5.46%
59	Surry	Virginia	29	356	11.96%	10.16%	450	5,599	7.95%	5.83%
39	Nine Mile Point	New York	13	139	19.32%	10.01%	211	4,908	-1.58%	0.25%
55	Watts Bar	Tennessee	5	417	27.53%	9.74%	178	6,455	18.59%	11.07%
11	Braidwood	Illinois	10	327	16.71%	9.31%	660	5,806	14.84%	11.04%
15	La Salle	Illinois	3	188	3.08%	9.23%	235	5,775	29.79%	28.31%
29	Callaway	Missouri	2	237	32.03%	8.54%	81	5,752	16.39%	9.84%
49	Three Mile Island	Pennsylvania	41	297	13.25%	7.92%	448	5,941	10.89%	8.53%
20	Waterford	Louisiana	18	285	9.50%	7.66%	487	6,269	7.39%	-6.87%
2	Farley	Alabama	5	358	5.15%	7.17%	94	6,642	8.41%	4.86%
47	Peach Bottom	Pennsylvania	10	317	12.96%	6.53%	1008	5,794	7.73%	7.24%
31	Fort Calhoun	Nebraska	5	173	13.60%	6.25%	269	5,587	11.30%	9.49%
17	Duane Arnold	Iowa	22	291	16.84%	6.13%	129	5,795	10.26%	6.52%
44	Perry	Ohio	22	141	10.27%	5.49%	531	4,905	1.28%	-4.98%
10	Vogtle	Georgia	2	325	19.61%	5.10%	139	6,248	13.39%	6.99%
51	Oconee	South Carolina	8	309	6.44%	5.08%	276	6,175	18.08%	11.79%
22	Pilgrim	Massachusetts	15	453	11.69%	5.08%	747	4,394	5.87%	1.68%
36	FitzPatrick	New York	55	292	5.34%	4.88%	212	4,938	-1.54%	0.26%
1	Browns Ferry	Alabama	10	302	6.57%	4.28%	207	6,353	13.49%	8.91%
27	Prairie Island	Minnesota	5	226	7.74%	3.97%	450	5,643	15.12%	6.71%
24	Fermi	Michigan	24	268	5.05%	3.76%	1312	4,552	1.18%	-3.85%
12	Byron	Illinois	5	300	16.04%	3.71%	199	5,627	9.47%	6.22%
48	Susquehanna	Pennsylvania	14	287	2.93%	3.37%	303	5,863	1.28%	1.77%
32	Seabrook	New Hampshire	30	270	15.59%	3.19%	868	4,195	6.29%	2.27%
19	River Bend	Louisiana	5	239	12.52%	2.59%	161	6,869	12.37%	10.44%
6	Millstone	Connecticut	32	266	-3.04%	2.45%	395	5,103	4.44%	4.92%
21	Calvert Cliffs	Maryland	10	258	44.17%	0.16%	438	6,082	9.93%	7.09%
23	D.C. Cook	Michigan	16	183	4.08%	-0.61%	227	5,853	6.46%	1.38%
43	Davis-Besse	Ohio	6	416	2.18%	-0.80%	363	4,476	0.78%	-0.33%
9	Hatch	Georgia	3	333	11.26%	-0.97%	72	6,586	16.75%	6.89%
37	Ginna	New York	9	374	-1.26%	-1.00%	264	4,515	3.35%	1.47%
38	Indian Point	New York	9	374	-1.26%	-1.00%	3760	5,810	8.03%	1.51%
16	Quad Cities	Illinois	7	272	-2.90%	-1.65%	154	5,824	1.50%	-0.59%
13	Clinton	Illinois	4	254	0.22%	-2.12%	154	5,851	5.33%	5.33%
18	Wolf Creek	Kansas	3	654	5.49%	-3.14%	39	5,973	5.39%	-1.80%
25	Palisades	Michigan	8	183	1.35%	-3.36%	226	6,785	11.05%	1.93%
30	Cooper	Nebraska	3	670	-7.88%	-4.26%	38	6,168	-1.30%	-4.10%
61	Point Beach	Wisconsin	5	192	-0.41%	-5.13%	114	5,150	10.38%	5.17%
45	Beaver Valley	Pennsylvania	33	279	-2.71%	-6.24%	808	5,647	-2.36%	-3.71%
57	South Texas	Texas	1	25	-11.91%	-6.81%	52	4,663	8.25%	0.31%
28	Grand Gulf	Mississippi	4	662	4.99%	-15.75%	56	6,076	1.99%	-3.82%
5	Diablo Canyon	California	7	358	3.29%	-31.32%	90	3,048	13.32%	9.46%
	All 61 plants in operation (0-50)		881	17,411	15.60%	14.00%	21625	286,714	9.75%	5.47%
	All 61 plants in operation-outside		21625	286,714	9.75%	5.47%	51432	3,510,028	14.64%	9.07%

Fig. 2.7 Percent change of the total population living in census tracts within a 10-mile radius and 50-mile radius, as measured between the census-tract centroid point and the nearest NPP, during the periods 1990–2000 and 2000–2010, sorted according to percent change within a 50-mile radius

NPP. The largest negative percent change, −7%, occurred among the communities within a 50-mile radius of Waterford Steam Electric Station, Louisiana.

Overall, the total population in areas within a 10-mile radius of one of the 61 US-based commercial NPPs increased 14% during the period 2000–2010 and 15% during the period 1990–2000, while the total population in all other places outside of a 50-mile radius of any NPP saw increases of 38 and 23%, respectively, during the periods 1990–2000 and 2000–2010, respectively (Figs. 2.6, 2.7). The total population of communities sited within a 50-mile radius of any one of the 61 NPPs increased 6 and 10%, respectively, during the periods 2000–2010 and 1999–2000, respectively, while the total population in all other areas of the country increased 9 and 14%, respectively, during the periods 2000–2010 and 1990–2000, respectively (Figs. 2.6, 2.7).

2.8 Conclusion

This chapter provides information on demographic characteristics of population living around the 61 nuclear power plants which are in operations in the USA. There were about 88 million people living in areas within a 50-mile radius surrounding the nuclear power plants in 2010. Among them, 71% were White, 17% Black, 4.4% Asian, 6.9% listed as Other, 13.18% Hispanic, and 36.32% People of Color. Within a 10-mile radius area, the ten NPPs with the largest population range from 258,462 (Limerick Generating Station, Pennsylvania) to 120,876 (Seabrook Station, New Hampshire). The three smallest communities range from 2,654 (South Texas Project, Texas) to 6,007 (Columbia Generating Station, Washington). Within 50-mile radius areas, the ten NPPs with the largest population range from 15,118,181 (Indian Point Nuclear Generating, New York) to 3,130,058 (Pilgrim Plant, Massachusetts). The three smallest communities range from 111,113 (Cooper Nuclear Station, Nebraska) to 194,963 (Arkansas Nuclear One, Arkansas). From a nuclear emergency management point of view, these data could be interpreted as challenges to effectively carry out various activities in response to any of the potential nuclear emergencies associated with the NPPs. For example, if an emergency occurs at Limerick Generating Station in the Pennsylvania plant, the challenge is to evacuate a population of 258,462. Protecting a quarter of a million people living within a 10-mile radius poses tremendous challenges for federal, state, and local governments. In the previous three historic nuclear power accidents (discussed in Chap. 1 under the topic 3.1), evacuation orders actually took place many hours after the core-meltdown accident occured. Therefore, a quarter million people could face potential risks of exposure to the nuclear radiation materials before they have been officially evacuated if an accident takes place at the Limerick Generating Station. When looking at the larger area evacuation of within 50-mile radius, the impacted population size could be as big as 15 million if the nuclear accident takes place at the Indian Point Nuclear Generating, New York. Evacuation comes with a price, as evident by the evacuation-related deaths in the Fukushima nuclear power accident discussion in Chap. 1. The location could be more complex, and the deaths related with evacuation could be high in areas such as New York because of the high-rise buildings in close proximity and in a condensed area. According to the Fukushima accident, the evacuation-related deaths were about 10% (about 1,600 deaths out of 16,000 evacuees, Chap. 1, topic 3.1). If this death rate applies to the potential accident in New York, it could be 1.5 million deaths. In addition, the most vulnerable groups when addressing evacuation planning and risks of exposure are women, seniors, and children who live in areas surrounding the NPPs. As depicted in Table 2.1, 45 million women, 8 million seniors who are 65 or older, and 1.5 million children who are younger than 5 years live within 50-mile radius areas. This includes exposure to day-to-day low radiation risks as well as potential exposure to high-level radiation during a nuclear emergency. Low-level radiation from day-to-day operation could cause leukemia and cancer to child (Yoshimoto et al. 2004). Similarly, elderly populations, within the age range of 50–74 years, were noted to be linked to higher likelihoods of leukemia and cancer mortality rates (Yoshimoto et al. 2004).

References

Bouges S, Daures JP, Hebrard M (1999) Incidence of leukemia, lymphoma and thyroid cancers in children under 15 years old in the vicinity of Marcoule nuclear plant, 1985–95. Revue D Epidemiologie et de Sante Publique 47(3):205–217. Retrieved from WOS:000081141800002

Canu IG, Ellis ED, Tirmarche M (2008) Cancer risk in nuclear workers occupationally exposed to uranium—emphasis on internal exposure. Health Phys 94(1):1–17. Retrieved from WOS:000251786800001

Cardis E, Gilbert ES, Carpenter L, Howe G, Kato I, Armstrong BK et al (1995) Effects of low-doses and low-dose rates of external ionizing-radiation—cancer mortality among nuclear industry workers in 3 countries. Rad Res 142(2):117–132. Retrieved from ISI: A1995QV23100001

Cardis E, Kesminiene A, Ivanov V, Malakhova I, Shibata Y, Khrouch V et al (2005) Risk of thyroid cancer after exposure to I-131 in childhood. J National Cancer Institute 97(10):724–732. Retrieved from WOS:000229856500011

Denton HR (1987) The causes and consequences of the Chernobyl nuclear accident and implications for the regulation of U.S. nuclear power plants. Ann Nuclear Energy 14(6):295–315. doi:10.1016/0306-4549(87)90132-0. Retrieved from http://www.sciencedirect.com/science/article/B6V1R-498W137-18/2/cbefde3b6b634a1626c992bb4369ba8f

GeoLytics (2012a) 2000 Long form in 2010 boundaries. Retrieved from GeoLytics, Inc. http://www.geolytics.com/USCensus,2000-Long-Form-in-2010-Boundaries,Products.asp

GeoLytics (2012b) Census 1990 in 2010 boundaries. Retrieved from GeoLytics, Inc. http://www.geolytics.com/USCensus

GeoLytics (2012c) Census 2010/American community survey (ACS). Retrieved from GeoLytics, Inc. http://www.geolytics.com/USCensus,Census2010ACS,Categories.asp

Green Peace (2010) Nuclear power: decommissioning risks. Retrieved from Green Peace USA. http://www.greenpeace.org/usa/en/media-center/reports/Nuclear-Power-Decommissioning-Risks/

Mangano JJ, Sherman JD (2013) Long-term local cancer reductions following nuclear power plant shutdown. Biomed Int 1(4). Retrieved from http://www.bmijournal.org/index.php/bmi/article/viewFile/115/82

Kyne D (2014) Environmental justice issues in communities hosting US nuclear power plants (Order No. 3617964). Available from ProQuest Dissertations & Theses Global. (1528556250). Retrieved from http://ezhost.utrgv.edu:2048/login?url=http://search.proquest.com.ezhost.utrgv.edu:2048/docview/1528556250?accountid=7119

Mariottee M (2006) The NRC's reactor licensing process: an overview. Nuclear Information and Resource Services (NIRS)

National Research Council (U.S.), & Committee on the Analysis of Cancer Risks in Populations near Nuclear Facilities (2012) Analysis of cancer risks in populations near nuclear facilities phase 1. Retrieved from National Academies Press

National Research Council (U.S.), & Committee on the Analysis of Cancer Risks in Populations near Nuclear Facilities (2014) Analysis of cancer risks in populations near nuclear facilities phase 2, pilot planning. Retrieved from http://www.nap.edu/catalog/18968/analysis-of-cancer-risks-in-populations-near-nuclear-facilities-phase

Nebraska Energy Quarterly (1997) State's first nuclear plant buried near Lincoln. Retrieved from Nebraska Energy Quarterly Newsletter

NIRS (2013) Pettition for rulemaking to improve emergency planning regulations (10.C.F.R. 50.47). Retrieved from http://www.nirs.org/reactorwatch/emergency/petitionforrulemaking22012.pdf

NRC (2012) Information Digest, 2011–2012. Retrieved from http://www.nrc.gov/reading-rm/doc-collections/nuregs/staff/sr1350/

Nuclear Heritage (2013) Anti-nuclear movement in the United States. Retrieved from Nuclear Heritage http://www.nuclear-heritage.net/index.php/Anti-nuclear_movement_in_the_United_States

Ottaviani J, Wehe DK (1989) Measured radiation fields in commercial nuclear power plants. Ann Nuclear Energy 16(2):91–99. doi:10.1016/0306-4549(89)90033-9. Retrieved from http://www.sciencedirect.com/science/article/B6V1R-498W1JD-18G/2/da96f79562075196ed555b4043b4af32

Ramana MV (2009) Nuclear power: economic, safety, health, and environmental issues of near-term technologies. Ann Rev Environ Resour 34:127–152. Retrieved from ISI:000272082000007

Sexton K, Adgate JL (1999) Looking at environmental justice from an environmental health perspective. J Exposure Anal Environ Epidemiol 9(1):3–8. Retrieved from WOS:000086103000001

The US Census Bureau (2010a) 2010 TIGER/Line shapefiles technical documentation. Retrieved from The US Census Bureau http://www.census.gov/geo/maps-data/data/pdfs/tiger/tgrshp2010/TGRSHP10SF1.pdf

The US Census Bureau (2010b) TIGER/Line® shapefiles pre-joined with demographic data: 2010 census summary file 1 demographic profile (DP1). Retrieved from The US Census Bureau http://www.census.gov/geo/maps-data/data/tiger-data.html

US EPA (2013) Environmental justice. Retrieved from US Environmental Protection Agency http://www.epa.gov/environmentaljustice/

US NRC (2004) Nuclear power plant licensing process. Retrieved from US Nuclear Regulatory Commission www.nrc.gov

US NRC (2011) Nuclear regulatory commission. Retrieved from US Nuclear Regulatory Commission www.nrc.gov

US NRC (2013a) Decommissioning nuclear power plants. Retrieved from US Nuclear Regulatory Commission http://www.nrc.gov/reading-rm/doc-collections/fact-sheets/decommissioning.html

US NRC (2013b) Emergency planning zones. Retrieved from US Nuclear Regulartory Commission

US NRC (2016) Emergency planning zones. Retrieved from US Nuclear Regulartory Commission http://www.nrc.gov/about-nrc/emerg-preparedness/about-emerg-preparedness/planning-zones.html

Welsome E (1999) The plutonium files America's secret medical experiments in the Cold War. Delacorte Press, New York, N.Y

Yoshimoto Y, Yoshinaga S, Yamamoto K, Fijimoto K, Nishizawa K, Sasaki Y (2004) Research on potential radiation risks in areas with nuclear power plants in Japan: leukaemia and malignant lymphoma mortality between 1972 and 1997 in 100 selected municipalities. J Radiol Protect 24(4):343–368. Retrieved from ISI:000226325000003

Chapter 3
Nuclear Power Emergencies and Their Management Process

3.1 Disaster and Emergency Management Process

People in various parts of the world see disaster events that bring causalities, injuries, deaths, and damages to property and environments, in many different scales. Their encounters may vary from one person to another; some might have direct experience in a disaster event, where others might have seen the events on any media or live broadcast from news channels. Disaster events are so frequent that people feel that they become a part of everyday life. For example, New York City was hard hit by Hurricane Sandy, which is also known as Superstorm Sandy, in October 29, 2012. The possibility for this event to take place was discovered on October 22, 2012, 1 week before the event actually occurred. The hurricane produced 147 deaths in the northeast USA, Canada, and the Caribbean; the dead toll was 48 in New York, 12 in New Jersey, 5 in Connecticut, 2 in Pennsylvania, and five in other states. The surge level was measured as 13.88 ft. at Battery Park, and the wave was measured at 32.5 ft. at New York Harbor; many important organizations were shut down including The New York Stock Exchange, US federal offices, schools, and United Nations headquarters in Manhattan. More than 11 million commuters were without access to transportation and about 7.9 million businesses and households without electric power in 15 states; 1,400 guardsmen are in Massachusetts, 6,700 National Guard are on active duty; and 110 homes burn in the Breezy Point neighborhood of Queens, New York. Approximately, 9,000 people in 13 states stayed in 71 Red Cross operated shelters (CNN 2015). The damage was estimated at $19 billion (Rosenzweig and Solecki 2014). According to the National Hurricane Center, Sandy was ranked as the second costliest tropical cyclone, after Hurricane Katrina in 2005 (CNN 2015). Hurricane Sandy demonstrates the extent to which a disaster event could bring about causalities, injuries, deaths, damages, and destruction. To have a closer look at managing natural disasters, it is imperative to understand the three key terms, namely disasters, hazard, and vulnerability.

D. Kyne, *Nuclear Power Plant Emergencies in the USA*,
DOI 10.1007/978-3-319-50343-1_3

Disasters are deadly, destructive, and disruptive events that occur when a hazard interacts with human vulnerability. Disasters are significant societal events that injure and kill people, damage infrastructure, and personal property, and complicate the routine activities people undertake on a daily basis (e.g., bathing, cooking, travelling, going to school, working, etc.). **Hazard** is the threat or trigger that initiates a disaster. Hazards include natural, technological, or anthropogenic (human-induced) agents like earthquakes, industrial explosions, and even terrorist attacks that negatively affect people or critical infrastructure. **Vulnerability**, on the other hand, refers to the proneness of people to disasters based on factors such as their geographic location, exposure of property, and level of income or other social variables. The ability of individuals, organizations, and communities to deal with disaster also determines the degree of vulnerability. Vulnerability is therefore closely related to the human element of disasters, while hazards may or may not always have a direct social cause (McEntire 2007).

To minimize the negative impacts from a disaster, effectively managing the disaster process is essential. The disaster management process consists of five phases: prediction, warning, emergency relief, rehabilitation (short-term), and reconstruction (long-term) (Moe and Pathranarakul 2006). In general, mitigation and preparedness, response, and recovery activities are conducted in the predication phase, rehabilitation and reconstruction phases, and warning and emergency relief phases, respectively (Moe and Pathranarakul 2006). The authors perceived that mitigation and preparedness activities are normally conducted before any disaster event actually takes place, while response activities are necessary during the event, and recovery activities are vital for rehabilitation and reconstruction to bring back the mental and physical damages of people and materials to their normalcy.

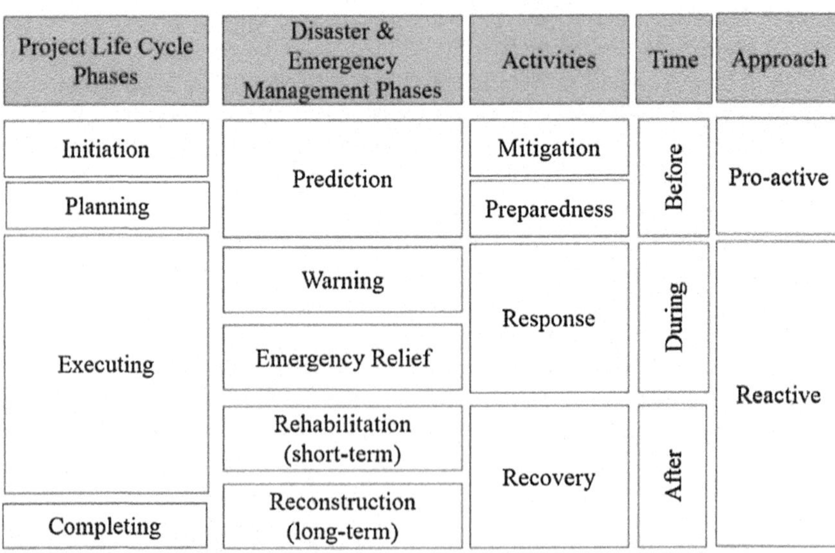

Fig. 3.1 An integrated approach to disaster management process. *Source* Adapted from Moe and Pathranarakul (2006) "Reprinted from Disaster Prevention and Management: An International Journal. Used with permission. All Rights Reserved."

Referring to the Hurricane Sandy event, the likelihood of the superstorm landing in New York was known a week before the event actually took place. Mitigation activities included closing of metro and subway system, a day prior to the storm. In addition, the National Guard was on standby, the preparing of shelters by Red Cross began, and other predisaster measures. An effective disaster management system requires the integration of both proactive approaches, which includes mitigation and preparedness activities and reactive approach which consists of response and recovery activities (Fig. 3.1).

3.2 Nuclear Power Emergency Management Process

As discussed in Chaps. 1 and 2, nuclear power plants are associated with inevitable risks. Chapter 1 discusses three historic nuclear power accidents, whereas Chap. 2 identifies populations that are at risk for potential nuclear power emergencies associated with 61 nuclear power plants in the USA. There are three nuclear power-induced disasters in nuclear power history, demonstrated as severe and unimaginable impacts that nuclear power plants might have on people living around the plant and environment. The most recent nuclear power-induced disaster, Fukushima Daiichi Nuclear Power Plant, which occurred in March 2011, caused severe damages, some irreversible and devastating. According to the Japanese science ministry, the accident left 11,580 square miles of land contaminated with the long-lived radioactive cesium and some 4,500 square miles (the size of Connecticut) had radiation levels higher than allowable exposure rate of 1 mSV (millisievert) per year (Starr 2016). In addition, some parts of the land are declared as “exclusion” zones because of high-level radiation for human habitation. The exclusion zones that were declared include 310 square miles of plant location and nearby areas where more than 159,128 people were evacuated. People living in the area received small amounts of moneys to cover their living costs, but they are being forced to still pay mortgages for their houses in the exclusion zones; the value of the exclusion zones could not be precisely estimated, but the rough estimated value reached to between \$250 and \$500. Radioactive cesium entered in the ecosystem, and it has been detected in various foods in Japan. While the chances to decontaminate the exclusion zones are very slim, there are additional areas that are contaminated, including homes and land due to rainwater flow. There are 733,000 curies of radioactive cesium, the largest discharge of radioactive materials pumped into the Pacific Ocean, after 15 months of the disaster. Reactor 4 building holds a spent fuel pool with 1,532 nuclear fuel assemblies, about 10 times more radioactive cesium than that was released by the Chernobyl disaster, a common pool containing 6,000 fuel assemblies located 50 ft. from building 4 (Starr 2016).

Unlike other disasters, nuclear power-induced disasters cause severe and irreversible damages. To minimize the negative impacts, effective nuclear power emergency management is essential. Like disaster management phases, nuclear power plant-induced disaster management includes four phases, namely prediction,

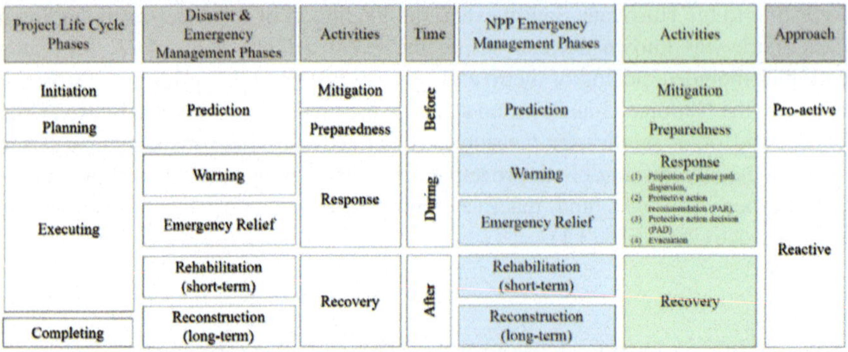

Fig. 3.2 An integrated approach to nuclear power plant-induced disaster management process. *Source* Kyne (2015) "Reprinted from Journal of Emergency Management. Used with permission. All Rights Reserved."

warning, emergency relief, rehabilitation, and reconstruction (Fig. 3.2) (Kyne 2015). Activities carried in nuclear power plant-induced disaster management are similar to those of disaster management phases, with the exception of response activities. There are four response activities, namely (1) projection of plume path dispersion, (2) protective action recommendation (PAR), (3) protective action decisions (PADs), and (4) evacuation.

Among the four phases, warning and emergency relief are very important, in order to save lives and minimize the negative impacts from nuclear power plant-induced disasters. This is due to the fact that there is no lead time in nuclear power emergencies. Hurricane Sandy, which was the most powerful storm, provided 7 days in advance of its likelihood of landing in New York City. This lead time of 7 days provided to do activities for mitigation and preparedness, before the disaster actually occurred. In nuclear emergencies, the response activities should begin soon after an incident is initiated, at the nuclear power plant. However, sometimes, an incident could go unnoticed for a period of time. For example, a crack in the pressure vessel in which nuclear fuel rods sit went unnoticed until the next fuel recharge cycle, which occurred every year and half at the Palo Verde Nuclear Power Generating Station (PVNGS), in October 2013 (Alltucker 2006).

According to the NRC (US NRC 2011a), US commercial nuclear power plants are required to notify the public when an accident occurs. For the purpose of notification, emergency levels are classified into four, namely (1) notification of unusual event (NOUE), (2) alert, (3) site area emergency (SAE), and (4) general emergency (GEm). The description of the four emergency classification and examples causing the four emergencies are discussed under the next subtitle.

An SAE is an event that requires protecting the public, whereas a GEm is an event such as a core-damage accident that requires protection of public. The release of radioactive materials from a SAE may not exceed the Environmental Protection

Agency (EPA) Protective Action Guides (PAGs), whereas the release resulting from a GEm may. In either event, the nuclear power plant is required to provide a protective action recommendation (PAR) to state, local, and tribal agencies within 15 min of the incident. To meet this requirement by EPA PAGs, the nuclear power plant must, first and foremost, project the radioactive plume path dispersion under the current and projected weather conditions, including wind speed and direction, humidity, and precipitation.

To accurately project and assess the source term dose and release pathway, the Nuclear Regulatory Commission utilizes the Radiological Assessment System for Consequence Analysis (RASCAL) Source Term to Dose (STDose) (RSICC 2013). The computer program was preinstalled with information on the type of reactor, the reactor's power output capacity, peak rod burn in the reactor, the discharge burn-up projected to occur in materials housed in spent fuel storage, and so on. However, the meteorological information, such as that on wind direction and speed, stability class, precipitation, and air temperature, must be entered manually. The RASCAL 4.3 program's method of calculating the source term is based on the methods documented in McKenna and Glitter (1988). The inhalation dose factors used in my calculations are based on the recommendations of the International Commission on Radiological Protection (ICRP 1977). Radiation dose was computed as a total effective dose equivalent (TEDE), which is defined as "the sum of the deep-dose equivalent (EDE) (for external exposures) and the committed effective dose equivalent (CEDE) (for internal exposures)" (US NRC 2011).

Once, the projection of the plume path dispersion and radioactivity was completed, the protective action recommendation (PAR) was generated. Then, the nuclear power plant sends the PAR to local radiological association, which conducts similar plume path dispersion analysis and generates PAR based on the study and recommendation done by the nuclear power plant. Then, the PAR is sent to local authorities to make protective action decisions (PADs). Based on the protective action decisions, evacuation activities are carried out accordingly. After the evaluation, activities to clean the radioactive contaminated areas will be followed. Cleanup activities pose a great challenge, and it is a time-consuming process.

3.3 Nuclear Power Emergency and Response

According to the NRC, there are four emergency events that could take place at a nuclear power plant (US NRC 2014a). An emergency classification is defined as "a set of names or titles established by the US Nuclear Regulatory Commission (NRC) for grouping off-normal events or conditions according to (1) potential or actual effects or consequences, and (2) resulting on-site and off-site response actions (US NRC 2014a)." The agency classified four levels of emergency in ascending order of severity:

(1) Notification of unusual event (NOUE)
(2) Alert
(3) Site area emergency (SAE)
(4) General emergency (GE)

(1) Notification of unusual event (NOUE or UE): "Events are in progress or have occurred which indicate a potential degradation of the level of safety of the plant or indicate a security threat to facility protection has been initiated. No releases of radioactive material requiring off-site response or monitoring are expected, unless further degradation of safety systems occurs (US NRC 2014a)." Some incidents causing this level of emergency are (1) the metal tubes holding uranium fuel pellets damage detected from reactor coolant samples, (2) low water level in spent fuel pool, (3) main turbine damage, (4) leaks from the reactor coolant system, and (5) loss of communication equipment (Table 3.1) (US GAO 2001). This is the lowest level of nuclear emergency, which requires all emergency and staff at the nuclear power plant to enter into a state of readiness, to handle systematically the current event.

(2) Alert: "Events are in progress or have occurred, which involve an actual or potential substantial degradation of the level of safety of the plant or a security event that involves probable life-threatening risk to site personnel or damage to site equipment because of what is considered 'hostile action'. Any releases are expected to be limited to small fractions of the Environmental Protection Agency (EPA) protective action guides (PAGs) (US NRC 2014a)." A few example situations causing this emergency event are (1) high radiation in primary containment, (2) a water leak more than 50 gallons per minute from the reactor coolant system, (3) high radiation level in vital areas, (4) main turbine damage, (5) unauthorized personnel entering the restrictive areas, and (6) an explosion taking place (Table 3.1) (US GAO 2001). This alert provides a warning to site emergency personnel to become ready to respond if the situation becomes more serious and informs current development to off-site authorities (US NRC 2014a).

(3) Site area emergency (SAE) : "Events are in progress or have occurred which involve actual or likely major failures of plant functions needed for protection of the public or hostile action that results in intentional damage or malicious acts; (1) toward site personnel or equipment that could lead to the likely failure of or; (2) that prevent effective access to, equipment needed for the protection of the public. Any releases are not expected to result in exposure levels which exceed EPA PAG exposure levels beyond the site boundary (US NRC 2014a)." Some examples of accidents causing this type of emergency are (1) failure in normal method of cooling the reactor, (2) alternative electrical power lost for more than 15 min, and (3) unauthorized personnel entering the restrictive areas (Table 3.1) (US GAO 2001). The goal of declaring this type of emergency is to ensure that response center, monitoring teams, and evaluation teams are ready to move forward if the situations get more serious (US NRC 2014a). During this type of emergency, radioactive materials could be released, but response action by an off-site organization is not expected (US NRC 2014a).

Table 3.1 Emergency action level and examples

Emergency action level and definition	Examples
Unusual event • An extremely low-level emergency that poses no threat to public safety but warrants increased awareness on the part of utility and off-site personnel	• Reactor coolant samples indicate measurable damage to the metal tubes that hold the uranium fuel pellets • The water level in the spent fuel pool is low • Water leaks from the reactor coolant system in excess of The Nuclear Regulatory Commission's allowed limits • The main turbine is severely damaged. • Loss of all on- and off-site communication equipment occurs
Alert • A low-level emergency that poses no threat to public safety but for which precautionary mobilization of certain emergency response functions is appropriate. Any radioactive release is expected to be limited	• High radiation readings occur inside the primary containment—a large concrete and steel structure that surrounds the reactor vessel and its coolant system • Water leaks from the reactor coolant system at a rate greater than 50 gallons per minute • Radiation levels in one or more vital areas (equipment necessary for the safe operation and shutdown of the plant) are high • Damage to the main turbine results in damage to vital equipment • Unauthorized personnel enter the protected area (area that includes vital plant structures and is surrounded by a security fence) • A fire occurs that could potentially affect safety systems, and an explosion occurs that damages permanent plant equipment • A toxic or flammable gas is released in or near a vital area
Site area emergency • Plant conditions degrade to a point where full activation of response functions is warranted. Any radioactive release is not expected to exceed the Environmental Protection Agency's exposure levels, except near the site's boundary	• The normal methods of cooling the reactor—feed water system, main steam system, or steam generators—do not function • All alternating current electrical power to vital busses (on-site network to supply electric power) is lost for more than 15 min • Unauthorized personnel enter a vital area of the plant
General emergency • Actual or imminent substantial degradation or melting of the reactor with the potential for a significant radioactive release to the environment beyond the plant's boundary occurs.	• Two of the three fission product barriers (fuel cladding, reactor coolant system, and containment) fail. • All alternating current electrical power (on- and off-site) is lost and not expected to be available for an extended period of time. • Unauthorized personnel take over the control room so that the utility loses the ability to safely operate or shut down the plant.

Source US GAO (2001)

(4) General emergency: "Events are in progress or have occurred which involve actual or imminent substantial core degradation, or melting with potential for loss of containment integrity, in addition to possible hostile action that results in an actual loss of physical control of the facility. Releases can be reasonably expected to exceed the EPA's PAG exposure levels off-site for more than the immediate site area (US NRC 2014a)." A few situations that might lead to this type of emergency are (1) fuel cladding, reactor coolant system failure, (2) electrical power system failure on- and off-site, and (3) unauthorized personnel entering into restrictive areas (Table 3.1) (US GAO 2001). The purpose of declaring this type of emergency is "to initiate predetermined protective actions for the public, to provide continuous assessment of information from the licensee and off-site organizational measurements, to initiate additional measures as indicated by actual or potential releases, to provide consultation with off-site authorities, and to provide updates for the public through government authorities (US NRC 2014a)." During this type of emergency, radioactive materials could be released and off-site response organizations are expected to take response actions.

3.4 Projection of Plume Path Dispersion

As discussed earlier, the emergency response activities are critically important to achieve the ultimate goal of minimizing negative impacts and saving lives for individuals living near nuclear power plants. There are two conditions that challenge the key stakeholders to effectively carry out their response activities. First, there is no lead time provided before an emergency event actually takes place. Second, there are rigorous guidelines to strictly follow during the emergency response. The radiological emergency response constitutes four steps, namely projection of plume path dispersion, protective action recommendation, protective action decisions, and evacuation.

3.4.1 RASCAL Computer Code

First and foremost, the projection of the plume path dispersion and radiation level requires a computer code, namely the Radiological Assessment System for Consequence Analysis (RASCAL) Source Term to Dose (STDose) (RSICC 2013). The Radiation Protection Computer Code Analysis and Maintenance Program (RAMP) NRC has developed the RASCAL code for dose and consequence projection tool for 25 years, with a primary purpose of making projection for independent dose and consequence in case of radiological incidents and emergencies (RAMP 2016a). The tool could be used by the Protective Measures Team in the US Nuclear Regulatory Commission's (NRC's) Operations Center and licensed facilities. The RASCAL estimates radioactive effluents from various sources including nuclear power plants, spent fuel storage pools and casks, fuel cycle facilities, and radioactive material handling facilities (RAMP 2016a).

3.4.2 *Obtaining RASCAL Computer Code*

The computer code could be obtained from the RAMP at a cost of $5000 per reactor site and $1000 one-time free for documentation including RASCAL code documents (Fig. 3.3). Users will have to complete a "non-disclosure agreement" form, including your payment information in order to request use of the computer code. Users may send back the completed from via an email attachment to ramp@nrc.gov

Who	Services	Cost	How
	U.S. Users		
U.S. Companies U.S. NRC Licensees/Vendors	∘ RASCAL Code ∘ Documentation ∘ RASCAL Forums ∘ RASCAL Support	5000 USD/year per Company or 5000 USD/year per Reactor Site*	**Step 1:** Fill out, sign and email this non-disclosure agreement to ramp@nrc.gov or fax to 1-888-573-9473. **Step 2:** The RAMP contractor will present an invoice and contract to the entities. **Step 3:** Upon payment, you will be given access by the NRC, for the services listed.
	∘ RASCAL Code ∘ Documentation ∘ Read-only access to RASCAL Forums ∘ Installation Assistance Only	1000 USD one-time fee, future upgrades not included	
U.S. Federal/State /Local Government entities ****** U.S. Universities using codes for Research Purposes	∘ RASCAL Code ∘ Documentation ∘ RASCAL Forums ∘ Limited RASCAL Support	Free	**Step 1:** Fill out, sign and email this non-disclosure agreement to ramp@nrc.gov or fax to 1-888-573-9473. **Step 2:** Upon receipt of the NDA, you will be notified via email that you have access for the services listed.
All Others	Contact RAMP@nrc.gov		
	International Users		
Non-Regulatory Authority, Companies, Universities, Hospitals	∘ RASCAL Code ∘ Documentation ∘ RASCAL Forums ∘ RASCAL Support ∘ Limited training may be negotiated	5000 USD/year	**Step 1:** Fill out, sign and email this non-disclosure agreement to ramp@nrc.gov or fax to 1-888-573-9473. **Step 2:** The RAMP contractor will present an invoice and contract to the entities. **Step 3:** Upon payment, you will be given access by the NRC, for the services listed.
All Others	Contact RAMP@nrc.gov		

***For Domestic Licensee with multiple sites and corporate offices, fees may be adjusted to align with the RASCAL code use and needs.**

Fig. 3.3 Obtaining a RASCAL computer code. *Source* RAMP (2016a)

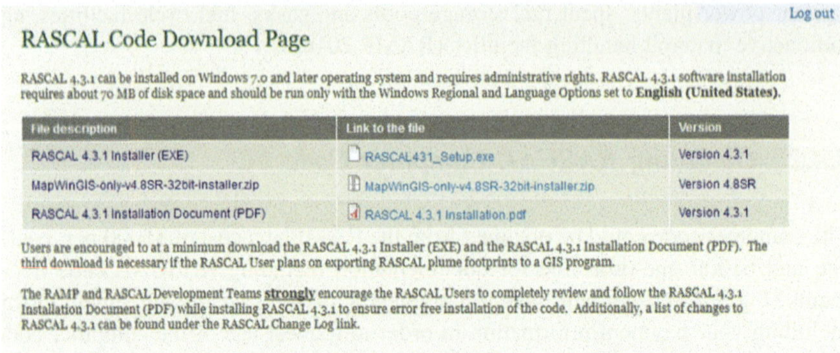

Log out

RASCAL Code Download Page

RASCAL 4.3.1 can be installed on Windows 7.0 and later operating system and requires administrative rights. RASCAL 4.3.1 software installation requires about 70 MB of disk space and should be run only with the Windows Regional and Language Options set to **English (United States)**.

File description	Link to the file	Version
RASCAL 4.3.1 Installer (EXE)	RASCAL431_Setup.exe	Version 4.3.1
MapWinGIS-only-v4.8SR-32bit-installer.zip	MapWinGIS-only-v4.8SR-32bit-installer.zip	Version 4.8SR
RASCAL 4.3.1 Installation Document (PDF)	RASCAL 4.3.1 Installation.pdf	Version 4.3.1

Users are encouraged to at a minimum download the RASCAL 4.3.1 Installer (EXE) and the RASCAL 4.3.1 Installation Document (PDF). The third download is necessary if the RASCAL User plans on exporting RASCAL plume footprints to a GIS program.

The RAMP and RASCAL Development Teams **strongly** encourage the RASCAL Users to completely review and follow the RASCAL 4.3.1 Installation Document (PDF) while installing RASCAL 4.3.1 to ensure error free installation of the code. Additionally, a list of changes to RASCAL 4.3.1 can be found under the RASCAL Change Log link.

Fig. 3.4 File download for RASCAL 4.3.1 program. *Source* RAMP (2016a)

or fax them to 1-888-573-9473. Additional information could be obtained at https://www.usnrc-ramp.com/RASCAL%20Obtaining%20Code.

3.4.3 Installing RASCAL

RAMP administrators process the application and will send a notification email when user's "user account" has been created. With the user account, users will have an access to RASCAL's most updated version and will be able to download (Fig. 3.4).

For the program installation, you will need at least three files to download:

(1) RASCAL431_Setup.exe
(2) MapWinGIS-only-v4.8SR-32bit-installer.zip
(3) RASCAL 4.3.1 Installation.pdf

The file with the name RASCAL431_Setup.exe is the program file user will need to run "step up" with administrative right. When user runs "setup," users will see instruction with which location to install. The default location is predetermined at C:\NRC\RASCAL43 (RAMP 2016b). The program requires Windows 7 or higher, with a disk space of 70 MB (RAMP 2016b). Users will need to unzip the file MapWinGIS-only-v4.8SR-32bit-installer.zip to install, and this file will help users export the map to ArcGIS shapefile.

While the program files are downloaded for installation, other technical document files (Fig. 3.5) should be downloaded for future reference. Among technical documents, there are two groups, namely models and methods and technical references.

Models and methods

1. NUREG-1940 Supplement 1—RASCAL 4.3 Description of Models and Methods
2. NUREG-1940—RASCAL 4 Description of Models and Methods

RASCAL Technical Documents Download Page

MODELS AND METHODS

File description	Link to the file
NUREG-1940 Supplement 1 - RASCAL 4.3 Description of Models and Methods	RASCAL 4.3 Description of Models and Methods.pdf
NUREG-1940 - RASCAL 4 Description of Models and Methods	NUREG-1940_RASCAL_4_Description_of_Models_and_Methods.pdf

TECHNICAL REFERENCES

File description	Link to the file
State-of-the-Art Reactor Consequence Analyses Project Volume 2: Surry Integrated Analysis	NUREG_CR-7110_Vol._2_Rev._1.zip
State-of-the-Art Reactor Consequence Analyses Project Volume 1: Peach Bottom Integrated Analysis	NUREG_CR-7110_Vol._1_Rev._1.pdf
State-of-the-Art Reactor Consequence Analyses (SOARCA) Report Part 2	NUREG-1935_Part2.zip
State-of-the-Art Reactor Consequence Analyses (SOARCA) Report Part 1	NUREG-1935_Part1.zip
Source Term Estimation Using MENU-TACT	NUREG_CR-4722 (ORNL-6314).pdf
Spent Fuel Pool Study SECY-13-0112	SECY-13-0112.pdf
Source Term Estimation During Incident Response to Severe Nuclear Power Plant Accidents	NUREG-1228.pdf
RTM Response Technical Manual 96 (NUREG/BR-0150, Vol. 1, Rev. 4)	NUREG-BR-0150_Vol.1_Rev.4.pdf
RASCAL 3.0.5 Workbook	NUREG-1889_RASCAL_3.0.5_Workbook.pdf
RASCAL 3.0.5: Description of Models and Methods	NUREG-1887_RASCAL_3.0.5_Description_of_Models_and_Methods.pdf
PLANNING BASIS FOR THE DEVELOPMENT OF STATE AND LOCAL GOVERNMENT RADIOLOGICAL EMERGENCY RESPONSE PLANS IN SUPPORT OF LIGHT WATER NUCLEAR POWER PLANTS	NUREG-0396_EPA_520_1-78-016.pdf
MANUAL OF PROTECTIVE ACTION GUIDES AND PROTECTIVE ACTIONS FOR NUCLEAR INCIDENTS	EPA_Manual_For_Nuclear_Incidents.pdf
Limiting Values of Radionuclide Intake And Air Concentration and Dose Conversion Factors For Inhalation, Submersion, And Ingestion - Federal Guidance Report No.11	EPA_Federal_Guidance_Report_11_EPA-520_1-88-02.pdf
Guidance for Industry KI in Radiation Emergencies — Questions and Answers (\\CDS029\CDERGUID\5386fnl.doc)	Guidance_for_Industry_KI_in_Radiation_Emergencies.pdf
FR Notice of FDA Guidance on Use of Potassium Iodide as a Thyroid Blocking Agent in Radiation Emergencies; Availability	FR_Vol.66_No.238_December_11_2001_pg_64046.pdf
Fission Product Removal in Engineered Safety Feature (ESF) Systems - Data Base Assessment and Suggested Experimental	NUREG-CR-3727_PNL-5050_R.pdf

Fig. 3.5 RASCAL technical documents download page. *Source* RAMP (2016a)

Technical references

1. State-of-the-Art Reactor Consequence Analyses Project Volume 2: Surry Integrated Analysis
2. State-of-the-Art Reactor Consequence Analyses Project Volume 1: Peach Bottom Integrated Analysis
3. State-of-the-Art Reactor Consequence Analyses (SOARCA) Report Part 2
4. State-of-the-Art Reactor Consequence Analyses (SOARCA) Report Part 1
5. Source Term Estimation Using MENU-TACT
6. Spent Fuel Pool Study SECY-13-0112

7. Source Term Estimation During Incident Response to Severe Nuclear Power Plant Accidents
8. RTM Response Technical Manual 96 (NUREG/BR-0150, Vol. 1, Rev. 4)
9. RASCAL 3.0.5 Workbook
10. RASCAL 3.0.5: Description of Models and Methods
11. Planning Basis for the Development of State and Local Government Radiological Emergency Response Plans in Support of Light-Water Nuclear Power Plants
12. Manual of Protective Action Guides and Protective Actions for Nuclear Incidents
13. Limiting Values of Radionuclide Intake And Air Concentration and Dose Conversion Factors For Inhalation, Submersion, And Ingestion—Federal Guidance Report No. 11
14. Guidance for Industry KI in Radiation Emergencies—Questions and Answers (\\CDS029\CDERGUID\5386fnl.doc)
15. FR Notice of FDA Guidance on Use of Potassium Iodide as a Thyroid Blocking Agent in Radiation Emergencies; Availability
16. Fission Product Removal in Engineered Safety Feature (ESF) Systems—Data Base Assessment and Suggested Experimental Program
17. Federal Guidance Report No. 12 External Exposure to Radionuclides in Air, Water, and Soil (EPA-402-R-93-081)
18. EPA AERSURFACE User's Guide
19. Criteria for Preparation and Evaluation of Radiological Emergency Response Plans and Preparedness in Support of Nuclear Power Plants
20. Consequence Study of a Beyond-Design-Basis Earthquake Affecting the Spent Fuel Pool for a U.S. Mark I Boiling Water Reactor
21. Cancer Risk Coefficients for Environmental Exposure To Radionuclides—Federal Guidance Report No. 13
22. Calculation of Releases of Radioactive Materials in Gaseous and Liquid Effluents from Pressurized Water Reactors (PWR-GALE Code)
23. Calculation of Releases of Radioactive Materials in Gaseous and Liquid Effluents from Boiling Water Reactors (GALE-BWR Code)
24. Accident Source Terms for Light-Water Nuclear Power Plants
25. 2013 EPA Protective Action Guides And Planning Guidance For Radiological Incidents

3.4.4 RASCAL Tools

The RASCAL constitutes two groups of tools, namely primary and additional (Fig. 3.6). The primary tools include four tools:

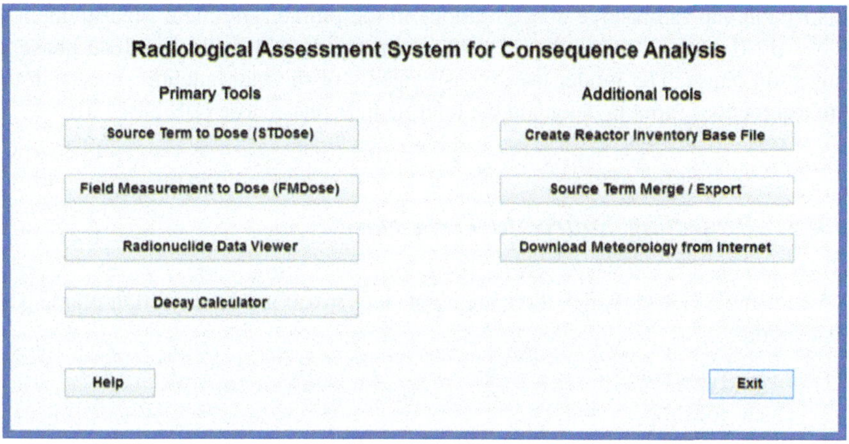

Fig. 3.6 Seven tools in the Radiological Assessment System for Consequence Analysis (RASCAL). *Source* (RAMP (2016a)

(1) Source Term to Dose (STDose)
(2) Field Measurement to Dose (FMDose)
(3) Radionuclide Data Viewer, and
(4) Decay Calculator.

The secondary tools include three tools:

(1) Create Reactor Inventory Base File,
(2) Source Term Merge/Export, and
(3) MetFetch

There are four phases in nuclear emergency, namely prerelease, plume (early), intermediate, and ingestion. The projection of plume dispersion for the first two phases must be done with the Source Term to Dose (STDose), whereas the latter two phases must be computed with the Field Measurement to Dose (Fig. 3.7) (FMDose) (RAMP 2016a)

When the radiological emergency is in either prerelease or plume (early), user uses the STDose tool to estimate radiation from a radioactive plume dispersed to the people downwind. The STDose first produces a time-dependent source term as a function of time for any release from the plant. This time-dependent radioactive release rate is used as input for an atmospheric dispersion and transport model to project plume

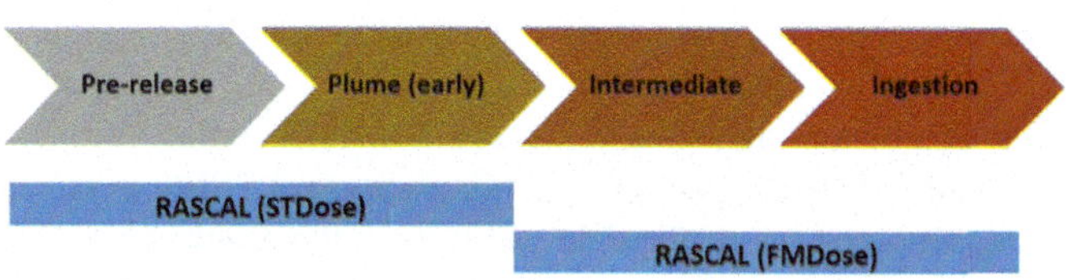

Fig. 3.7 Plume dispersion phases in a nuclear emergency. *Source* RAMP (2016a)

dispersion and radioactive concentrations in the plume. The Field Measurement to Dose (FMDose) is normally used when the accident is in intermediate phase or ingestion phase. The model uses measurements of the radionuclide activity in the environment as input to calculate the intermediate phase dose (US NRC 2015).

3.4.5 *Projecting Source to Term Dose*

When using STDose model, there are eight steps involving in projecting plume path dispersion:

(1) Event type,
(2) Event location,
(3) Source term,
(4) Release path,
(5) Metrology,
(6) Calculate doses,
(7) Detailed results, and
(8) Save case

To demonstrate how to project plume path dispersion and radioactive release in the plume downwind, a hypothetical core-meltdown accident at reactor Unit 1 at the Palo Verde Nuclear Generating Station (PVNG) is used as an example.

(1) Event Type: First, user selects the type of event. There are four event types preinstalled:

(1) Nuclear power plant,
(2) Spent fuel,
(3) Fuel cycle/UF6/criticality,
(4) Other radioactive material releases (Fig. 3.8).

In this case, nuclear power plant was selected.

(2) Event Location: Second, user determines the event location. In this exercise, Plao Verde Unit 1 was selected (Fig. 3.9).

(3) Source Term: Third, user selects the appropriate source term. In the RASCAL program, there are two groups of conditions, namely source term based on reactor conditions and source term based on nuclide specific data. RASCAL has two sequences for estimating releases from core-melt accidents. The long-term station blackout (LTSBO) sequence due to gradual decay of the reactor coolant as heat evaporates the coolant. The temperature increases in the core after the coolant evaporates. The LOCA sequence is based on an assumption that the accident is due to rapid loss of coolant due to a large break in the cooling system line. The temperature rapidly increases the core. In this exercise, LTSBO is selected and the user selects LTSBO and inputs the date and time of the accident with "No" to core recovered (Fig. 3.10).

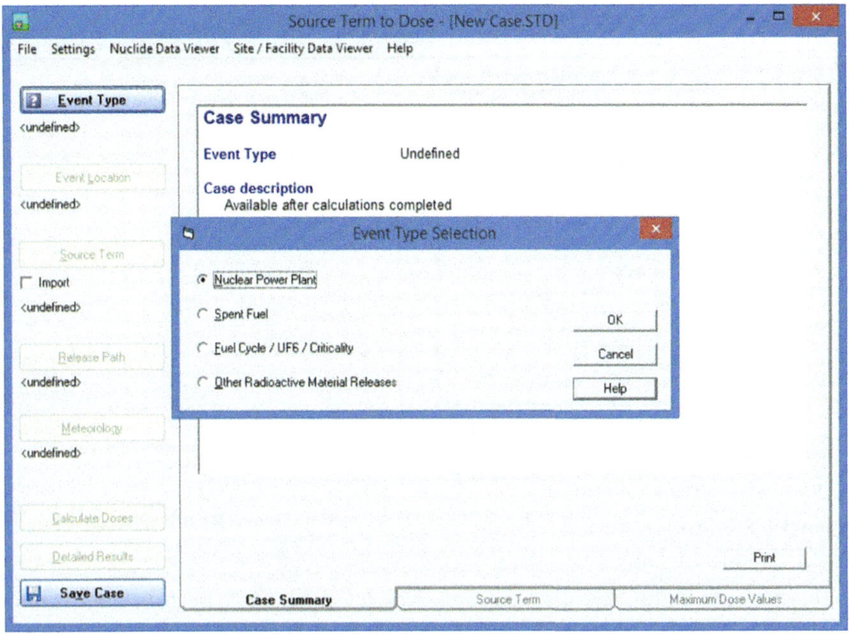

Fig. 3.8 Event type selection in RASCAL STDose model. *Source* RAMP (2016a)

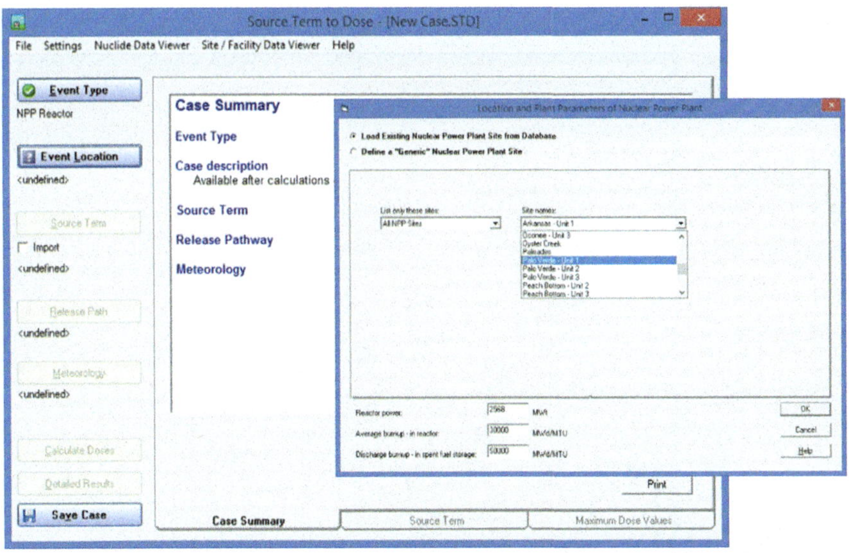

Fig. 3.9 Event location selection in RASCAL STDose model. *Source* RAMP (2016a)

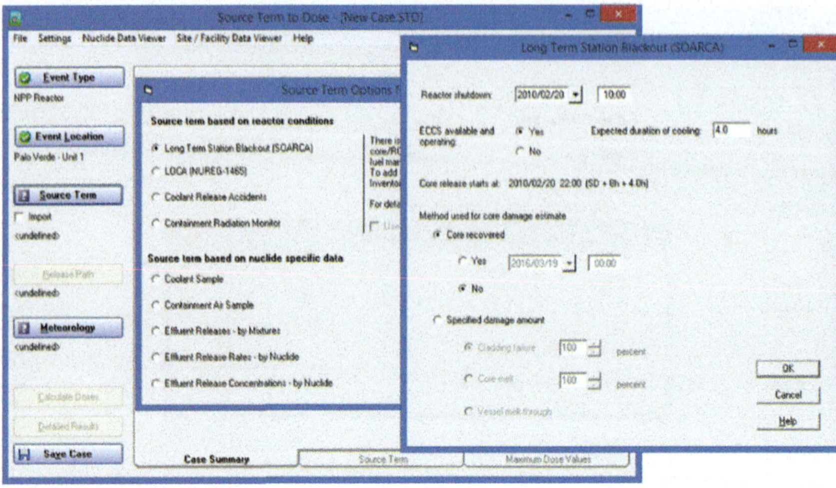

Fig. 3.10 Source term selection in RASCAL STDose model. *Source* RAMP (2016a)

(4) **Release Path**: The next step involves selecting a release path out of preinstalled three pathways, namely containment leakage/failure, steam generator tube rupture, and containment bypass. In this exercise, containment leakage/failure was selected (Fig. 3.11).

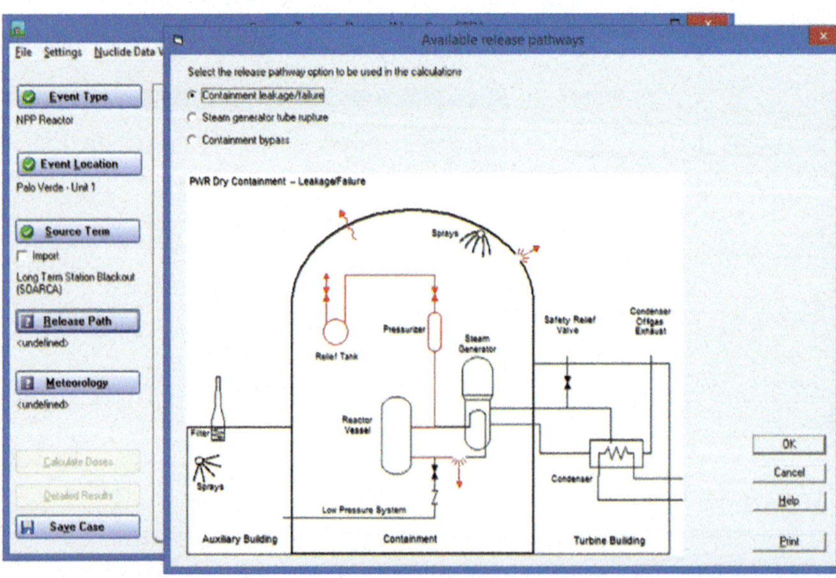

Fig. 3.11 Release path selection in RASCAL STDose model. *Source* RAMP (2016a)

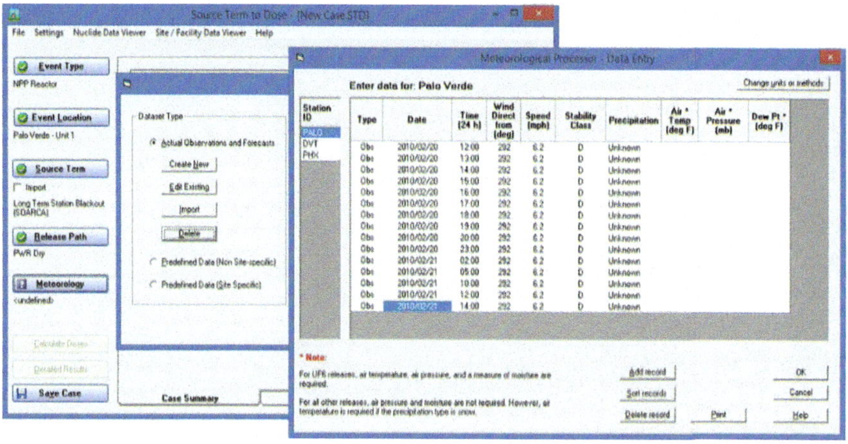

Fig. 3.12 Meteorology data selection in RASCAL STDose model. *Source* RAMP (2016a)

(5) Meteorology: Fifth, user selects meteorology dataset options: create new, edit existing, import, and delete. In this exercise, user selects create new and enters data as shown in Fig. 3.12. The meteorological data were selected from the joint frequency distribution (JFD) tables published in the *2010 Annual Radioactive Effluent Release Report* for Palo Verde Nuclear Generating Station: Units 1, 2, and 3 (Kutner 2010). Metrological information includes time, wind direction (degree), wind speed (mph), stability class, precipitation, and air temperature (degree F).

(6) Calculate Doses: Sixth, calculations are carried out for "close-in + out to 100 miles," and 24 h is set for "start of release to atmosphere plus." ICRP 26/30 was used for inhabitation dose factors in the calculations (Fig. 3.13).

(7) Detailed Results: Seventh, detailed results can be viewed in two types of display formats: from 100-mile calculation and from close-in calculation. Each of them can be viewed in two formats: footprint and numeric table (Fig. 3.14).

The results of "from close-in calculation" are displayed in footprint form (Fig. 3.15).

The results of "from 100-mile calculation" are displayed in footprint form (Fig. 3.16).

The source term results are displayed in a tab source term (Fig. 3.17).

The maximum values can be viewed for both close-in calculation and 100-mile calculation. Figure 3.18 shows a close-in dose. Underlined values exceed the recommended thresholds. TEDE refers "to the sum of inhalation dose and the external doses from cloudshine and 4-days of groundshine (US NRC 2013)." "Although, the PAG is expressed as a range of 1–5 REM, it is emphasized that, under normal conditions and evacuation of members of the general population should be initiated for most incidents at a projected dose of 1 REM (US EPA 2013)."

The maximum dose is viewed for 100-mile calculation (Fig. 3.19).

The case summary could be viewed in a case summary tab (Fig. 3.20).

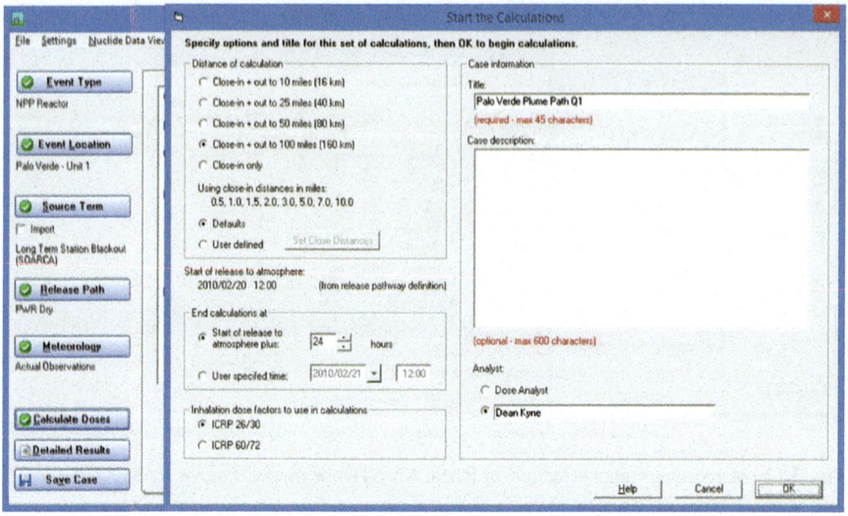

Fig. 3.13 Calculating doses in RASCAL STDose model. *Source* RAMP (2016a)

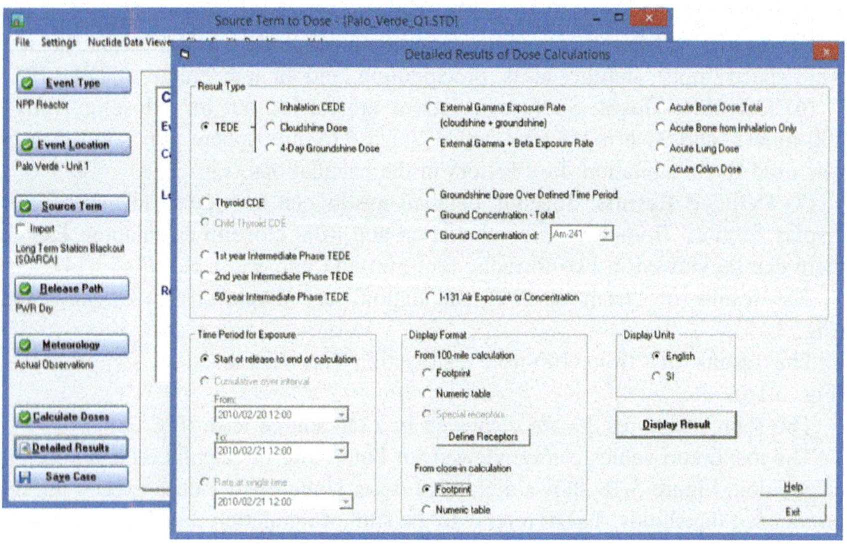

Fig. 3.14 Detailed results for close-in calculation to be displayed in footprint format. *Source* RAMP (2016a)

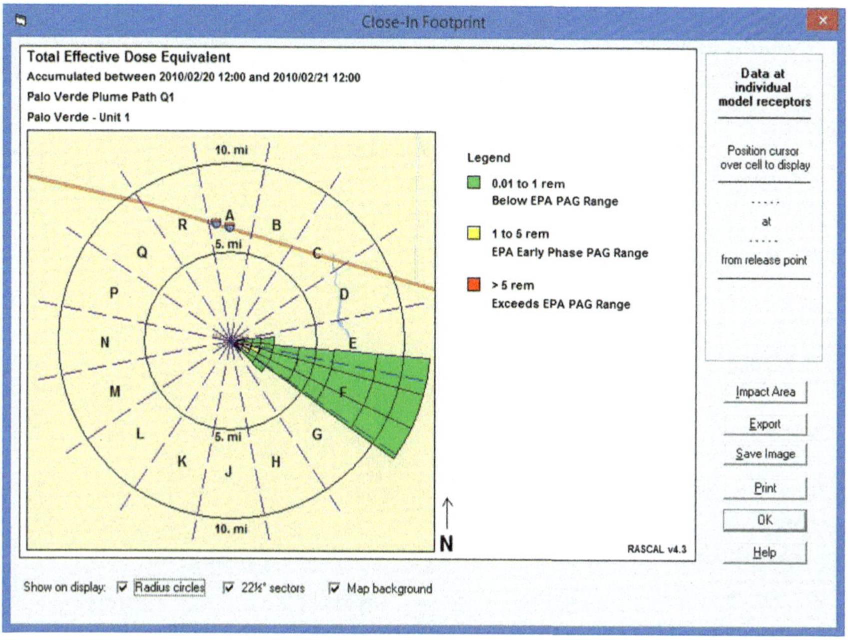

Fig. 3.15 Detailed results for close-in displayed in footprint format. *Source* RAMP (2016a)

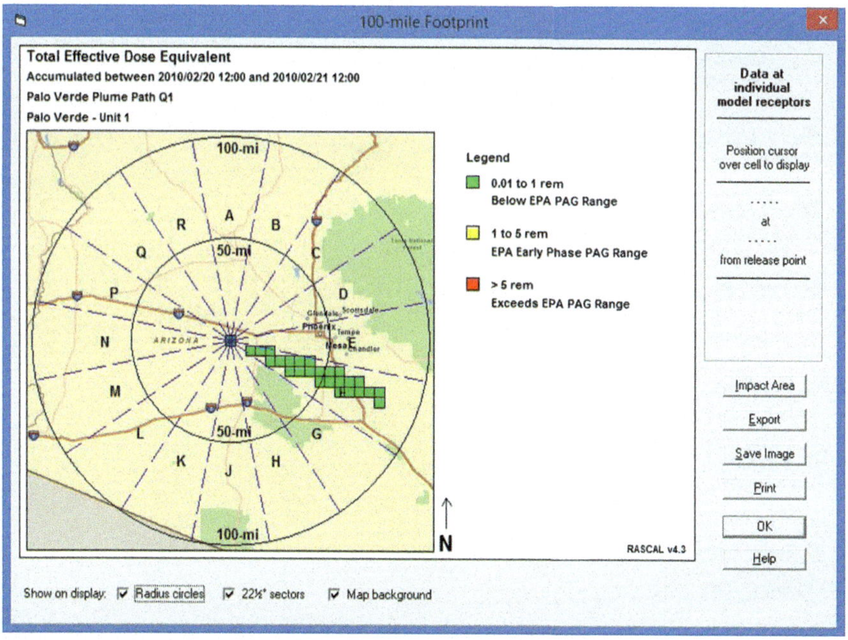

Fig. 3.16 Detailed results for 100-mile calculation to be displayed in footprint format. *Source* RAMP (2016a)

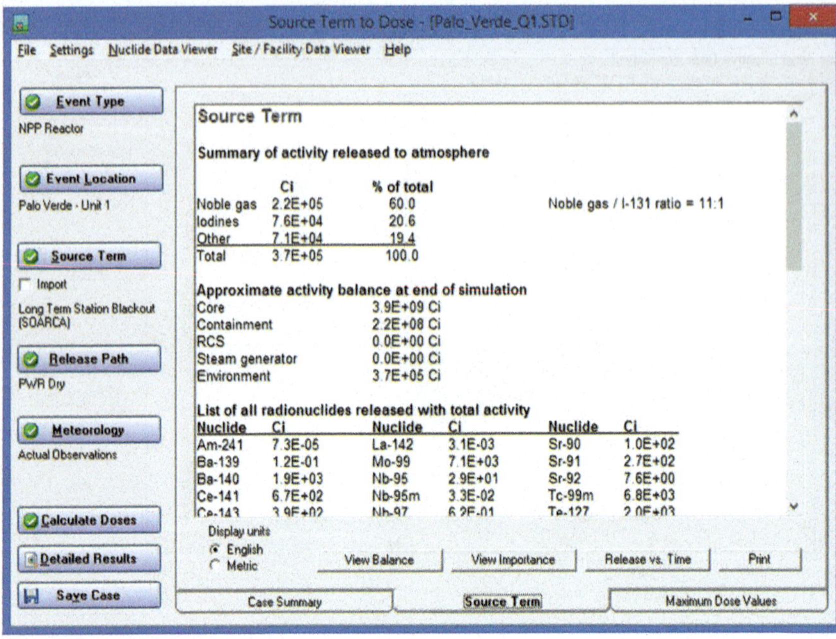

Fig. 3.17 Detailed results for source term displayed in source term tab under detailed results. *Source* RAMP (2016a)

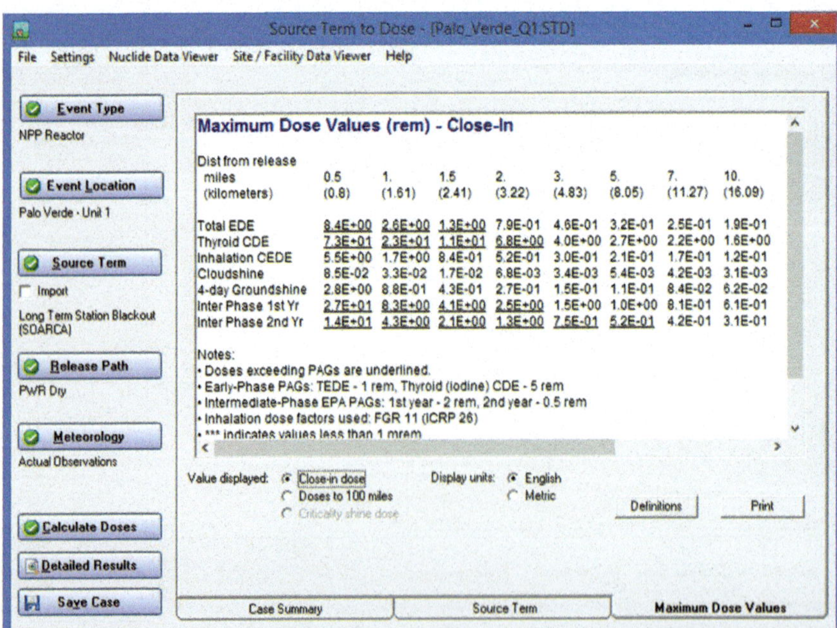

Fig. 3.18 Detailed results for maximum dose values for close-in values displayed in maximum dose value tab under detailed results tab. *Source* RAMP (2016a)

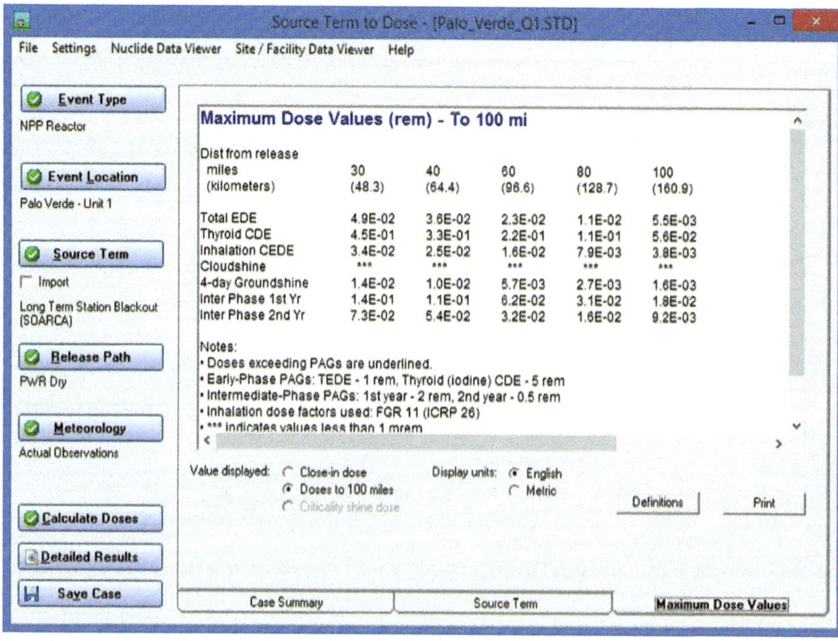

Fig. 3.19 Detailed results for maximum dose values for dose to 100-mile values displayed in maximum dose value tab under detailed results tab. *Source* RAMP (2016a)

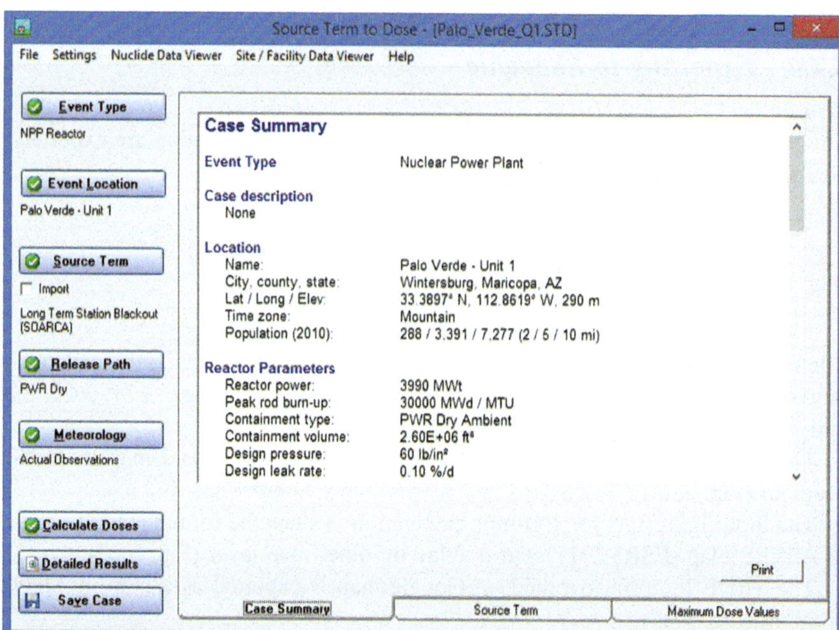

Fig. 3.20 Detailed results for case summary displayed in case summary tab under detailed results tab. *Source* RAMP (2016a)

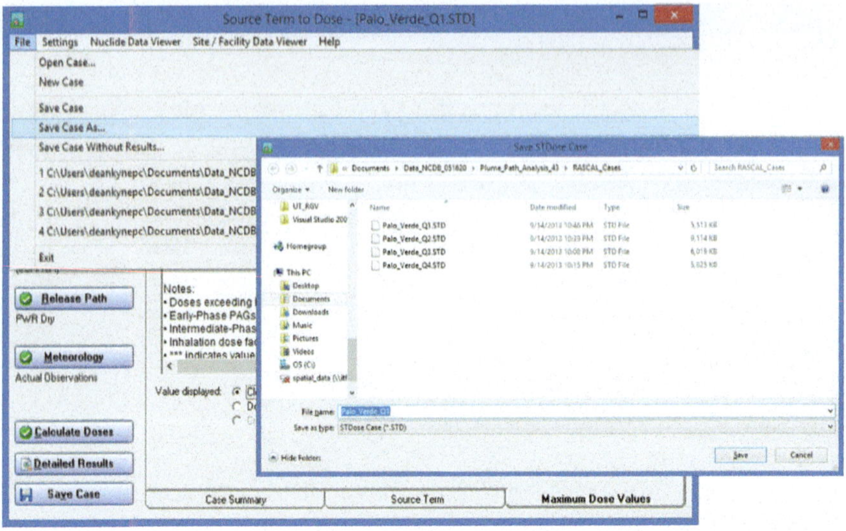

Fig. 3.21 Saving a case file in.STD extension in a preferred location. *Source* RAMP (2016a)

(8) Save Case: The calculations could be saved in a file with an extension ".STD" and could be recalled for future calculation or modification to compute (Fig. 3.21).

3.4.6 *Exporting to Shapefile*

The footprint files for close-in calculation and 100-mile calculation are exported to GIS shapefile (Fig. 3.22).

3.4.7 *Overlaying on Other Map Layers*

The footprint results for a close-in calculation in shapefile format could be added to ArcGIS map (ESRI 2011) and overlay on other map layer (Figs. 3.21, 3.22, 3.23 and 3.24).layer (Figs. 3.21, 3.22, 3.23 and 3.24).

The TEDE footprint overlaid on Google map is exported as a map in ArcGIS program (Fig. 3.24)

The footprint results for 100-mile calculation in shapefile format could be added to ArcGIS map (ESRI 2011) and overlay on other map layer (Fig. 3.25).

The TEDE footprint overlaid on Google map is exported as a map in ArcGIS program (Fig. 3.26)

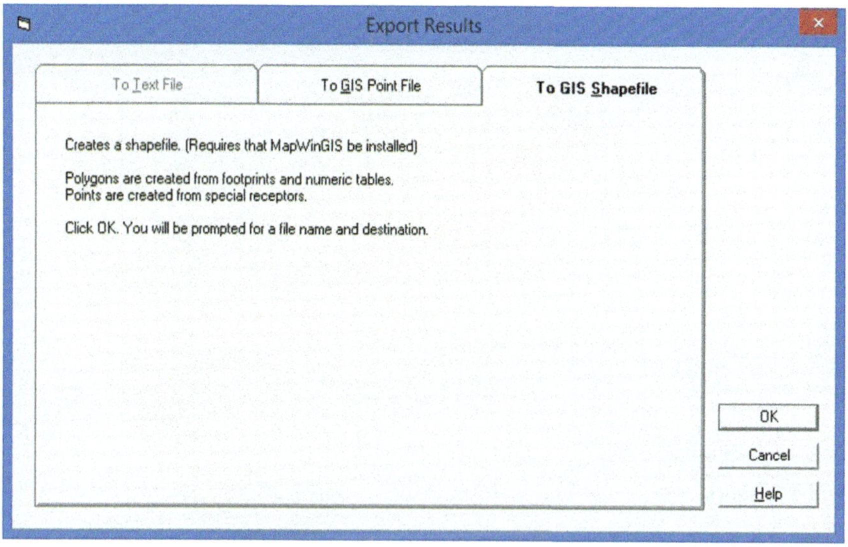

Fig. 3.22 Exporting the footprints into GIS shapefile. *Source* RAMP (2016a)

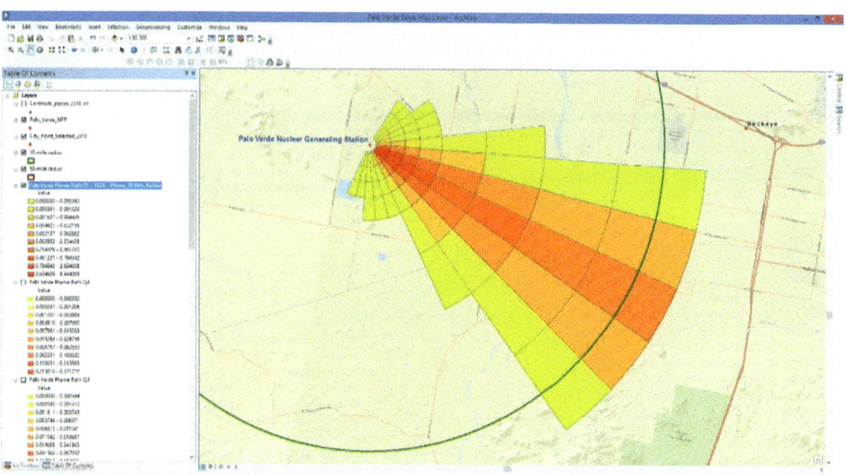

Fig. 3.23 Over laying close-in dose footprint GIS shapefile on Google Base map in ArcGIS program

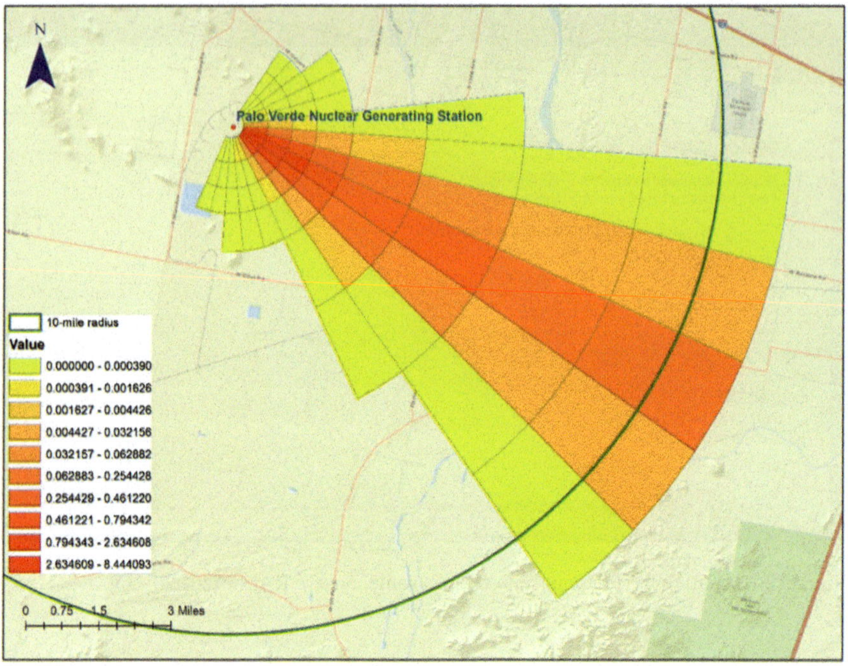

Fig. 3.24 A GIS map displaying projected close-in plume pathway overlaid on the Google street map

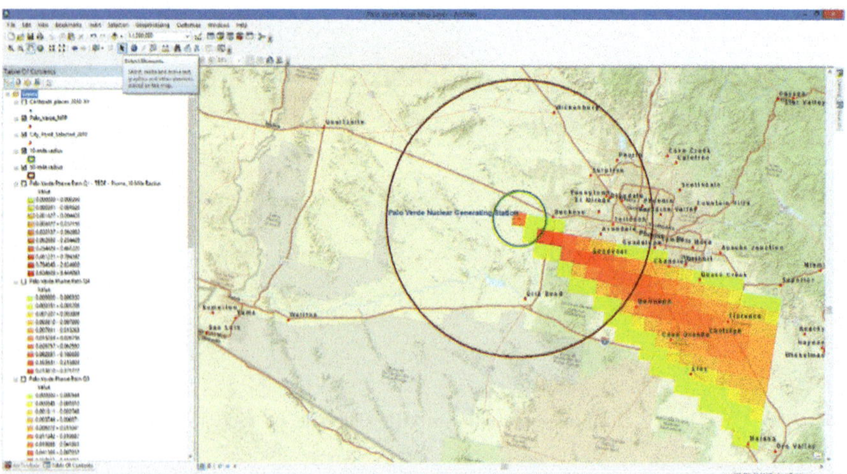

Fig. 3.25 Overlaying close-in dose footprint GIS shapefile on Google Base map in ArcGIS program

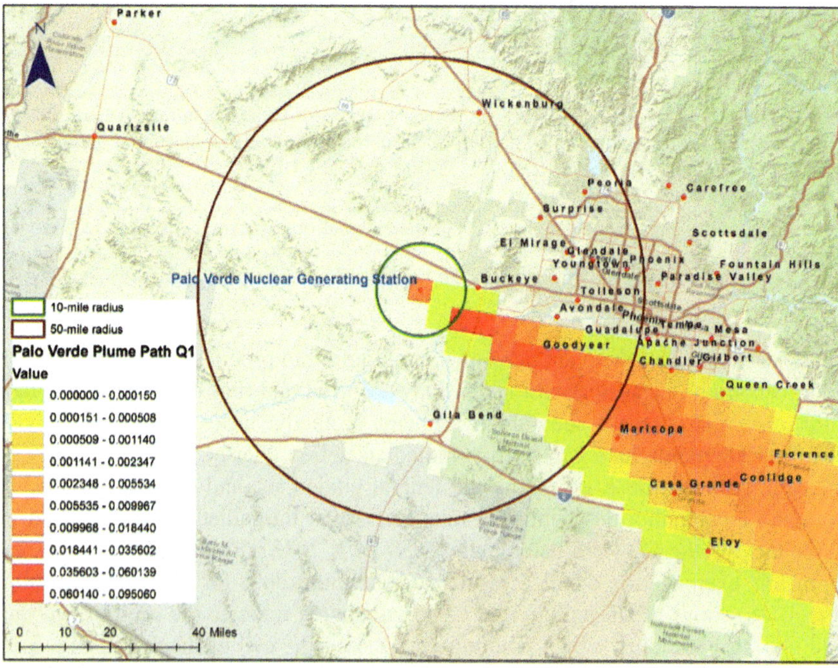

Fig. 3.26 A GIS map displaying projected 100-mile plume pathway overlaid on the Google street map

3.5 Protective Action Recommendation (PAR)

According to NRC (2014b), the requirement of the nuclear power plant personnel to produce the protective action recommendation (PAR) to state and local government agencies are stated clearly as follows:

> Immediately upon becoming aware that an incident has occurred that may result in a radiation dose that exceeds federal government protective action guides, responsible nuclear power plant personnel evaluate plant conditions and then make protective action recommendations (PARs) to the state and local government agencies on how to protect the population. Nuclear power plant personnel are required to report the PARs to the state or local government agencies (within 15 min) (US NRC 2014b).

The protection action recommendation (PAR) could be generated based on the projection of plume dispersion in the RASCAL. However, it must be noted that RASCAL does not provide protective action recommendations. The nuclear power plant personnel must read the detailed results including source term (Fig. 3.17), maximum dose value including close-in (Fig. 3.18) and 100-mile (Fig. 3.19), and plume footprint including close-in (Fig. 3.16) and 100-mile (Fig. 3.16).

Table 3.2 Protective actions recommended by the EPA's PAGs

Protective action	PAG (Projected dose)	Comments
Evacuation (or) Sheltering	TEDE 1–5 REM thyroid CDEThyroid CDE 5–25 REM	Evacuation (or for some situations, sheltering) should normally be initiated at 1 REM TEDE or 5 REM thyroid CDE

Note TEDE is the sum of inhalation dose and the external dose from cloudshine and 4 days of groundshine
Source US NRC (2013), US EPA (2013)

As stated clearly, the plant personnel are required to produce PAR "immediately upon becoming aware that an incident has occurred and that may result in a radiation dose that exceeds federal government protective action guides (US NRC 2014b)." There are three key words: immediate action after becoming aware of the incident, radiation dose exceeding the federal guidelines, and sending the PAR to state and local authorities within 15 min. Regarding the radiation dose, the federal guidelines by the EPA are listed below (US EPA 2013) (Table 3.2).

In this exercise, the detailed results of maximum dose for close-in calculation (Fig. 3.18) show TEDE and thyroid CDE calculation indicate the need to evacuate within a 3-mile radius. When pointing cursor over the cell or display in Fig. 3.15, the data at individual model receptors are shown in (Fig. 3.27). This information

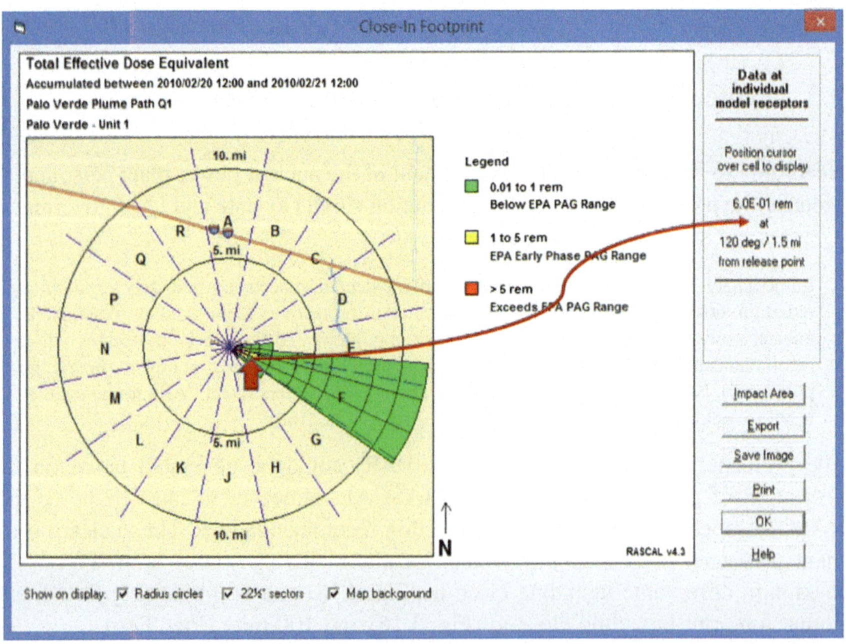

Fig. 3.27 A close-in dose calculation displayed in footprint map with data at individual model receptors when mouse pointer hovering on a selected model receptor. *Source* RAMP (2016a)

provides a dose rate at the wind direction and distance from the release point. This information could be incorporated within the maps overlaid on Google maps (Figs. 3.24 and 3.26) to make protective action recommendation.

3.6 Protective Action Decisions (PADs)

According to NRC, the state and local authorities must make a decision on the protective action recommendation (PAR) submitted by the plant.

> State and local officials make the final decision on what protective action is necessary to protect public health and safety, and then relay these decisions to the public in a timely manner (normally within approximately 15 min) (US NRC 2014b).

The 15-min time frame is another challenge for state and local agencies to make a certain decision on the protective action decisions. In addition, the agencies must generate a PADs, which must be signed by a State Governor. For example, in case of such emergency in Arizona, the PADs is signed by the Governor of the State of Arizona; in the absence of the Governor, it must be signed by a Maricopa County Judge (Kyne 2015). Once the PADs has been signed, the joint information committee must be formed and disseminate the PADs information to the public and emergency management agencies and responders.

3.7 Evacuation

The emergency agencies and personnel must carry out evacuation activities. The PADs is based on PAR which was generated by the plant personnel. Thus, plant personnel as well as state and local agencies must be aware of EPA's protective action guidelines. There are a few items that should be aware of before making PAR and PADs which will have an impact on evacuation.

First, the TEDE's range is between 1 REM and 5 REM, and the projected dose of 1 REM is suggested to evacuate the general population. The EPA's PAG states as follows:

> Although the PAG is expressed as a range of 1-5 REM, it is emphasized that, under normal conditions, evacuation of members of the general population should be initiated for most incidents at a projected dose of 1 REM (US EPA 2013). (pp. 2–5)

Second, the sheltering is more preferred for some group of general public. For example, the elderly, disabled persons, children, and other vulnerable populations may face difficulty in evacuating. It is justified that those groups of people may be at the projected doses of up to 5 REM. In addition, under very hazardous conditions, the group could even be sheltered at the projected dose rate of 10 REM.

> Sheltering may be preferable to evacuation as a protective action in some situations. Because of the higher risks associated with evacuation of some special groups in the population (e.g. those who are not readily mobile), sheltering may be the preferred alternative for such groups as a protective action at projected doses up to 5 REM. In addition, under unusually hazardous environmental conditions, use of sheltering at projected doses up to 5 rem to the general population (and up to 10 REM to special groups) may become justified (US EPA 2013). (pp. 2–5)

Third, the specific situations that are not in favor for evacuation of the specific population at the projected dose of 1 REM are severe weather conditions, competing disasters, institutionalized persons, and local physical factors (US EPA 2013).

> Illustrative examples of situations or groups for which evacuation may not be appropriate at 1 REM include: a) the presence of severe weather, b) competing disasters, c) institutionalized persons who are not readily mobile, and d) local physical factors which impede evacuation (US EPA 2013). (pp. 2–6)

Fourth, the EPA's PAG highlighted that there is no minimum radiation level to initiate the sheltering. The advantage of sheltering is that it is low-cost and low-risk protective action to protect the general public (US EPA 2013).

> No specific minimum level is established for initiation of sheltering. Sheltering in place is a low-cost, low-risk protective action that can provide protection…(US EPA 2013) (pp. 2–7)

Fifth, the plant personnel and state and local agencies must be aware that the EPA's PAG radiation dose levels are projected for a four-day period, whereas the PAR and PADs are based on early stages of prerelease and plume (early) (Fig. 3.7). Therefore, the plant personnel must project the dose with FMDose to reevaluate the radiation dose level. Based on the projected dose levels generated by the FMDose, PAR and PADs should be updated and necessary changes in evacuation must be carried out.

> The projected dose comparison to the early phase PAGs is normally calculated for exposure during the first four days following the projected (or actual) start of a release. The objective is to encompass the entire period of exposure to the plume and to deposited material prior to implementation of any further, longer-term protective action, such as relocation. 4 days is chosen here as the duration of exposure to deposited materials during the early phase because, for planning purposes; it is a reasonable estimate of the time needed to make measurements, reach decisions, and prepare to implement relocation (US EPA 2013). (pp. 5–6)

Finally, the PAR, PADs, and evacuation activities must be carried by local and state agencies. The NRC does not involve in any of the activities.

> The NRC monitors the actions of nuclear power plants to ensure the protective actions taken or recommended by the nuclear power plant personnel are appropriate. Additionally, state and local agencies may independently assess the situation to ensure that the correct protective action decisions are made. Independent assessments performed during an accidental radiological release from a nuclear power plant ensure that the best possible action is taken (US NRC 2014b).

3.8 Conclusions

This chapter explains four types of nuclear power emergencies: notification of unusual event (NOUE), alert, site area emergency (SAE), and general emergency classified by the US Nuclear Regulatory Commission (US NRC). Among them, the latter two emergencies require projection of plume dispersion and carrying out emergency management activities including mitigation, preparedness, response, and recovery. Among the four stages in nuclear emergency management process, the response to nuclear power emergency is different from other emergency responses. For example, responding to a fire emergency, a fire station could go to the fire accident place as soon as the station was notified of the fire. However, responding to a nuclear power emergency requires special technical skills or technical-oriented response. In the response stage in the nuclear power emergency management, there are four activities to carry out, including projection of radioactive plume path dispersion, protective action decisions (PADs), protective action recommendation (PAR), and evacuation. The projection of radioactive plume path dispersion requires a special software, RASCAL, and weather information. Only a trained technician could do projection of the radioactive plume path dispersion. This chapter provides detailed explanation on the process of projecting plume path dispersion with RASCAL. In the most recent Fukushima nuclear power emergency (Chap. 1, Sect. 1.3), evacuation was delayed several hours after the nuclear power accident was initiated. In addition, the determination for evacuation zones was also delayed a couple of days to evacuate larger areas. This plume projection with RASCAL could help accelerate the response process and could help expedite the evaluation.

References

Alltucker K (2006) Palo Verde's costly slump: customers may pay for nuke plant's struggles. Retrieved from AZ Central News http://www.azcentral.com/12news/news/articles/0212paloverde12-CP.html#

CNN (2015) Hurricane sandy fast facts. Retrieved from CNN http://www.cnn.com/2013/07/13/world/americas/hurricane-sandy-fast-facts/

ESRI (2011) ArcGIS desktop: Release 10.1 Redlands. Environmental Systems Research Institute, CA. Retrieved from http://www.esri.com/software/arcgis/arcgis10/whats-coming/

ICRP (1977) Recommendations of the ICRP. ICRP Publication 26. Ann ICRP 1(3). Retrieved from http://www.icrp.org/publication.asp?id=ICRP%20Publication%2026. Retrieved from International Commission on Radiological Protection

Kutner K (2010) 2010 annual radioactive effluent release Report. Retrieved from Palo Verde Nuclear Generating Station http://pbadupws.nrc.gov/docs/ML1113/ML11136A042.pdf

Kyne D (2015) Managing nuclear power plant induced disasters. J Emergency Manag (Weston, Mass.) 13(5):417–429. doi:10.5055/jem.2015.0252. Retrieved from Prime National Pub. Corp

McEntire DA (2007) Disciplines, disasters, and emergency management the convergence and divergence of concepts, issues and trends from the research literature. Charles C Thomas, Springfield, Ill

McKenna TJ, Glitter JG (1988) Source term estimation during incident response to severe nuclear power plant accidents (NUREG-1228). Nuclear Regulatory Commission, Washington, DC (USA). Office for Analysis and Evaluation of Operational Data. Retrieved from http://www.osti.gov/scitech/biblio/6822946

Moe TL, Pathranarakul P (2006) An integrated approach to natural disaster management. Disaster Prevent Manag Int J 15(3):396–413. doi:10.1108/09653560610669882. Retrieved from http://www.emeraldinsight.com/doi/abs/10.1108/09653560610669882. Retrieved from Emerald

RAMP (2016a) Radiological Assessment System for Consequence AnaLysis (RASCAL) Overview. Online: US Nuclear Regulatory Commission, Radiation Protection Computer Code Analysis and Maintenance Program. Retrieved from https://www.usnrc-ramp.com/RASCAL%20Overview

RAMP (2016b) RASCAL 4.3.1 Installation. US Nuclear Regulatory Commission, Radiation Protection Computer Code Analysis and Maintenance Program

Rosenzweig C, Solecki W (2014) Hurricane sandy and adaptation pathways in New York: lessons from a first-responder city. Global Environ Change-Human Policy Dimen 28:395–408. Retrieved from WOS:000343839100034

RSICC (2013) RASCAL 4.3: radiological assessment systems for consequence analysis reactor and nuclear systems division (RNSD), Oak Ridge National Laboratory, The Radiation Safety Information Computational Center (RSICC). Retrieved from https://rsicc.ornl.gov/PackageDetail.aspx?p=RASCAL%204.3&id=C00783&cpu=PCX86&v=02&t=Radiological%20Assessment%20Systems%20for%20Consequence%20AnaLysis

Starr S (2016) Costs and consequences of the Fukushima Daiichi Disaster. In: PSR's Environmental Health Policy Institute. Retrieved from http://www.psr.org/environment-and-health/environmental-health-policy-institute/responses/costs-and-consequences-of-fukushima.html

US EPA (2013) Manual of protective action guides and protective action for nuclear incidents. Office of Radiations Program, United States Environmental Protection Agency, Washington DC 20460. Retrieved from: http://www.epa.gov/sites/production/files/2015–06/documents/pag-manual-interim-public-comment-4-2-2013.pdf

US GAO (2001) Nuclear regulations: progress made in emergency preparedness at indian point 2, but additional improvements needed (GAO-01-605). United States Government Accountable Office (US GAO). Retrieved from http://www.gao.gov/assets/240/231903.pdf

US NRC (2011) Nuclear Regulatory Commission. Retrieved from US Nuclear Regulatory Commission www.nrc.gov

US NRC (2013) RASCAL 4.3 Workbook Washington, DC 20555-0001: US Nuclear Regulatory Commission. Retrieved from http://pbadupws.nrc.gov/docs/ML1328/ML13281A475.pdf

US NRC (2014a) Emergency classification U.S. nuclear regulatory commission. Retrieved from http://www.nrc.gov/about-nrc/emerg-preparedness/about-emerg-preparedness/emerg-classification.html

US NRC (2014b) Protective action recommendations. Retrieved from US Nuclear Regulatory Commission http://www.nrc.gov/about-nrc/emerg-preparedness/about-emerg-preparedness/protective-action-recommendations.html

US NRC (2015) RASCAL 4.3: User's Guide. Washington, DC 20555-0001: US Nuclear Regulatory Commission. Retrieved from http://pbadupws.nrc.gov/docs/ML1328/ML13281A475.pdf

Chapter 4
Simulation of Nuclear Power Plant Core-Meltdown Accidents

4.1 Nuclear Power Plant Core-Meltdown Accidents

Historically, there has only been one plant, Three Mile Island, in Pennsylvania, that experienced a partial nuclear core-meltdown, in U.S. commercial nuclear power operations, which occurred in 1997. As discussed in the previous chapters, the impacts of the nuclear power core-meltdown accidents are unimaginable. The most destructive power came from the tons of nuclear fuel assemblies containing many hundreds of nuclear fuel rods, sitting inside the core of the nuclear reactor in the containment building. When the fuel rods are uncovered and they enter into the melting stage, there was high–level toxic radioactive plume released into the air. Gradually, the fuel rods are melted to the bottom of the reactor or breach the reactor. The entire containment room could quickly fill up with the radioactive plume and accumulate pressure from it. When the 4-feet thick containment wall no longer resists the enormous pressure, the dome-shaped containment building erupts.

4.2 Radioactive Plume Dispersion

Radiation from the toxic radioactive plume is considered to be the most immediate threat from nuclear power core-meltdown accidents. The radioactive plume will move from the source depending on the wind speed, wind direction, and other weather conditions. People living near nuclear power plants need to be evacuated immediately when an accident takes place. Therefore, the projection of the radioactive plume dispersion is a necessary first step to evacuate the individuals living around the nuclear power plant. This chapter will conduct a plume dispersion projection at two nuclear power plants: Palo Verde Nuclear Power Generating Station and Indian Point Nuclear Power Station.

D. Kyne, *Nuclear Power Plant Emergencies in the USA*,
DOI 10.1007/978-3-319-50343-1_4

4.3 A Scenario of a Core-Meltdown Accident at the Palo Verde Nuclear Power Plant

The Palo Verde Nuclear Generating Station (PVNGS) is located 25 miles west of Phoenix, Arizona and has three currently operating reactors. There are two reasons for selecting this nuclear power plant as a hypothetical scenario for a core meltdown: First, unlike other reactors in the U.S., the PVNGS is not located next to a body of water, such as a river or ocean. The plant is located in the desert, next to a pond, where the water is stored for cooling purposes. The risk associated with this power plant is that in case of a core-meltdown accident, the power plant may face challenges in cooling down the melting nuclear fuel rods with water. In Japan, when the Fukushima four reactors experienced partial and full core-meltdown, water must have been poured over the melted cores of unit 1, 2 and 3, to keep the fuel from overheating and melting again. The water seeping into ground water and some of the contaminated water are stored in 10-meter tall steel tanks holding 750,000 tons of water (Normile 2016). The nuclear power plant is located close by the fifth largest city in the U.S., with a population of more than six million, and more than 2.2 million living within a 50-mile radius from the nuclear power plant.

4.3.1 Palo Verde Nuclear Generation Station (PVNGS)

There are nearly 2.2 million people living in the host communities located within a 50-mile radius from the PVNGS, whereas about four million of their counterparts live in the "outside areas" of the plant (Tables 4.1, 4.2, and 4.3). Outside areas are defined as areas that fall outside of a 50-mile radius from the nearest nuclear power plant, in this case the PVNGS. The racial demographic characteristics in the communities within a 50-mile radius are observed as 77% White, 6% Black, 37% Hispanic, and 3% Asian. The demographic characteristics in the communities located in what qualify as outside areas are observed as 78% White, 3% Black, 24% Hispanic and 3% Asian (Kyne 2014). Overall, more People of Color (49%) live in the host communities than their counterparts (36%) in the outside areas (Table 4.3, Fig. 4.1).

4.3.2 Pressurized Water Reactor (PWR) Core-Damage Accident

In this exercise, the objective is to evaluate the simulated dispersal of a radioactive effluent plume path, in a scenario in which only one of the NPP's three PWRs experiences a core-damage accident, at the PVNGS in Arizona. The Radiological Assessment System for Consequence Analysis (RASCAL) Source Term to Dose

Table 4.1 Demographic composition of population, as sorted by distance from Palo Verde Nuclear Generating Station, Arizona in 1990; *Source* (Kyne 2014)

	0–10	11–20	21–30	31–40	41–50	0–50	Outside
Year 1990							
Tract	1	8	39	192	294	534	992
Tract area (sq. mile)	120	1,658	672	2,661	1,831	6,942	107,049
Total population	1,365	10,128	38,556	379,343	866,410	1,295,802	2,369,419
White	1,090	8,112	28,273	299,286	714,771	1,051,532	1,916,149
Black	0	328	2,073	15,153	42,671	60,225	49,833
Asian	0	44	563	5,254	14,728	20,589	33,539
Native American	72	209	496	6,985	18,992	26,754	177,831
Others	203	1,440	7,144	52,651	75,260	136,698	192,071
Hispanic	203	2,580	11,418	85,971	162,063	262,235	418,401
Color	275	3,137	14,272	112,206	234,538	364,428	668,883
White (%)	79.85	80.09	73.33	78.90	82.50	81.15	80.87
Black (%)	0.00	3.24	5.38	3.99	4.93	4.65	2.10
Asian (%)	0.00	0.43	1.46	1.39	1.70	1.59	1.42
Native American (%)	5.27	2.06	1.29	1.84	2.19	2.06	7.51
Others (%)	14.87	14.22	18.53	13.88	8.69	10.55	8.11
Hispanic (%)	14.87	25.47	29.61	22.66	18.71	20.24	17.66
Color (%)	20.15	30.97	37.02	29.58	27.07	28.12	28.23
Female (%)	48.21	49.81	47.30	51.82	50.43	50.74	50.64
Old (65+ years) (%)	9.23	8.18	7.12	19.38	10.08	12.70	13.19
Kid (<5 years) (%)	6.23	11.33	11.99	9.84	9.80	9.89	9.30
Native-born (%)	92.89	90.32	91.61	92.68	91.30	91.71	92.80
Renter housing units (%)	20.21	22.27	34.78	23.84	37.08	32.95	27.74
Education (%)	14.12	8.44	12.64	13.97	20.77	18.45	21.31
Unemployment (%)	12.74	10.17	8.92	7.56	6.49	6.85	7.36
Poverty (%)	26.30	22.84	20.52	13.05	14.66	14.42	16.47
Mean household income ($)	38,481	48,753	54,608	54,603	60,623	58,632	57,882

(STDose) model is used to project plume dispersion, in case of one of the three reactors experiencing a core-meltdown accident (RSICC 2013). The results of the dispersal projections were imported in shapefile format to ESRI's ArcMap program (US NRC 2013). The resulting TEDE is measured in Roentgen Equivalent Man (REM) and is equal to the absorbed dose (in RADS) multiplied by the quality factor of the type of radiation (US NRC 2011).

Table 4.2 Demographic composition of population, as sorted by distance from Palo Verde Nuclear Generating StationPalo Verde Nuclear Generating Station, Arizona in 2000; *Source* (Kyne 2014)

	0–10	11–20	21–30	31–40	41–50	0–50	Outside
Year 2000							
Total population	2,112	15,826	62,558	570,281	1,149,152	1,799,929	3,330,703
White	1,615	11,767	45,916	398,763	841,429	1,299,490	2,572,225
Black	15	583	2,661	26,280	55,123	84,662	69,654
Asian	10	99	922	8,189	28,148	37,368	60,021
Native American	49	309	703	9,584	25,715	36,360	217,182
Others	423	3,073	12,357	127,445	198,728	342,026	411,644
Hispanic	712	5,260	20,013	208,105	347,341	581,431	713,886
Color	779	6,310	24,968	258,156	469,773	759,986	1,098,581
White (%)	76.47	74.35	73.40	69.92	73.22	72.20	77.23
Black (%)	0.71	3.68	4.25	4.61	4.80	4.70	2.09
Asian (%)	0.47	0.63	1.47	1.44	2.45	2.08	1.80
Native American (%)	2.32	1.95	1.12	1.68	2.24	2.02	6.52
Others (%)	20.03	19.42	19.75	22.35	17.29	19.00	12.36
Hispanic (%)	33.71	33.24	31.99	36.49	30.23	32.30	21.43
Color (%)	36.88	39.87	39.91	45.27	40.88	42.22	32.98
Female (%)	48.48	43.41	48.87	51.28	49.04	49.69	50.43
Old (65+ years) (%)	10.18	7.92	8.98	17.07	8.55	11.26	13.96
Kid (<5 years) (%)	9.14	8.72	9.78	10.24	9.79	9.92	8.34
Native-born (%)	81.53	86.71	87.52	83.72	81.68	82.57	89.72
Renter housing units (%)	23.38	26.40	24.29	23.10	38.25	32.89	25.37
College degree or higher (%)	7.02	8.83	17.09	15.40	23.71	20.68	25.00
Unemployment (%)	6.13	5.84	4.55	5.80	5.62	5.64	5.63
Poverty (%)	21.02	17.93	12.74	13.00	15.61	14.70	13.48
Mean household income ($)	53,204	59,309	71,249	60,305	68,888	66,197	67,691

4.3.3 *Projected Plume Path for Quarter 1*

The plume dispersion after the NPP encounters a core- meltdown accident is largely dependent by the weather condition. To project the plume path, the actual and forecast weather information on the day the event is taking place are required to input into the RASCAL model. In this exercise, a typifying weather information for quarter 1 (January to March) was based on the meteorological data, reported in the

Table 4.3 Demographic composition of population, as sorted by distance from Palo Verde Nuclear Generating Station, Arizona in 2010; *Source* (Kyne 2014)

	0–10	11–20	21–30	31–40	41–50	0–50	Outside
Year 2010							
Total population	3,090	41,543	164,444	805,703	1,193,109	2,207,889	4,038,927
White	2,446	32,437	132,810	612,665	929,449	1,709,807	3,173,799
Black	224	2,368	9,639	49,776	70,875	132,882	107,773
Asian	10	852	4,670	19,844	41,677	67,053	111,648
Native American	0	761	1,769	13,084	26,470	42,084	236,252
Others	410	5,125	15,556	110,334	124,638	256,063	409,455
Hispanic	1,066	16,096	49,966	357,283	402,723	827,134	987,540
Color	1,319	20,479	67,230	446,487	555,346	1,090,861	1,488,924
White (%)	79.16	78.08	80.76	76.04	77.90	77.44	78.58
Black (%)	7.25	5.70	5.86	6.18	5.94	6.02	2.67
Asian (%)	0.32	2.05	2.84	2.46	3.49	3.04	2.76
Native American (%)	0.00	1.83	1.08	1.62	2.22	1.91	5.85
Others (%)	13.27	12.34	9.46	13.69	10.45	11.60	10.14
Hispanic (%)	34.50	38.75	30.38	44.34	33.75	37.46	24.45
Color (%)	42.69	49.30	40.88	55.42	46.55	49.41	36.86
Female (%)	49.94	47.17	51.24	51.15	49.38	50.13	50.31
Old (65+ years) (%)	9.00	5.97	8.00	14.46	8.65	10.67	14.80
Kid (<5 years) (%)	6.89	11.41	9.96	9.04	7.76	8.46	6.84
Native-born (%)	87.57	88.74	88.10	79.95	80.63	81.10	88.43
Renter housing units (%)	21.78	22.18	20.49	27.12	36.01	31.61	25.23
College degree or higher (%)	7.27	15.58	24.73	18.38	26.31	23.13	27.96
Unemployment (%)	10.90	7.32	7.31	8.89	7.15	7.75	7.60
Poverty (%)	21.23	12.79	9.81	16.41	18.84	17.18	14.21
Mean household income ($)	66,449	67,996	79,633	58,958	67,297	65,194	68,577

2011 Radioactive Effluents Report by the PVNGS. With a prevailing wind direction of west-northwest, at a speed of 6.2 miles per hour, the model provides a projected plume path which dispersed to the east-south, travelling a distance greater than 100 miles in a 24-h period (Fig. 4.2). The plume path covers an area of approximately 3,477 square miles, dispersing over an estimated population of 663,543 (Kyne 2015). Under the plume pathway, different groups of individuals are exposed to different levels of TEDE ranging from 0.000 to 0.095 REM, grouped

Fig. 4.1 Spatial distribution of percent Color among the populations surrounding Palo Verde Nuclear Generating Station (PVNGS)

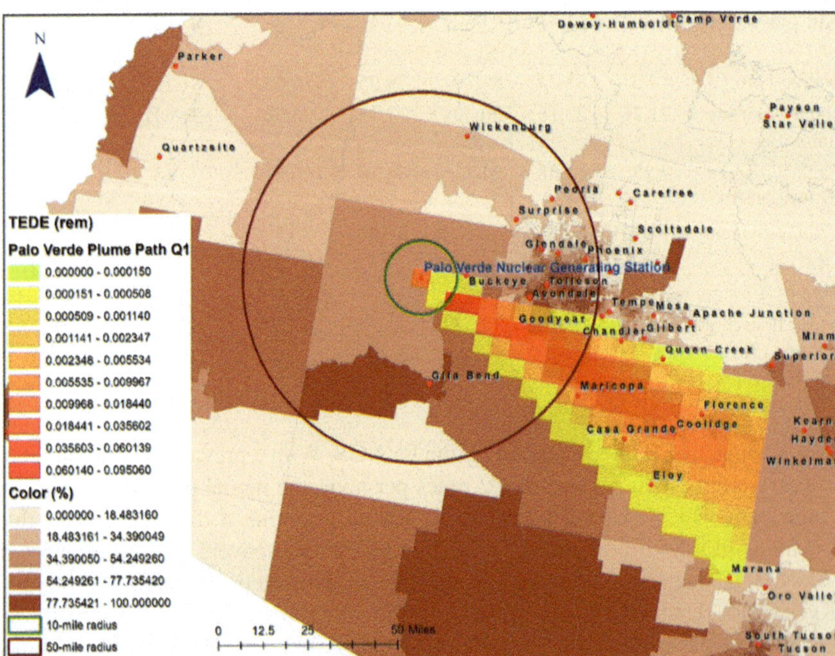

Fig. 4.2 The projected plume path emanating from Palo Verde NPP after a simulated core-damage accident conducted under meteorological conditions typifying Quarter 1; *Source* Adapted from (Kyne 2014)

Table 4.4 Demographic composition in areas around Palo Verde NPP located beneath the projected plume path with meteorological conditions typifying Quarter 1, *Source* (Kyne 2015) "Reprinted from Journal of Emergency Management. Used with permission. All Rights Reserved"; *Source* (Kyne 2014)

Demographics/ (TEDE) rem)	(0.000–0.0003)	(0.0003–0.002)	(0.002–0.005)	(0.005–0.01)	(0.01–0.4)	Total	Outside
Tracts	213	145	50	73	68	549	977
Tract area (sq. miles)	1,201	726	425	475	650	3,477	110,513
Total population	305,529	207,069	41,606	75,125	34,214	663,543	5,583,273
White	235,302	157,724	31,010	57,221	21,834	503,090	4,380,516
Black	16,143	9,151	2,337	3,266	1,574	32,471	208,184
Asian	16,948	12,539	1,094	2,214	581	33,375	145,326
Native American	6,267	4,629	1,855	4,149	7,049	23,949	254,387
Others	30,870	23,025	5,310	8,276	3,177	70,658	594,860
Hispanic	66,773	45,665	10,932	18,591	6,745	148,705	1,665,969
Color	110,872	75,803	16,869	29,130	16,337	249,011	2,330,774
White (%)	77.01	76.17	74.53	76.17	63.82	75.82	78.46
Black (%)	5.28	4.42	5.62	4.35	4.60	4.89	3.73
Asian (%)	5.55	6.06	2.63	2.95	1.70	5.03	2.60
Native American (%)	2.05	2.24	4.46	5.52	20.60	3.61	4.56
Others (%)	10.10	11.12	12.76	11.02	9.29	10.65	10.65
Hispanic (%)	21.85	22.05	26.27	24.75	19.71	22.41	29.84
Color (%)	36.29	36.61	40.55	38.77	47.75	37.53	41.75
Renter-occupied housing units (%)	26.70	20.73	15.66	16.42	18.38	22.57	27.81
College degree or higher (%)	37.20	34.26	19.32	23.37	18.08	32.43	25.60
Unemployed (%)	6.18	6.25	6.28	8.07	8.08	6.47	7.80
Below poverty line (%)	8.95	7.00	8.90	10.68	15.65	8.86	16.01
Mean household income ($)	86,360	85,695	80,214	69,205	59,475	82,421	65,767

Source (Kyne 2015)

in Table 4.4. EPA recommends evacuation when the dose level reaches 1 REM (US EPA 2013; Kyne 2015). In short, in the event under the typifying Quarter 1 weather conditions, minority populations—including People of Color, Black and Asian populations who live in owner-occupied housing units and who live under the poverty line, are likely to be exposed to high doses of radiation.

4.3.4 Projected Plume Path for Quarter 2

This projection exercise is carried out under meteorological conditions typifying Quarter 2 (April to June) with the PVNGS, as subject of a core-meltdown accident. The weather information for Quarter 2 includes a northeasterly wind direction at a speed of 7.6 miles per hour, and the resulting plume pathway covers an estimated area of 6,004 square miles in which about 1 million people (964,774) exposed to radiation level ranging between 0.00 and 0.023 REM (which is still below the 1 REM requirement, to call for an evacuation or sheltering according to the NRC's protection action guidelines) within a period of 24 h (Table 4.5) (Kyne 2014). The population beneath the projected plume pathway under meteorological conditions

Table 4.5 Demographic composition in areas around Palo Verde NPP located beneath the projected plume path with meteorological conditions typifying Quarter 2; *Source* (Kyne 2014)

Demographics/ (TEDE) rem)	(0.000–0.0003)	(0.0003–0.002)	(0.002–0.005)	(0.005–0.01)	(0.01–0.4)	Total	Outside
Tracts	421	185	131	55	12	804	722
Tract area (sq. miles)	2,276	1,801	1,251	525	150	6,004	107,987
Total population	643,380	189,204	97,495	30,819	3,876	964,774	5,282,042
White	547,531	156,891	86,769	27,471	3,124	821,787	4,061,819
Black	21,935	6,386	3,131	786	78	32,316	208,339
Asian	21,031	4,340	2,210	762	129	28,472	150,229
Native American	6,682	2,352	683	239	11	9,967	268,369
Others	46,201	19,235	4,701	1,561	534	72,232	593,286
Hispanic	122,483	35,082	10,238	2,987	832	171,623	1,643,052
Color	180,171	50,293	17,322	5,090	1,190	254,066	2,325,719
White (%)	85.10	82.92	89.00	89.14	80.60	85.18	76.90
Black (%)	3.41	3.38	3.21	2.55	2.01	3.35	3.94
Asian (%)	3.27	2.29	2.27	2.47	3.32	2.95	2.84
Native American (%)	1.04	1.24	0.70	0.78	0.28	1.03	5.08
Others (%)	7.18	10.17	4.82	5.06	13.79	7.49	11.23
Hispanic (%)	19.04	18.54	10.50	9.69	21.48	17.79	31.11
Color (%)	28.00	26.58	17.77	16.51	30.71	26.33	44.03
Renter-occupied housing units (%)	22.65	17.10	11.54	8.98	12.47	19.75	28.70
College degree or higher (%)	27.26	27.77	29.40	32.87	18.33	27.77	26.04
Unemployed (%)	6.31	6.65	6.38	7.32	7.25	6.41	7.88
Below poverty line (%)	9.21	9.25	5.87	5.70	10.69	8.77	16.47
Mean household income ($)	74,049	77,654	71,533	76,691	68,980	74,527	66,108

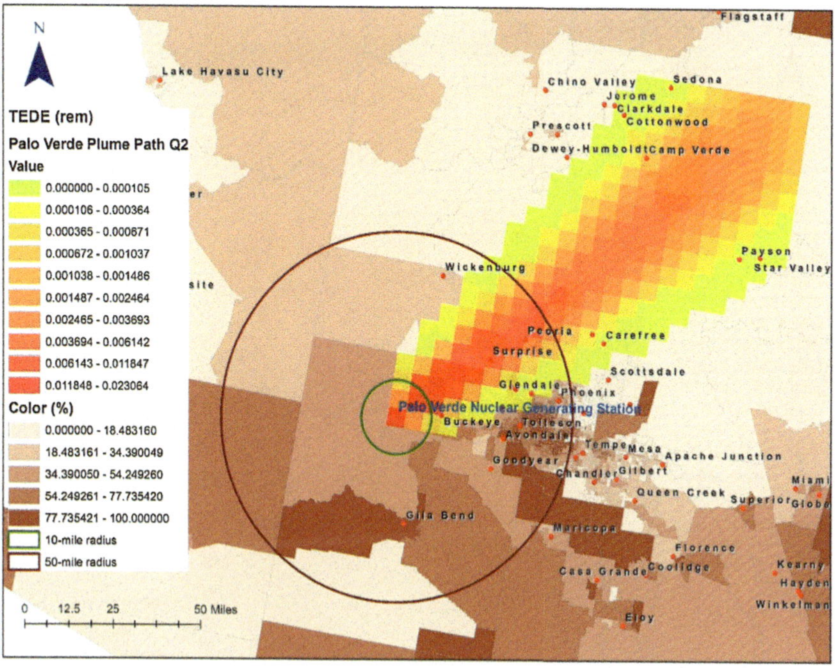

Fig. 4.3 The projected plume path emanating from Palo Verde NPP after a simulated core-damage accident conducted under meteorological conditions typifying Quarter 2; *Source*: Adapted from (Kyne 2014)

typifying Quarter 2 consists of a racial demographic with of White (85%) and Hispanic (18%); 10% of them live under poverty and are exposed to higher levels of radiation doses (Kyne 2014). In short, this projection provides that an estimated one million people would be exposed to the radioactive plume.

4.3.5 Projected Plume Path for Quarter 3

In this exercise, the projection of the radioactive plume was carried out with the meteorological conditions typifying Quarter 3 (July to September)—prevailing winds toward west-southwest (248°) and an average speed of 7.3 miles per hour. The findings show that the west-southwest wind pushed the radioactive plume to the east-northeast of the NNP (Fig. 4.4) covering an area of more than 3,702 square miles (Table 4.6). Approximately 3.5 million people are observed to live beneath the plume pathway (Kyne 2014). The populations exposed include a racial demographic makeup of a higher percentages of Black, Hispanic, Asian and People

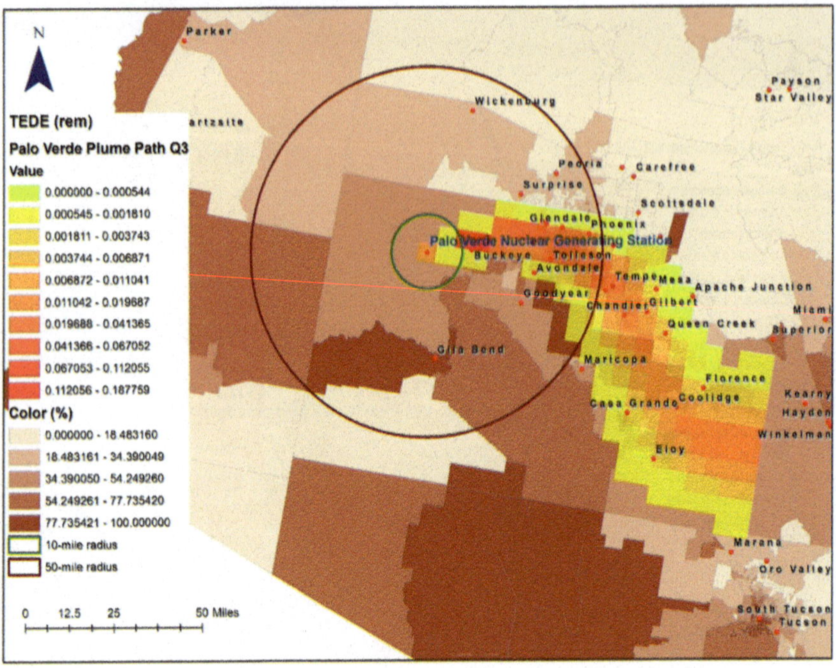

Fig. 4.4 The projected plume path emanating from Palo Verde NPP after a simulated core-damage accident conducted under meteorological conditions typifying Quarter 3; *Source* Adapted from (Kyne 2014)

of Color, among the population demographics, while their exposure to radiation levels was estimated between 0.00 and 0.188 REM, which is below the 1 REM TEDE required to initiate either an evacuation or shelter-in-place order. The estimated radiation level must be validated with field data collected from the affected area at the time of the accident (Kyne 2014). The projected radioactive plume would likely be dispersed over populated areas including: Buckeye, Maricopa, Guadalupe, Chandler, Queen Creek, Florence, Coolidge, Casa Grande, Glendale, Litchfield Park, Phoenix, Tolleson, Avondale, Gilbert, Tempe, Apache Junction, and Paradise Valley (Fig. 4.4). There are 12 cities that would be exposed to the higher TEDEs, shown in red and brown colors (Fig. 4.4). They include: Buckeye, Guadalupe, Chandler, Queen Creek, Florence, Phoenix, Tolleson, Avondale, Gilbert, Tempe, Apache Junction, and Paradise Valley (Kyne 2014). This projection estimates about 55% (3.5 million out of 6.25 million total) of the total population of Arizona would be likely to be beneath the projected radioactive plume pathway. This is cause for concern to many key stakeholders who are involved in radiological emergency planning, to effectively plan and response in case of the core-meltdown accident under similar circumstances and impacts.

Table 4.6 Demographic composition in areas around Palo Verde NPP located beneath the projected plume path with meteorological conditions typifying Quarter 3; *Source* (Kyne 2014)

Demographics/ (TEDE) rem)	(0.000–0.0003)	(0.0003–0.002)	(0.002–0.005)	(0.005–0.01)	(0.01–0.4)	Total	Outside
Tracts	484	168	97	172	589	1,510	16
Tract area (sq. miles)	1,426	550	325	575	826	3,702	110,288
Total population	1,024,487	335,466	162,860	344,957	1,601,288	3,469,058	2,777,758
White	856,746	261,947	131,867	278,370	1,195,778	2,724,708	2,158,899
Black	33,563	17,001	7,129	13,314	107,419	178,427	62,228
Asian	29,532	9,657	6,904	11,525	62,536	120,154	58,547
Native American	16,039	9,150	2,425	13,270	35,586	76,470	201,866
Others	88,608	37,711	14,534	28,478	199,968	369,299	296,219
Hispanic	207,529	93,733	25,983	92,706	685,726	1,105,677	708,997
Color	300,148	133,971	44,709	134,774	907,417	1,521,017	1,058,768
White (%)	83.63	78.08	80.97	80.70	74.68	78.54	77.72
Black (%)	3.28	5.07	4.38	3.86	6.71	5.14	2.24
Asian (%)	2.88	2.88	4.24	3.34	3.91	3.46	2.11
Native American (%)	1.57	2.73	1.49	3.85	2.22	2.20	7.27
Others (%)	8.65	11.24	8.92	8.26	12.49	10.65	10.66
Hispanic (%)	20.26	27.94	15.95	26.87	42.82	31.87	25.52
Color (%)	29.30	39.94	27.45	39.07	56.67	43.85	38.12
Renter-occupied housing units (%)	23.29	24.85	19.24	34.99	37.84	30.84	23.41
College degree or higher (%)	27.30	23.11	37.66	31.16	24.13	26.39	26.23
Unemployed (%)	6.58	6.80	5.90	7.11	8.20	7.37	8.03
Below poverty line (%)	10.22	11.92	6.28	13.17	20.34	15.16	15.41
Mean household income ($)	73,210	68,029	102,708	68,579	62,674	68,917	65,685

4.3.6 *Projected Plume Path for Quarter 4*

This projection which was conducted with the meteorological conditions typifying Quarter 4—a prevailing wind direction to the west (270°) and a wind speed of 5.6 miles per hour provides an estimated plume pathway moving to the east of the Palo Verde NNP (Fig. 4.5), covering an area of 3,327 square miles (Table 4.7) (Kyne 2014). There are about 2.3 million people living beneath the radioactive plume pathway and they are estimated to be exposed to radiation levels ranging from 0.00 and 0.371 REM (1 REM is a threshold to call for either an evacuation or shelter-in-place order) (Kyne 2014). The presence of minority groups, including People of Color, Black, Hispanic and Asian (estimated in percentages) is greater

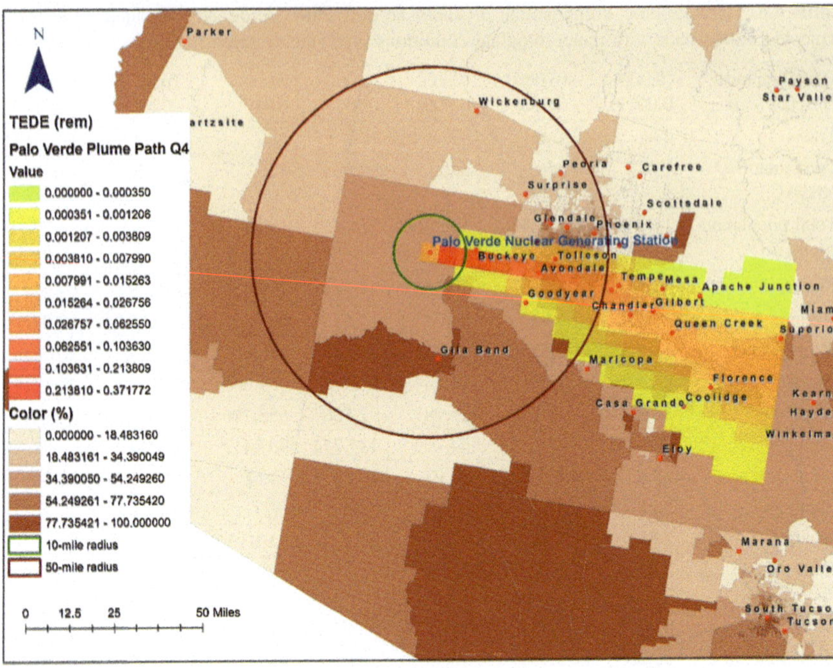

Fig. 4.5 The projected plume path emanating from Palo Verde NPP after a simulated core-damage accident conducted under meteorological conditions typifying Quarter 4; *Source* Adapted from (Kyne 2014)

than the populations living outside of the projected plume path (Kyne 2014). The cities which are observed to be beneath the plume pathway are: Buckeye, Goodyear, Guadalupe, Chandler, Queen Creek, Florence, Coolidge, Litchfield Park, Tolleson, Avondale, Gilbert, Tempe, Apache Junction, and Paradise Valley (Fig. 4.5); while the cities exposed to the higher TEDEs, shown in red (Fig. 4.5) include: Buckeye, Guadalupe, Chandler, Queen Creek, Litchfield Park, Tolleson, Avondale, Gilbert, Tempe, Apache Junction, and Paradise Valley (Kyne 2014).

The findings from the four projections for the four different quarters in a calendar year provide an understanding that the impacted areas could be between 3,327 and 6,004 square miles, while the impacted population size could be between 0.66 million (11% of the total 6.25 million Arizonians) and 3.5 million (55% of the total 6.25 million Arizonians). The differences in weather conditions including wind direction and wind speed are mainly accountable for the variations in affected areas and populations. The 10-mile emergency planning zone is not sufficient for the radiological emergency planning because the radioactive plume could travel a

Table 4.7 Demographic composition in areas around Palo Verde NPP located beneath the projected plume path with meteorological conditions typifying Quarter 4; *Source* (Kyne 2014)

Demographics/ (TEDE rem)	(0.000–0.0003)	(0.0003–0.002)	(0.002–0.005)	(0.005–0.01)	(0.01–0.4)	Total	Outside
Tracts	142	159	199	229	356	1,085	441
Tract area (sq. miles)	1,126	525	400	550	725	3,327	110,663
Total population	158,798	336,474	461,824	580,558	798,994	2,336,649	3,910,167
White	131,442	281,968	368,090	442,544	603,861	1,827,905	3,055,701
Black	4,680	11,963	19,428	32,461	58,381	126,912	113,743
Asian	4,034	6,342	10,642	20,528	41,548	83,094	95,607
Native American	7,414	9,795	13,343	11,251	15,471	57,273	221,063
Others	11,228	26,406	50,322	73,774	79,734	241,465	424,053
Hispanic	27,843	90,260	163,568	260,908	251,684	794,263	1,020,411
Color	46,200	122,338	211,141	331,486	378,173	1,089,338	1,490,448
White (%)	82.77	83.80	79.70	76.23	75.58	78.23	78.15
Black (%)	2.95	3.56	4.21	5.59	7.31	5.43	2.91
Asian (%)	2.54	1.88	2.30	3.54	5.20	3.56	2.45
Native American (%)	4.67	2.91	2.89	1.94	1.94	2.45	5.65
Others (%)	7.07	7.85	10.90	12.71	9.98	10.33	10.84
Hispanic (%)	17.53	26.83	35.42	44.94	31.50	33.99	26.10
Color (%)	29.09	36.36	45.72	57.10	47.33	46.62	38.12
Renter-occupied housing units (%)	17.55	29.99	35.66	36.67	27.17	30.85	25.45
College degree or higher (%)	27.40	24.34	21.51	22.42	31.96	26.12	26.43
Unemployed (%)	5.63	6.66	7.86	8.59	6.64	7.29	7.88
Below poverty line (%)	9.12	17.57	18.80	18.99	11.58	15.51	15.13
Mean household income ($)	80,123	58,299	57,641	63,297	80,402	68,100	67,071

distance longer than 100 miles within a 24-hour period. The findings from the projections do not provide an alarming radiation level measured by TEDE, but it must be verified with the filed data collected in the day of the event. In addition, this projection is assumed that one out of three reactors will encounter a core-meltdown accident. The four projections of plume dispersion demonstrate that nuclear power emergencies such as nuclear power plant core-meltdown accident, are different from other emergencies because of its potential severity and impacts and fundamental requirements for mass evacuations and timely responses.

4.4 A Scenario of Core-Meltdown Accident at the Indian Point Nuclear Power Plant

4.4.1 Indian Point Nuclear Power Plant

There are close to 15 million people living in the host communities located within a 50-mile radius from the IPNG, whereas nearly 16 million of their counterparts live in the "outside areas" (Tables 4.8, 4.9, and 4.10) (Kyne 2014). In other words, more than five out of every ten people (15 million out of 31.7 million total population in New York, Connecticut, and New Jersey) live within a 50-mile radius of an NPP, according to the 2010 U.S. Census. The racial demographic composition in the communities within a 50-mile radius are observed as 65% White, 22% Black, 18% Hispanic and 5% Asian, whereas the demographic characteristics in the communities located in the outside areas are observed as 88% White, 8% Black, 4% Hispanic and 2% Asian (Kyne 2014). Overall, more People of Color (43%) live in the host communities than their counterparts (14%) in the outside areas (Table 4.10).

The spatial distribution of higher percent of Color living around the IPNG was depicted in Fig. 4.2. It is evident that many of the census tracts that consist of below 20% were mainly scattered in the north, the east and the west, whereas census tracts with greater than 20% were located in the south (Fig. 4.6).

4.4.2 Projected Plume Path for Quarter 1

Assuming that one of the two reactor units at the IPNG experience a core-meltdown accident, under the meteorological conditions typifying Quarter 1 (January to March), with a prevailing wind direction of north-east (23°), at a wind speed of 3 miles per hour, the projection results show a plume path traveling toward a south-west direction, more than 100 miles, within 48-h after the core-meltdown accident took place (Kyne 2016). The plume footprint is estimated to cover an area as large as 9,612 square miles, dispersing above an estimated population of 20 million (19,855,006) likely to be exposed to radiation level ranging from 0.000 to 1.103 REM, which is above the recommended evaluation dose level of 1 REM (Table 4.11) (Kyne 2016). Interestingly, there are about 4.25 out of 20 million people who are likely to be exposed to a high level of radiation, exceeding the threshold required for an evacuation (Table 4.11). Among the populations living beneath the projected radioactive plume footprint, Blacks and Asians were living in the pathway area at higher percentages than other races. While fewer of them live below the federal poverty line than compared with the counterpart population living outside of the plume's footprint (Kyne 2016).

Table 4.8 Demographic composition of population, as sorted by distance from Indian Point Nuclear Generating, New York, in 1990; *Source* (Kyne 2014)

	0–10	11–20	21–30	31–40	41–50	0-50	Outside
Year 1990							
Tract	55	187	596	1,252	1,670	3,760	4,002
Tract area (sq. miles)	292	1,005	1,487	1,919	1,107	5,810	63,011
Total population	229,882	749,366	2,453,297	4,911,005	5,442,748	13,786,298	15,221,250
White	201,501	647,849	1,855,732	3,162,924	3,096,381	8,964,387	13,431,768
Black	16,429	64,693	368,460	861,202	1,657,783	2,968,567	1,200,971
Asian	5,847	23,174	91,373	317,468	303,274	741,136	266,893
Native American	425	1,661	4,420	12,869	13,867	33,242	47,283
Others	5,669	11,994	133,340	556,475	371,457	1,078,935	274,366
Hispanic	18,731	43,014	324,710	1,214,297	825,975	2,426,727	648,859
Color	39,637	129,807	757,725	2,266,527	2,700,268	5,893,964	2,127,483
White (%)	87.65	86.45	75.64	64.40	56.89	65.02	88.24
Black (%)	7.15	8.63	15.02	17.54	30.46	21.53	7.89
Asian (%)	2.54	3.09	3.72	6.46	5.57	5.38	1.75
Native American (%)	0.18	0.22	0.18	0.26	0.25	0.24	0.31
Others (%)	2.47	1.60	5.44	11.33	6.82	7.83	1.80
Hispanic (%)	8.15	5.74	13.24	24.73	15.18	17.60	4.26
Color (%)	17.24	17.32	30.89	46.15	49.61	42.75	13.98
Female (%)	49.70	51.09	52.43	52.56	52.72	52.48	51.44
Old (65+ years) (%)	10.37	11.00	13.76	13.15	13.02	13.04	13.40
Kid (<5 years) (%)	8.35	8.53	8.11	7.60	8.36	8.05	8.44
Native-born (%)	89.09	87.42	81.67	74.57	76.13	77.39	93.56
Renter housing units (%)	25.75	26.01	41.22	60.66	53.40	52.34	28.04
Education (%)	32.51	34.60	29.01	28.94	20.61	26.10	22.16

(continued)

Table 4.8 (continued)

	0–10	11–20	21–30	31–40	41–50	0-50	Outside
Unemployment (%)	3.96	4.35	5.61	7.72	8.14	7.23	5.62
Poverty (%)	5.02	6.38	7.89	16.16	15.07	13.55	8.41
Mean household income ($)	103,352	113,938	97,399	82,528	70,724	82,538	72,323

Table 4.9 Demographic composition of population, as sorted by distance from Indian Point Nuclear Generating, New York, in 2000; *Source* (Kyne 2014)

	0–10	11–20	21–30	31–40	41–50	0–50	Outside
Year 2000							
Total population	242,165	826,902	2,627,286	5,352,294	5,845,359	14,894,006	15,902,366
White	196,235	673,908	1,759,588	3,040,855	2,825,575	8,496,161	13,272,190
Black	17,902	70,385	410,274	800,710	1,750,064	3,049,335	1,370,075
Asian	8,780	34,826	130,263	500,998	486,427	1,161,294	459,169
Native American	431	2,370	7,084	23,711	18,208	51,804	54,916
Others	18,804	45,422	320,086	985,956	765,109	2,135,377	746,051
Hispanic	30,788	77,568	495,739	1,592,978	1,054,119	3,251,192	1,048,920
Color	61,284	195,385	1,073,881	2,944,270	3,443,913	7,718,733	3,123,957
White (%)	81.03	81.50	66.97	56.81	48.34	57.04	83.46
Black (%)	7.39	8.51	15.62	14.96	29.94	20.47	8.62
Asian (%)	3.63	4.21	4.96	9.36	8.32	7.80	2.89
Native American (%)	0.18	0.29	0.27	0.44	0.31	0.35	0.35
Others (%)	7.76	5.49	12.18	18.42	13.09	14.34	4.69
Hispanic (%)	12.71	9.38	18.87	29.76	18.03	21.83	6.60
Color (%)	25.31	23.63	40.87	55.01	58.92	51.82	19.64
Female (%)	50.78	50.92	52.31	52.01	52.58	52.20	51.32
Old (65 + years) (%)	11.52	11.93	13.10	12.21	12.12	12.31	13.84
Kid (< 5 years) (%)	8.73	8.92	8.43	7.62	8.39	8.16	7.68
Native-born (%)	85.06	84.71	76.46	68.28	68.34	70.93	91.29
Renter housing units (%)	25.56	26.47	41.05	60.63	53.30	52.23	27.18
College degree or higher (%)	38.04	39.90	33.57	33.57	25.04	30.72	26.40
Unemployment (%)	3.56	4.01	5.68	8.10	8.42	7.47	5.67
Poverty (%)	6.01	7.92	10.23	17.31	17.28	15.35	9.17
Mean household income ($)	109,993	123,688	103,111	89,616	74,011	88,163	77,552

Table 4.10 Demographic composition of population, as sorted by distance from Indian Point Nuclear Generating, New York, in 2010; *Source* (Kyne 2014)

	0–10	11–20	21–30	31–40	41–50	0–50	Outside
Year 2010							
Total population	253,977	866,039	2,685,808	5,416,986	5,895,371	15,118,181	16,378,985
White	196,989	679,057	1,697,802	2,981,212	2,799,672	8,354,732	13,279,606
Black	21,031	75,492	428,486	765,664	1,716,532	3,007,205	1,502,391
Asian	12,594	45,390	166,660	646,636	662,158	1,533,438	695,603
Native American	261	1,937	5,135	17,716	16,853	41,902	52,616
Others	23,102	64,163	387,725	1,005,758	700,156	2,180,904	848,769
Hispanic	48,788	114,365	622,117	1,732,926	1,182,057	3,700,253	1,503,914
Color	83,856	244,517	1,238,828	3,138,954	3,625,473	8,331,628	3,940,985
White (%)	77.56	78.41	63.21	55.03	47.49	55.26	81.08
Black (%)	8.28	8.72	15.95	14.13	29.12	19.89	9.17
Asian (%)	4.96	5.24	6.21	11.94	11.23	10.14	4.25
Native American (%)	0.10	0.22	0.19	0.33	0.29	0.28	0.32
Others (%)	9.10	7.41	14.44	18.57	11.88	14.43	5.18
Hispanic (%)	19.21	13.21	23.16	31.99	20.05	24.48	9.18
Color (%)	33.02	28.23	46.12	57.95	61.50	55.11	24.06
Female (%)	49.24	51.22	51.94	51.90	52.29	51.98	51.10
Old (65+ years) (%)	12.80	13.01	13.16	12.55	12.00	12.47	14.14
Kid (<5 years) (%)	5.96	6.82	6.38	6.07	6.62	6.38	5.78
Native-born (%)	80.58	82.49	74.88	66.85	66.66	69.33	89.09
Renter housing units (%)	22.67	24.79	38.10	56.15	49.51	48.41	25.33
College degree or higher (%)	40.49	43.98	37.58	39.18	30.93	36.01	30.52
Unemployment (%)	6.00	5.55	7.45	8.29	8.35	7.97	7.26
Poverty (%)	7.08	8.73	10.30	16.31	15.63	14.39	10.14
Mean household income ($)	111,702	124,939	105,091	93,828	78,343	91,901	79,841

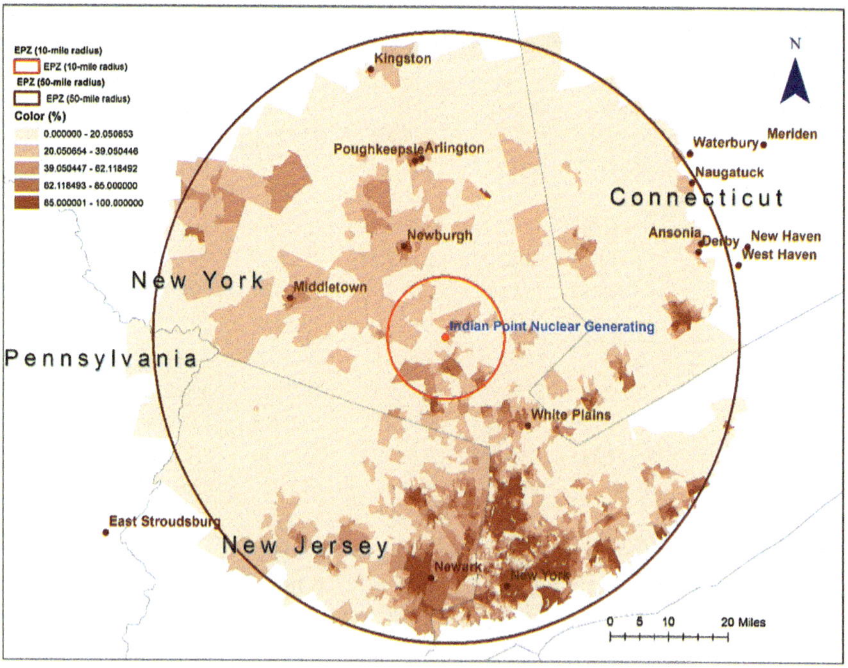

Fig. 4.6 Spatial distribution of percent Color among the populations surrounding Indian Point Nuclear Generating

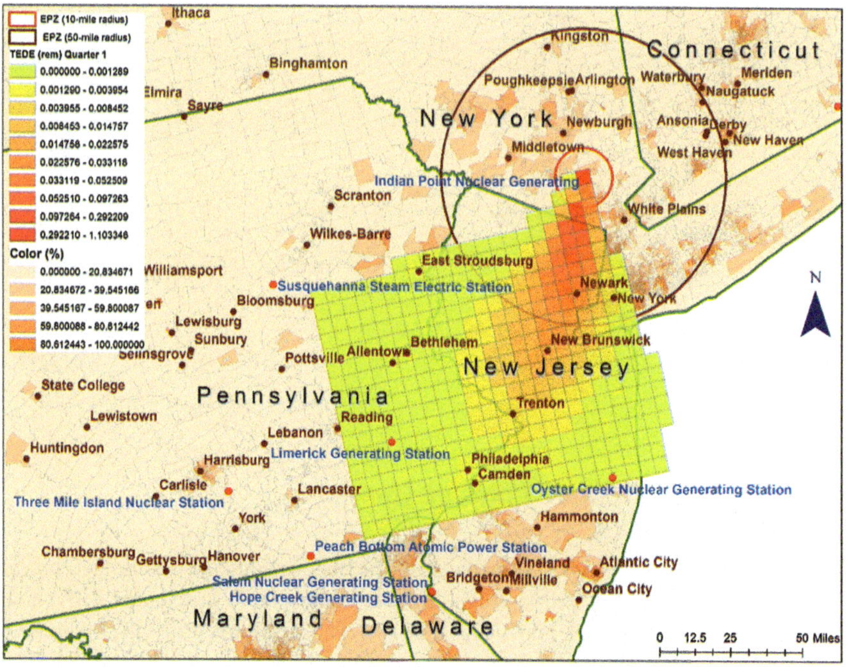

Fig. 4.7 The projected plume path emanating from Indian Point Nuclear Generating after a simulated core-damage accident conducted under meteorological conditions typifying Quarter 1; *Source* (Kyne 2016)

Table 4.11 Demographic composition in areas around Indian Point Nuclear Generating located beneath the projected plume path with meteorological conditions typifying Quarter 1; *Source* (Kyne 2016)

Demographics/ (TEDE) rem)	(0.000–0.0003)	(0.0003–0.002)	(0.002–0.005)	(0.005–0.01)	(0.01–2.00)	Total
Plume pathway area (sq. miles)	4,633	2,252	951	650	1,126	9,612
Total population	6,406,282	4,043,251	3,798,295	1,350,646	4,256,533	19,855,006
White	3,931,013	2,688,114	2,087,612	916,557	2,664,943	12,288,238
Black	1,199,355	753,589	793,762	137,920	694,945	3,579,571
Asian	486,509	319,271	336,838	201,967	397,238	1,741,823
Native American	16,001	8,610	12,024	3,481	8,674	48,789
Others	773,403	273,667	568,060	90,721	490,733	2,196,585
Hispanic	1,241,120	444,956	950,273	197,915	1,001,888	3,836,151
Color	2,977,345	1,556,422	2,075,321	551,010	2,129,061	9,289,159
White (%)	61.36	66.48	54.96	67.86	62.61	61.89
Black (%)	18.72	18.64	20.90	10.21	16.33	18.03
Asian (%)	7.59	7.90	8.87	14.95	9.33	8.77
Native American (%)	0.25	0.21	0.32	0.26	0.20	0.25
Others (%)	12.07	6.77	14.96	6.72	11.53	11.06
Hispanic (%)	19.37	11.00	25.02	14.65	23.54	19.32
Color (%)	46.48	38.49	54.64	40.80	50.02	46.78
Female (%)	51.91	51.80	52.46	51.67	51.30	51.85
Old (65+ years) (%)	13.18	13.78	11.78	12.93	12.37	12.84
Kid (<5 years) (%)	6.29	6.54	5.86	6.04	6.65	6.32
Native-born (%)	79.48	79.82	72.16	74.25	72.42	76.28
Renter housing units (%)	38.60	34.65	60.10	35.06	38.35	42.16
College degree or higher (%)	28.83	29.77	44.22	43.17	35.39	34.44
Unemployment (%)	8.29	8.33	8.26	6.66	7.64	8.04
Poverty (%)	14.95	13.53	18.04	8.78	10.55	13.89
Mean household income ($)	72,709	75,906	98,518	105,781	94,536	85,420

4.4.3 *Projected Plume Path for Quarter 2*

The projection exercise was conducted with the meteorological conditions typifying Quarter 2 (April to June) with a prevailing wind direction to the South (180°) and a wind speed of 3 miles per hour (Kyne 2016). The resulting projected radioactive

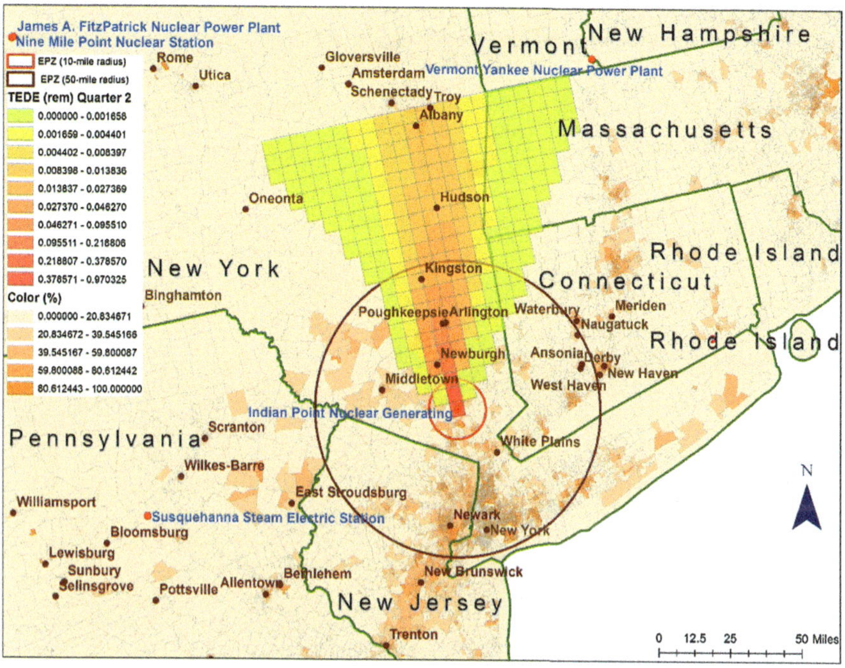

Fig. 4.8 The projected plume path emanating from Indian Point Nuclear Generating after a simulated core-damage accident conducted under meteorological conditions typifying Quarter 2; *Source* (Kyne 2016)

plume dispersion footprint shows that the radioactive plume was pushed toward the northerly direction covering an estimated area of 6,654 square miles and there was an estimated population of 1.4 million people (1,390,699) exposing to the radiation level ranging from between 0.00 and 0.97 REM, which is about the 1 REM that requires either evacuation or sheltering by the NRC's protection action guidelines (Fig. 4.8, Table 4.12) (Kyne 2016). Among the populations exposing to the radiation underneath the plume footprint, the presence of White (83%) is observed as predominant, whereas minorities consist of Black (8%) and Hispanic (8%); about 10% of the total impacted people are estimated to live under the poverty (Table 4.12) (Kyne 2016).

4.4.4 *Projected Plume Path for Quarter 3*

This resulting projected radioactive plume dispersion under meteorological conditions typifying Quarter 3 (July to September) with a prevailing wind direction of north-east (45°) and an average wind speed of 3 miles per hour shows a radioactive plume footprint covering an area of 8,957square miles (Table 4.13) and an

Table 4.12 Demographic composition in areas around indian point nuclear generating located beneath the projected plume path with meteorological conditions typifying Quarter 2; *Source* (Kyne 2016)

Demographics/ (TEDE) rem)	(0.000–0.0003)	(0.0003–0.002)	(0.002–0.005)	(0.005–0.01)	(0.01–2.00)	Total
Plume pathway area (sq. miles)	2,276	1,176	825	1,126	1,251	6,654
Total population	205,832	172,200	76,904	414,501	521,261	1,390,699
White	186,572	156,525	71,071	337,736	410,013	1,161,917
Black	6,656	6,315	2,219	42,175	56,533	113,899
Asian	3,322	2,445	1,058	15,974	14,175	36,974
Native American	425	320	88	908	897	2,639
Others	8,857	6,595	2,468	17,709	39,642	75,270
Hispanic	15,235	8,822	3,716	18,931	65,882	112,586
Color	27,459	20,457	8,383	86,031	143,661	285,990
White (%)	90.64	90.90	92.42	81.48	78.66	83.55
Black (%)	3.23	3.67	2.89	10.17	10.85	8.19
Asian (%)	1.61	1.42	1.38	3.85	2.72	2.66
Native American (%)	0.21	0.19	0.11	0.22	0.17	0.19
Others (%)	4.30	3.83	3.21	4.27	7.61	5.41
Hispanic (%)	7.40	5.12	4.83	4.57	12.64	8.10
Color (%)	13.34	11.88	10.90	20.76	27.56	20.56
Female (%)	49.36	51.25	49.36	51.29	50.61	50.64
Old (65+ years) (%)	13.65	17.10	14.78	13.98	13.54	14.20
Kid (<5 years) (%)	4.97	4.90	4.35	5.24	5.59	5.24
Native-born (%)	93.11	94.33	94.20	91.80	88.63	91.25
Renter housing units (%)	15.21	23.84	17.07	34.12	30.59	27.63
College degree or higher (%)	30.86	29.20	35.30	35.28	28.94	31.49
Unemployment (%)	6.55	7.19	5.64	6.58	6.98	6.75
Poverty (%)	7.19	10.62	7.20	12.44	10.55	10.44
Mean household income ($)	85,701	70,260	82,767	73,346	77,576	76,703

approximately 12.4 million people living beneath the projected plume dispersed areas (Kyne 2016). The impacted population is anticipated to be exposed to radiation levels ranging from 0.00 and 1.565 REM, which is above the 1 REM TEDE required to initiate either an evacuation or shelter-in-place order (Table 4.13) (Kyne 2016). The projected radioactive plume would cover areas in three states: New York, New Jersey, and Pennsylvania, placing the cities and their surrounding areas,

Table 4.13 Demographic composition in areas around Indian Point Nuclear Generating located beneath the projected plume path with meteorological conditions typifying Quarter 3; *Source* (Kyne 2016)

Demographics/ (TEDE) rem)	(0.000–0.0003)	(0.0003–0.002)	(0.002–0.005)	(0.005–0.01)	(0.01–2.00)	Total
Plume pathway area (sq. miles)	3,603	2,527	1,151	651	1,026	8,957
Total population	3,454,294	5,215,402	1,482,878	1,047,547	1,236,696	12,436,816
White	2,594,096	3,437,311	974,204	622,578	1,016,798	8,644,986
Black	414,021	1,050,103	208,426	249,870	62,291	1,984,710
Asian	223,345	318,739	170,130	55,775	97,536	865,525
Native American	6,541	10,406	2,319	2,021	2,378	23,666
Others	216,292	398,843	127,799	117,302	57,692	917,928
Hispanic	459,556	697,099	229,760	213,461	121,487	1,721,363
Color	1,132,178	2,120,398	626,265	529,505	294,388	4,702,735
White (%)	75.10	65.91	65.70	59.43	82.22	69.51
Black (%)	11.99	20.13	14.06	23.85	5.04	15.96
Asian (%)	6.47	6.11	11.47	5.32	7.89	6.96
Native American (%)	0.19	0.20	0.16	0.19	0.19	0.19
Others (%)	6.26	7.65	8.62	11.20	4.67	7.38
Hispanic (%)	13.30	13.37	15.49	20.38	9.82	13.84
Color (%)	32.78	40.66	42.23	50.55	23.80	37.81
Female (%)	51.40	51.67	51.25	51.62	51.22	51.50
Old (65+ years) (%)	13.26	13.38	12.14	11.57	13.66	13.08
Kid (<5 years) (%)	6.31	6.13	6.54	6.67	6.27	6.29
Native-born (%)	84.46	85.45	76.71	77.30	82.03	83.11
Renter housing units (%)	29.62	32.14	31.05	35.24	20.71	30.51
College degree or higher (%)	33.17	30.89	38.11	34.21	48.92	34.46
Unemployment (%)	7.53	8.40	7.21	8.16	5.69	7.72
Poverty (%)	9.96	12.87	8.38	10.74	5.37	10.59
Mean household income ($)	87,696	77,504	95,126	95,031	133,381	89,125

primarily Newark, East Stroudsburg, New Brunswick, Bethlehem, Trenton, and Allentown, to higher levels of TEDEs, shown in red and brown colors (Fig. 4.9) (Kyne 2016). The impacted populations include higher presence of minority groups including People of Color, Black, Hispanic, and Asian, most living under poverty (Table 4.13).

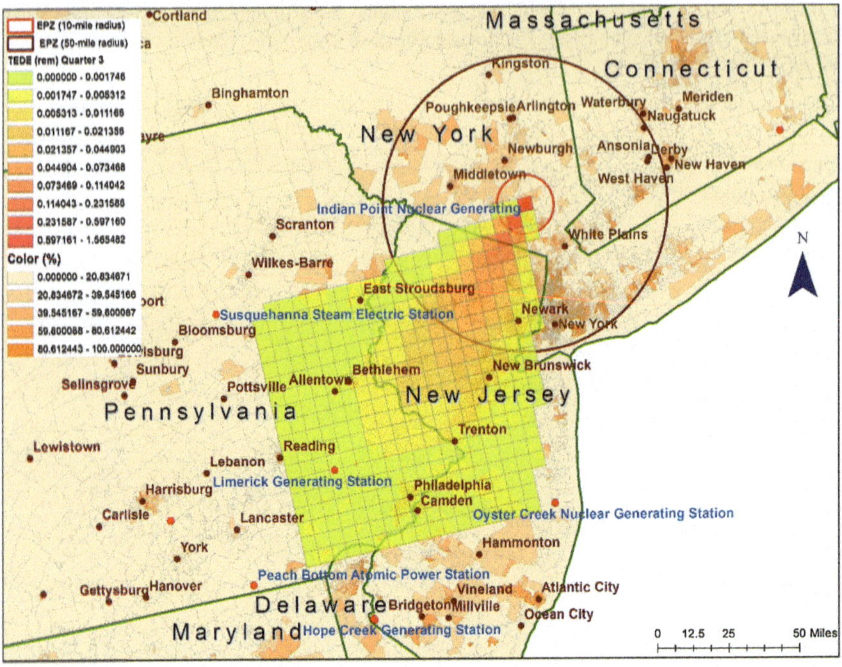

Fig. 4.9 The projected plume path emanating from Indian Point Nuclear Generating after a simulated core-damage accident conducted under meteorological conditions typifying Quarter 3; *Source* (Kyne 2016)

4.4.5 *Projected Plume Path for Quarter 4*

This projection of radioactive plume dispersion was conducted inputting the meteorological conditions typifying Quarter 4 (October to December), with a prevailing wind direction to the north-east (23°) and a wind speed of 6 miles per hour (Fig. 4.10). The resulting plume footprint covers an area as large as 5,054 square miles (Table 4.14) and an estimated 10.9 million people expose to the radiation carried by the plume ranging from 0.00 and 0.134 REM, which is lower than the 1 REM threshold to call for either an evacuation or shelter-in-place order (Kyne 2016). The impacted populations include a higher presence of Black, Hispanic, and Asian (Table 4.14). The projected radioactive plume pathway placed the six cities and their surrounding areas, namely, Newark, New York, New Brunswick, Trenton, Philadelphia, and Camden under the higher TEDEs level (Fig. 4.10) (Kyne 2016).

The four projection exercises provide estimated populations living in the impacted areas and their exposure to radiation from the dispersed radioactive plume

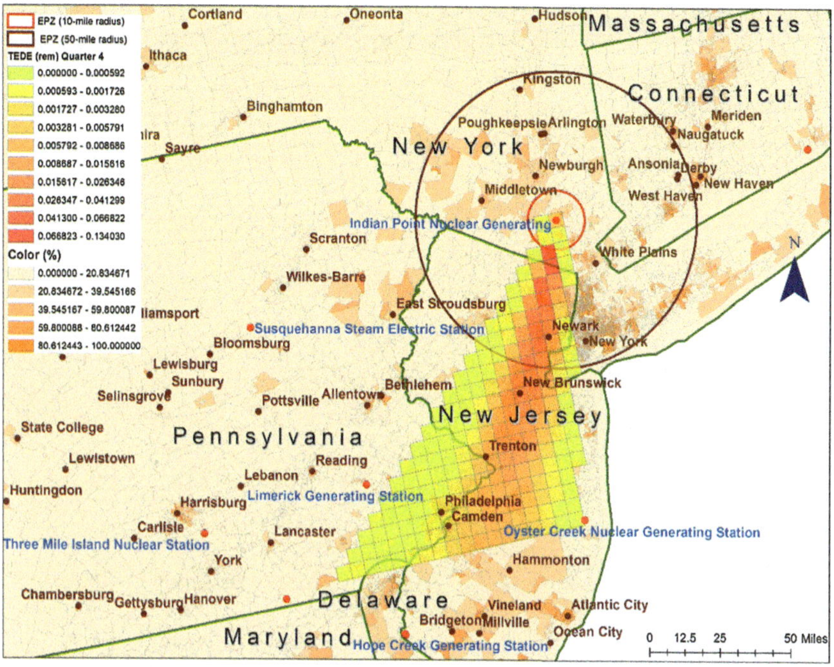

Fig. 4.10 The projected plume path emanating from Indian Point Nuclear Generating after a simulated core-damage accident conducted under meteorological conditions typifying Quarter 4; *Source* (Kyne 2016)

scenario. In the projection with meteorological conditions typifying Quarter 1, the resulting plume pathway placed an estimated population of 7.2 million people, 7.68 million, 4.89 million, and 0.06 million population living beneath the radioactive plume pathway areas located in the four states, namely New York, New Jersey, Pennsylvania and Delaware, respectively (Fig. 4.11) (Kyne 2016). Among the four states, New Jersey hosts the largest population that could be exposed to dangerous levels of radiation. Similar findings are observed in projection results with meteorological conditions typifying Quarter 4 (October to December) (Kyne 2016). The projection results with typifying weather conditions in Quarter 3 (July to September) provide estimates in populations of1 million people in New York, about 6.78 million in New Jersey, 4.91 million in Pennsylvania, and 0.06 million in Delaware; indicating that the New Jersey host the largest population that would be exposed to radiation. The projection results with the typifying weather conditions in quarter 2 (April to June) show that 1.23 million out of the total estimated 1.39 million living in impacted areas live in New York (Fig. 4.11) (Kyne 2016).

Table 4.14 Demographic composition in areas around Indian Point Nuclear Generating located beneath the projected plume path with meteorological conditions typifying Quarter 4; *Source* (Kyne 2016)

Demographics/ (TEDE) rem)	(0.000–0.0003)	(0.0003–0.002)	(0.002–0.005)	(0.005–0.01)	(0.01–2.00)	Total
Plume pathway area (sq. miles)	1,926	901	676	701	851	5,054
Total population	2,866,750	2,831,498	1,171,610	1,015,330	3,029,595	10,914,783
White	2,117,502	1,755,476	852,117	722,686	1,789,422	7,237,203
Black	396,992	635,352	162,029	142,431	588,037	1,924,840
Asian	226,263	196,489	79,354	70,270	300,291	872,667
Native American	4,572	6,976	2,674	1,816	5,751	21,790
Others	121,422	237,204	75,438	78,126	346,093	858,283
Hispanic	242,294	461,169	123,837	159,521	642,261	1,629,084
Color	895,341	1,316,486	381,473	385,785	1,560,135	4,539,219
White (%)	73.86	62.00	72.73	71.18	59.06	66.31
Black (%)	13.85	22.44	13.83	14.03	19.41	17.64
Asian (%)	7.89	6.94	6.77	6.92	9.91	8.00
Native American (%)	0.16	0.25	0.23	0.18	0.19	0.20
Others (%)	4.24	8.38	6.44	7.69	11.42	7.86
Hispanic (%)	8.45	16.29	10.57	15.71	21.20	14.93
Color (%)	31.23	46.49	32.56	38.00	51.50	41.59
Female (%)	51.79	51.90	51.64	50.75	51.47	51.62
Old (65+ years) (%)	14.71	12.71	13.05	13.01	12.24	13.17
Kid (<5 years) (%)	5.98	6.42	6.10	6.05	6.79	6.34
Native-born (%)	85.01	83.31	86.06	79.83	73.42	80.98
Renter housing units (%)	28.70	35.48	26.68	30.34	36.70	32.55
College degree or higher (%)	42.45	28.91	32.55	33.45	35.16	35.07
Unemployment (%)	6.46	9.72	7.63	7.16	7.91	7.88
Poverty (%)	8.75	16.77	8.45	8.62	10.50	11.27
Mean household income ($)	101,091	73,630	90,017	92,233	93,212	89,745

4.5 Projection Limitations

The RASCAL computer code is not free from limitations in carrying out estimating radioactive plume dispersion (Kyne 2014). The projection is highly dependent on prevailing weather conditions on the day the event takes place (Ramsdell et al.

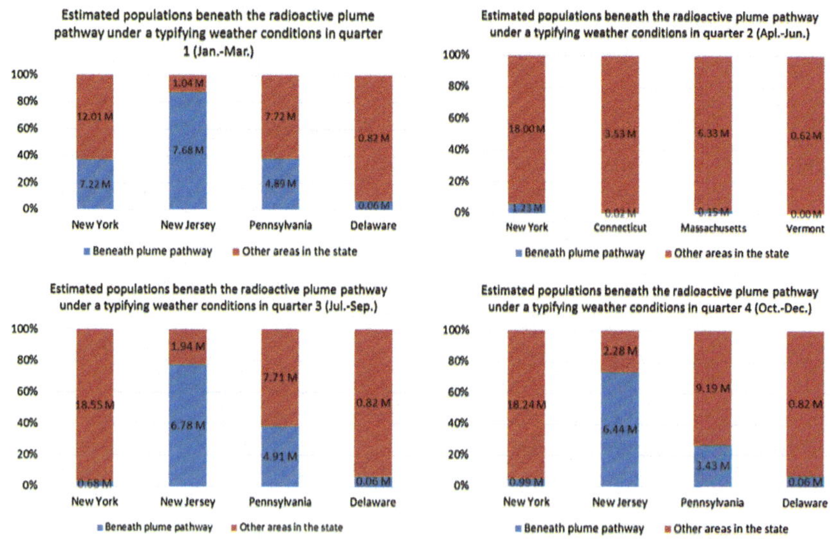

Fig. 4.11 A summary of total populations in areas around Indian Point Nuclear Generating located beneath the projected plume path with meteorological conditions typifying four quarters; *Source* (Kyne 2016)

2013). It should be noted that the eight projections conducted in this chapter may not provide the actual plume footprint that would occur in the event actual accident, on any given day, during any calendar quarter because the assumed typifying weather conditions could be different from those of the actual conditions on the day the event takes place (Kyne 2014). Second, it is not advisable to use the this stimulation data in briefings delivered to decision markers making critical decisions for evacuation during an actual emergency (Ramsdell et al. 2013). The resulting projected stimulation data must be incorporated with the current meteorological data for a few hours prior to the event (Kyne 2014). Third, extreme caution must be taken when using the radiation level of TEDE estimated by the STDose model in the RASCAL computer code (Kyne 2014). During the emergency, the actual radiation does must be recomputed with the data collected on site by the EPA's Radiological Emergency Response Team (RERT); the verification and computation of the TEDE could be done using the Field Measurement to Dose model in the RASCAL computer code (RASCAL Class Training July 11 and 12, 2013). In the beginning, the projection provides estimated TEDE levels, and either evacuation or shelter-in-place suggestions are provided based on the EPA's *Manual of protective action guides and protective actions for nuclear incidents* (US EPA 1992) while a does of 1 REM is recommended by the NRC to evacuate in most incidents and the agency did not specify minimum level for a shelter-in-place order (Ramsdell et al. 2013; Kyne 2014).

4.6 Conclusion

This chapter demonstrates the significant potential and scale of a nuclear power emergency and its impacts, providing tools necessary to understand where the plume path will disperse if a nuclear core-melt down accident occurs. The immediate threat after the nuclear core-melt down accident is the release of radioactive toxic waste to the environment. Radiation doses could be lethal. During the Fukushima accident, the radiation dose was measured to be nearly 12,000 $\mu Sv\ h^{-1}$ near the main gate of the plant (Thielen 2012). The main radionuclide releases were Iodine-131 which has a half-life of 8 days, Caseium-137 which has a 30-year half-life, and Caesium-134 which has a two-year half-life (World Nuclear Association 2016). One of the four activities that needs to be carried out in the response stage of the response management process is to project radioactive plume path dispersion in order to estimate the radiation does level and to determine the evacuation zones. This chapter also demonstrates the two simulation exercises to project radiative plume path dispersion at two nuclear power plants: Palo Verde Nuclear Generating Station (PVNG) in Arizona and Indian Point Nuclear Generation (IPNG) in New York. The impact of a nuclear power accident depends on weather conditions on the day the event takes place. In the simulation exercise for PVNG, the estimated population underneath the radioactive plume path could be 0.66, 0.96, 3.5, and 2.3 million in typifying weather conditions, in quarters one through four, respectively. (Tables 4.4, 4.5, 4.6, and 4.7). In the similar simulation for IPNG, the estimated population underneath the radioactive plume path could be 19.86, 1.39, 12.49, and 10.91 million, in typifying weather conditions in quarters one through four respectively (Tables 4.11, 4.12, 4.13, and 4.14). With such population sizes being vulnerable to exposure of the radioactive plume pathway, carrying out emergency management process effectively is critical for the lives of nearly 3.5 million people, and 19.86 million people in the simulated worst case scenarios at PVNG and IPNG respectively. Any delay in the four activities in response stage, namely projection of radioactive plume dispersion, protective action recommendation, protective action decision, and evacuation will place people living near the power plant in danger of exposure to the high level radioactive material releases. The previous nuclear power accidents including Three Mile Island, Chernobyl, and Fukushima Daiichi power plants share a common issue, which is that of a delay in evacuation of people living close to the nuclear power plants. The delayed evacuation might have caused an increase in the unnecessary exposure to the radioactive materials for populations that were not able to evacuate in a timely manner. To save millions of people from unnecessary and dangerous exposure to radiation, addressing the problem with the delay in evacuation is imperative. To solve this issue, two preventive actions are necessary: administering potassium (KI) pills and informing the emergency alert to the residents before the nuclear core-melt down accident takes place.

References

Kyne D (2014) *Environmental justice issues in communities hosting US nuclear power plants* (Order No. 3617964). Available from ProQuest Dissertations & Theses Global. (1528556250). Retrieved from http://ezhost.utrgv.edu:2048/login?url=http://search.proquest.com.ezhost.utrgv.edu:2048/docview/1528556250?accountid=7119

Kyne D (2015) Managing nuclear power plant induced disasters. J Emerg Manag (Weston, Mass.) 13(5):417–429. doi:10.5055/jem.2015.0252. Retrieved from Prime National Pub. Corp

Kyne D (2016) A Fukushima-like disaster in New York city and possible issues associated with the extreme event management system. J Extreme Events 3(2). doi:10.1142/S234573761650007X

Normile D (2016) Five years after the meltdown, is it safe to live near Fukushima? Retrieved from Science http://www.sciencemag.org/news/2016/03/five-years-after-meltdown-it-safe-live-near-fukushima

Ramsdell JV Jr, Athey GF, Rishel JP (2013) RASCAL 4.3 user's guide (Draft). Retrieved from U.S. Nuclear Regulatory Commission http://pbadupws.nrc.gov/docs/ML1328/ML13281A701.pdf

RSICC (2013) RASCAL 4.3: radiological assessment systems for consequence analysis Reactor and Nuclear Systems Division (RNSD), Oak Ridge National Laboratory, The Radiation Safety Information Computational Center (RSICC). Retrieved from https://rsicc.ornl.gov/PackageDetail.aspx?p=RASCAL%204.3&id=C00783&cpu=PCX86&v=02&t=Radiological%20Assessment%20Systems%20for%20Consequence%20AnaLysis

Thielen H (2012) The Fukushima Daiichi nuclear accident-an overview. Health Physics 103(2): 169–174. Retrieved from WOS:000306109900009

US EPA (1992) Manual of protective action guides and protective actions for nuclear incidents (EPA-400-R-92-001). Office of Radiation Programs, United States Environmental Protection Agency, Washington, DC. Retrieved from: http://www.epa.gov/radiation/docs/er/400-r-92-001.pdf

US EPA (2013) Manual of protective action guides and protective action for nuclear incidents. Office of Radiations Program, United States Environmental Protection Agency, Washington DC 20460. Retrieved from: http://www.epa.gov/sites/production/files/2015-06/documents/pag-manual-interim-public-comment-4-2-2013.pdf

US NRC (2011) Nuclear Regulatory Commission. Retrieved from US Nuclear Regulatory Commission www.nrc.gov

US NRC (2013) RASCAL 4.3 Workbook. US Nuclear Regulartory Commission, Washington, DC 20555-0001. Retrieved from http://pbadupws.nrc.gov/docs/ML1328/ML13281A475.pdf

World Nuclear Association (2016) Fukushima accident. Retrieved from World Nuclear Association http://www.world-nuclear.org/information-library/safety-and-security/safety-of-plants/fukushima-accident.aspx

References

Chapter 5
Issues Associated with the Nuclear Power Emergency

5.1 Policy Issues Associated with Nuclear Power Emergency Response

The nature of the nuclear emergency management process requires urgency and timeliness in carrying out activities. Why does the process require "urgency" and "timeliness" actions? The powerful destructive nuclear fuel assemblies containing hundreds of fuel rods may uncover and may enter into a meltdown state. In this stage, high-level radioactive plume is released from the reactor and could disperse into the environment. Soon after a nuclear power plant encounters the core-meltdown accident, the immediate threat is the emission of radioactive plume. How fast the highly toxic radioactive plume could disperse or travel will depend on the wind speed, wind direction, precipitation, temperature, and weather conditions on that particular day. For example, if the wind is at 10 miles per hour, then the radioactive plume could travel a distance of 10 miles per hour. Thus, this requires urgency and timeliness in carrying out the activities related to the emergency management process. If there is a delay, the plume path could even travel longer distances, placing more people at risk of exposure to high levels of radioactive plume.

Understanding the nature of the nuclear power emergency management response system, the NRC has set rigorous steps to follow in carrying out the response activities. The NRC sets the policy to reflect the urgency and timeliness nature of the nuclear emergency management by allocating 15 min each for the first three activities (US NRC 2014). The Federal Emergency Management Agency (FEMA) also plays an important role in assuring that the emergency planning in EPZ is in order. The facility owners must follow the NRC guidelines in preparing the emergency plan for EPZ while they need approval by FEMA, which also has the authority to set the minimum time required to alert and notify in order to completely cover 100% of the plume path exposure zone within a recommended goal of 45 min (US FEMA 2016).

D. Kyne, *Nuclear Power Plant Emergencies in the USA*,
DOI 10.1007/978-3-319-50343-1_5

The NRC's policy seems to promote the effectiveness of successfully carrying out the activities in the response phase of the emergency management process that reflects the nature of urgency and timeliness. To recall response activities discussed earlier, there are four steps in the response activities:

(1) projection of plume path dispersion,
(2) protective action recommendation (PAR),
(3) protective action decision (PAD), and
(4) evacuation.

To evaluate the list above, NRC's policy toward four response activities, the practicality of 15 min for each activity should be investigated. Looking at the practicality of the 15-min allocated time, there are two research questions that arise:

(1) Is this 15-min time doable in practice? and
(2) What are some potential issues associated with the policy toward response activities?

In search of answers to the questions, it is imperative to look into a projected case of a core-meltdown accident at the PVNGS. The response activities to be carried out are depicted in Fig. 5.1.

Step (1 & 2) Project plume path dispersion and provide a PAR
As depicted in the flowchart in Fig. 5.1, the first two steps involve two parties, the PVNGS and the ARRA. The two steps are closely related and they could be studied together.

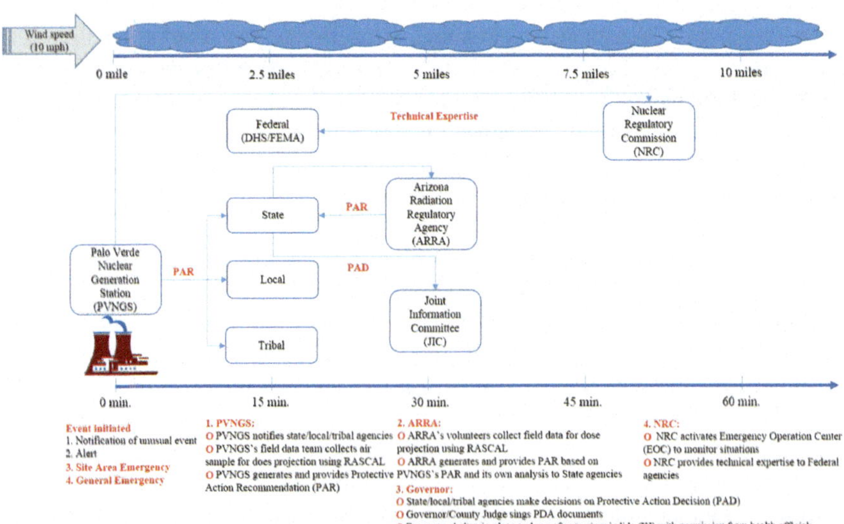

Fig. 5.1 A tentative timeline for preparing a protective action decision (PAD) in the case of a GEm at PVNGS, *Source* (Kyne 2015) "Reprinted from Journal of Emergency Management. Used with permission. All Rights Reserved"

Palo Verde Nuclear Generating Station (PVNGS): The first activity to be carried out by the plant personnel is to provide a protective action recommendation (PAR) to the state, local, and tribal agencies, within 15 min from the time of the initial incident. To do so, the plant personnel must follow the following steps:

(1) become aware of the initiation of an incident,
(2) gather meteorology information for projection,
(3) conduct a projection of radioactive plume path,
(4) have a team meeting to generate a PAR based on the projection results,
(5) deploy staff to collect the air sample beneath the radioactive plume pathway,
(6) update the projection with the changes in weather conditions,
(7) validate the projection using the collected air sample data,
(8) standby to answer to any questions other agencies may arise, and
(9) coordinate with other agencies for further actions.

Among the above-mentioned nine steps, the first four steps are mandatory to generate a PAR at the plant. A closer look at these first four steps has been taken into consideration, in order to estimate the time requirement.

The first challenge the plant personnel must face is becoming aware of the incident as soon as it is initiated. There are mechanism and devices that help indicate any potential destruction or failure of any system or parts. It could be fairly stated that the plant personnel will be notified with the assistance of devices and mechanism about any incident. In the past, the plant experienced two incidents which went unnoticed for an unknown period of time. One incident was that one of the three reactors experienced a water leak from the reactor vessel which contains hundreds of nuclear fuel rods (Randazzo 2013). The leak was discovered in reactor unit 3, on October 5, 2013, when the plant underwent regular re-fueling every 18–24 months. The leak, which releases radioactive fluid at a rate of 0.01 gallons per minute was located within the reactor unit housing, and did not reach the outside environment. The humidity detector inside the reactor's containment dome did not detect the leak (Kyne 2015). The other incident was that the plant had to shut down because there was a problem with a value on one of the reactor's two steam generators, which occurred on November 11, 2013 (The Associated Press 2013). That incident was recorded as a near-miss event by the NRC due to the risks of damage to the reactor core (Lochbaum 2012). Given the past experience, it is highly likely that the plant management may not notice the incident as soon as it is initiated and it is highly likely that the plant management encounters an unknown amount of delay in preparing a PAR, within 15 minutes of a core-meltdown accident (Kyne 2015).

Next, once the plant personnel become aware of the incident, they must project a radioactive plume pathway and radiation dose. The plant personnel must utilize the Radiological Assessment System for Consequence Analysis (RASCAL) (RSICC 2013). The computer code will allow the plant personnel to project an estimated radiation dose. The radiation dose was computed as a total effective dose equivalent (TEDE), which is defined as "the sum of the deep-dose equivalent (EDE) (for external exposures) and the committed effective dose equivalent (CEDE) (for internal exposures)" (US NRC 2011). How long will it take to conduct a plume projection?

(1) To start, the plant personnel must switch on the computer and open the RASCAL code program. This will take about 5 min.
(2) There is preinstalled information in the computer code about the meteorology data such as wind speed, wind direction, temperature, precipitation, and stability condition at the plant and other stations. The plant meteorology data could be available, but the plant personnel may need to type them in. Other weather information at other weather stations that were included in the projection could be available to download directly from online systems, using direct download function in the RASCAL program. This step may require at least 30 min to project a radioactive plume dispersion and estimated dose.
(3) Once the projected plume dispersion pathway and the estimated radioactive dose are ready, it is necessary to call for a meeting with team members who are responsible for making decisions for certain conclusions to be included in the PAR. Then, a PAR must be formally prepared, which is a process that could require an additional 15 min.
(4) When the PAR is ready, it could be transmitted via a fax or email to the ARRA. In addition, a telephone call must be made in order to let the agency officials know that the PAR is transmitted and to assure they received the documents successfully. In some cases, the radioactive plume projection file must be saved as "case file" with an extension ".STD." This file will also help ARRA in checking the input information, including meteorology data, and reuse this information for projection at their end. These steps may also require at least another 5 min of time.

In practice, the process of generating PAR at the PVNGS would require at least 55 min. This 55-min period excludes the time the incident may have gone unnoticed. If there is a period of time when the incident goes unnoticed, the process would take even longer than 55 min. However, the completion of generating and delivering the PAR does not stop PVNGS's activities to an end and PVNGS has to carry out other activities. It has to deploy staff or its team members to collect the air sample beneath the radioactive plume pathway. The weather conditions could be very fluid depending on the day and month of the year. It is also necessary to monitor the conditions by collecting the up-to-date meteorology data and updating the plume path dispersion with the changes in weather conditions. As discussed in Chap. 3, there are four phases in nuclear emergency: prerelease, plume (early), intermediate, and ingestion. The projection of plume dispersion for the first two phases must be done with the Source Term to Dose (STDose) , whereas the latter two phases must be computed with the Field Measurement to Dose. To project the plume dispersion in the latter two phases, the PVNGS must collect air sample data beneath the plume pathway after the radioactive plume has started dispersion. Collection of field data at 10-min intervals along the central axis requires a certain amount of time. The latter two phases require the use the Field Measurement to Dose model to project plume dispersion. Therefore, the PVNGS must deploy its team member to collect air samples. Next, the PVNGS personnel must standby to answer any questions the other agencies may have after they have sent out their PAR especially from the ARRA agency. In addition, the plant personnel must coordinate with other agencies in carrying out other activities.

Arizona Radiation Regulatory Agency (ARRA): After ARRA receives PAR information, this step requires ARRA to project a plume dispersion based on the PAR prepared by the plant personnel at PVNGS (Kyne 2015). The preceding step is preparing PAR by PVNGS. In other words, ARRA could begin this step only after the agency receives the PAR report from the plant personnel at the PVNGS. How could ARRA prepare a PAR at their end? The answer to this question is to follow certain steps:

(1) receive a PAR prepared by the PVNGS,
(2) gather meteorology information for projection,
(3) conduct a projection of radioactive plume path,
(4) have a team meeting to generate a PAR based on the projection results,
(5) deploy staff to collect air samples beneath the radioactive plume pathway,
(6) update the projection with the changes in weather conditions,
(7) validate the projection using the collected air sample data,
(8) standby to answer to any questions other agencies may arise, and
(9) coordinate with other agencies for further actions.

In order to begin with the response activities at the ARRA end, the first step is to receive the PAR prepared by the PVNGS. ARRA's primary goal in this exercise is to validate the PAR and the projection of radioactive plume dispersion generated by the PVNGS. ARRA will use the same computer code RASCAL that PVNGS used in projecting plume dispersion. Therefore, ARRA might find it useful to use the case file saved in a file with an extension ".STD." How long will it take to conduct a plume projection?

(1) First, ARRA personnel must switch on the computer and open the RASCAL code program. This will take about 5 min.
(2) There is preinstalled information in the computer code, about the meteorology data such as wind speed, wind direction, temperature, precipitation, and stability condition at the plant and other stations. The plant meteorology data could be available, but the plant personnel may need to type them in. Other weather information at other weather stations that were included in the projection could be available to download directly from online using direct download function in the RASCAL program. This step may require at least 30 min to project a radioactive plume dispersion and estimated dose.

(3) Once the projected plume dispersion pathway and the estimated radioactive dose are ready, it is necessary to call for a meeting with team members who are responsible for making decisions for certain authorizations to be included in the PAR. In the decision-making process, ARRA must make a comparison between their results and the results obtained by the PVNGS. Then, a PAR must be formally prepared. The PAR of ARRA and PVNGS may or may not be the same. This process could require another 20 min.
(4) When the PAR is ready, it could be transmitted via a fax or an email to the state, local, and tribal offices. In addition, a telephone call must be made in order to let the agency officials know that the PAR is transmitted and they received them successfully. This informing process many also require at least another 15 min.

In practice, the process of generating PAR and validating the PAR of PVNGS at the ARRA office would require at least 70 min. However, upon completion of the tasks of generating, validating, and delivering the PAR to the state, local, and tribal offices does not bring ARRA activities to an end. ARRA has to carry out other activities as well. ARRA is responsible to lead a group of volunteers to collect field data, including radiation surveys, and environmental and foodstuff samples, at the PVNG plant site. ARRA must provide training courses, including a week-long training and day-long annual refresher course, to a group of volunteers. In addition, ARRA is also responsible for collecting up-to-date meteorology data in order to update the projection of plume path dispersion. It must also project the plume dispersion using the Field Measurement to Dose model in order to estimate the radiation level, in the radioactive plume dispersed in the intermediate, and ingestion phases, utilizing the up-to-date meteorology data, and air and food sample data collected by its volunteers. ARRA officials must prepare and remain on standby to answer any questions from the state, federal, local, and tribal agencies on the PAR provided. Finally, ARRA officials are responsible for coordinating with other federal, local, state, and tribal agencies in order to carry out other related activities (Kyne 2014).

Step (3) Making a protective action decision (PAD)
State, local, and tribal agencies: In this step, the state, local, and tribal agencies have been provided with the PAR. Their goal is that the state, local, and tribal agencies review the PAR provided by ARRA in order to prepare a protection action decision (PAD).

(1) receive a PAR prepared by the ARRA,
(2) meet responsible parties to generate a PAD including instructions for evacuation, sheltering-in-place, and the administering of potassium iodide (KI),
(3) sign the PAD by the Governor or the County Judge,
(4) send the PAD to all concerned agencies,
(5) standby to answer any questions or concerns that may arise, and
(6) coordinate with other agencies for further actions.

The PAD must be prepared based on the information contained in the PAR provided by ARRA. Therefore, the concerned agencies at the state, local, and tribal levels must wait for the PAR from ARRA. Once the agencies receive the PAR, then responsible parties must be called for an urgent meeting to generate a PAD. The PAD must pertain information instructions for evacuation, sheltering-in-place, and the administering of potassium iodide (KI). Once a PAD is prepared, it must be signed by either the Governor of the State of Arizona or by a Maricopa County Judge and is then sent out to all concerned agencies. This entire process with the final product of signed PAD could take at least 45 min. Then, the agencies must send the PAD to all other concerned agencies (Kyne 2015).

Step (4) Evacuation

As soon as the PAD has been signed, the next step is to form a joint information committee (JIC) near the incident site. The exact location is not clearly described, however, it is indicated that the JIC would be near the incident site. A total of 48 outdoor warning sirens, located within the 10-mile radius of PVNGS's emergency planning zone are to be turned on as soon as the event is initiated (Maricopa County Emergency Management 2013; Kyne 2014).

The findings reveal that it could take at least 170 min for this to take place. The ARRA office would require at least 70 min, and the PVNGS would require at least 55 min, thus producing signed PAD could take at least 45 min. The NRC's policy is to complete the three steps within 45 min. The highly radioactive toxic plume might have traveled at least about 30 miles at a wind speed of 10 miles per hour. This does not include the possibility of unnoticing the initiation of the incident at the PVNGS. As discussed earlier, the nuclear power plant encountered several experiences in which the initiation of the incident went unnoticed. If this possibility is not ruled out, the radioactive plume might have traveled even a longer distance than 30 miles.

The lessons learned from this PVNGS case study are as follows:

(1) The population in the EPZ zone within a 10-mile radius is likely to be evacuated after they have unnecessarily been exposed to the radioactive plume, dispersed from the accident at the PVNGS.
(2) It would be too late to take KI pills which could protect their thyroid gland because evaluation information could reach them after they have been exposed to the radioactive plume or they have been beneath the radioactive plume pathway.
(3) During the 170 min, under the changing weather conditions the radioactive plume might have traveled various locations within the 10-mile-radius areas or in the EPZ zone. With these possibilities, many people might have become exposed to the high-level radioactive plume before the evacuation efforts take place.

5.2 Issues Associated with the Process

The nuclear emergency management process constitutes the four stages, namely mitigation, preparedness, response, and recovery. The nuclear emergency management and the disaster management processes are similar, but the response stage is different between the two (Fig. 3.2, Chap. 3). Unlike the response stage in the disaster management process, the response stage in the nuclear emergency management process includes four activities that are namely projection of plume path dispersion, protective action recommendation (PAR), protective action decision (PAD), and evacuation. These activities are very specific and technical in nature and require special training and specific skills in order to carry out the activities. For instance, projection of plume path requires a special training to use the RASCAL computer code and special knowledge about nuclear reactors and radiation. In disaster management, this response stage does not require much specialized skill. For example, to respond to a storm-hit area, almost anyone could carry out response activities, searching people who are displaced, assisting in accounting for those who have survived or may be injured.

In the nuclear emergency management process, the NRC is aware of the urgency and timeliness nature of response stage. With the awareness, the NRC has set clear policy guidelines on the flow of activities and the time allocated for each activities. A delay in one activity may cause another delay in the subsequent activity. However, as discussed earlier in this chapter, it is evident that the time allocated is not sufficient to carry out specified activities. In every step, there is a high likelihood of encountering a delay rather than accomplishing the specified tasks. In other words, the allocated time of 15 min does not relate to the practical aspects of accomplishing the tasks within an allocated time frame. According to Perry and Lindell (2003), to formulate an effective emergency preparedness plan, planning, training, and existing written plans must relate to one another. The degree of effectiveness always depends on the degree of relation among the plan, preparedness, training, and existing written plans. What is evident here is that the NRC's policy and guidelines provided for the response stage are not related to executing the specified activities.

One issue associated with the nuclear power emergency process is that there is a redundant work performed by two different parties in response activities. As discussed in the activities, projection of radioactive plume dispersion was done by the two parties, the personnel at PVNGS and the officials at ARRA. They used the same computer code, namely RASCAL to conduct the same study, a projection of plume dispersion. To solve redundancy issue, taking the advantage of the state-of-the-art technology, a virtual coordination system could be established between the two parties. The system will provide opportunity to work together without a requirement of traveling. In case of emergency at the nuclear power plant, both parties could be connected through the virtual system. Then, projection of the plume dispersion could be carried out by both parties at the same time. Both parties could discuss and make certain decisions on generating a PAR.

Another issue related to the nuclear power emergency management process is the lengthy process which could take up to an estimated 170 min prior to the PAD being officially signed and evacuation activities starting. The negative consequence of this delay is the individuals living near the nuclear power plant exposed to the radioactive plume before they could be evacuated. The prevention such as taking KI pills could not be done before the individuals are exposed to the high-level radioactive plume. There are some alternatives to reduce the processing time before evaluation. One possible alternative is that immediately after the incident is initiated at the nuclear power plant, the JIC should be established and notify the information to the public living within 50 miles. The center should provide instructions and guidelines such as taking KI pills and proper shelter placements. The JIC could also inform the public that there are further instructions and guidelines to be provided, including the detailed information on plume path dispersion, direction, and estimated dose level, once the PAD has been officially signed. In taking these steps, the public does not need to wait the estimated processing time for projecting plume dispersion and radiation levels.

In short, the findings uncover the potential issues in relation to the nuclear emergency management process and policy, the redundancy, and a lengthy process of nuclear power emergency management. These three issues need urgent attention from the NRC policy makers.

5.3 Issues Associated with the Priorities

This section looks into priorities specified in the radiological emergency planning.

In the discussion above, it is evident that there are issues with the nuclear power emergency management process. This section intends to search for and identify any issues associated with such priorities. There are certain priorities that are specified in the national emergency planning guidelines for all hazards. On December 17, 2003, the President signed a directive, namely Homeland Security Presidential Directive-8 (HSPD-8) of "*National Preparedness Guidelines*" to the Secretary of Homeland Security to develop a national domestic all-hazards preparedness goal. The Homeland Security Agency (US FEMA 2007) has set priorities in developing emergency planning for all hazards.

The agency's priorities for emergency planning include the following:

- Expand regional collaboration,
- Implement the National Incident Management System and National Response Plan,
- Implement the National Infrastructure Protection Plan,
- Strengthen Information Sharing and Collaboration Capabilities,
- Strengthen Interoperable and Operable Communications Capabilities,

- Strengthen CBRNE Detection, Response, and Decontamination Capabilities,
- Strengthen Medical Surge and Mass Prophylaxis Capabilities, and
- Strengthen Planning and Citizen Preparedness Capabilities.

The priority areas are highlighted in order to build "a nation prepared with coordinated capabilities to prevent, protect against, respond to, and recover from all hazards in a way that balances risk with resources and need (US FEMA 2007)." The areas that are placed under a spotlight in the HSPD-8's guidelines vary from collaboration, planning, preparedness, and participation. Priorities are those activities that are important to incorporate into emergency planning in order to respond with the sufficient resources to a particular risk.

The risks from nuclear power plants are different from other risks associated with other power facilities. The catastrophic nature of the nuclear power plant is very powerful in destructing the individuals who are living around the nuclear power plant. Priorities should lead to the ultimate goal of timely evacuation in order to save lives and minimize the impacts associated with such disasters. Reflecting the HSPD-8 directives, the priorities in nuclear power emergency planning should include (1) collaboration, (2) information sharing, (3) planning, (4) preparedness, and (5) citizen participation.

First, it is critical to have effective collaboration among the key stakeholders who are involved in the nuclear emergency management process. The two facts with nuclear power incidents are that the incident could go unnoticed and the radioactive plume could travel at the wind speed and direction on the day of the event taking place. It is vital to expand the collaboration among the key stakeholders who are involved in the emergency management process for these reasons. For instance, the plant personnel, such as the ARRA personnel, and the NRC officials need to perform the plume projection separately in case of emergency, such as in the case study of Palo Verde nuclear power plant accident. The three parties should collaborate and work together on plume path projection and utilize the resources to collect the air sample instead of gathering the air sample data separately, in different groups. Thus, they can project radioactive plume dispersion together so that the redundancy work could be eliminated and the process could be completed within a shorter time frame.

Second, information sharing should also be one of the priorities in emergency planning. There are two areas of information sharing that receive a worthy discussion in this section. On the one hand, the information sharing among the key stakeholders is not clearly outlined. For example, the information sharing between the plant personnel and the ARRA personnel could be carried out since the incident was initiated and the plume projection and PAR formulation could be shared. Second, the information sharing between the authorities and the public is not clear. The information sharing is regularly done with the residents who live within the 10-mile-radius or within the plume exposure pathway EPZ. However, the information about the plant potential risks and evacuation is not shared with the residents who live in the radiation ingestion pathway EPZ or within the 11–50-mile radius from the nuclear power plant. The reason is that in case of nuclear power emergency, the possibilities of evacuating the residents are not only within a 10-mile radius. As shown in the case study of the Palo Verde, the distance the radioactive plume could travel is far beyond the 50-mile-radius zones. Therefore, the information sharing must be regularly carried out with all residents at large, including those who live further than the 10-mile radius.

Third, another issue could be planning for a radiological emergency. The National Preparedness Guidelines highlight the importance of planning, in the priorities list. One issue associated with nuclear power emergency planning is lack of focus on mitigation before an incident takes place, but it rather focuses on the response to emergency events. An emergency management plan consists of four phases: mitigation, preparedness, response and recovery (Choi 2008). The first two phases take place before an emergency has occurred, whereas, the latter two phases focus on activities carried out after the emergency has already occurred. To formulate any effective emergency management plan, it is vital that every effort must be made for both mitigation and preparedness. The overall goal for the two phases is to minimize negative impacts from the emergency event. To do so, optimizing lead time is very important in the two phases of mitigation and preparedness. As an example, in case of a tsunami event, an early-warning could be issued anywhere from a few minutes to hours before the actual tsunami would hit the impacted areas. Such warning provides local people ample lead time to mitigate the risks and relocate themselves to a designated safe haven. However, a nuclear power emergency plan does not seem to embrace the preciousness of the lead time. The emergency plan clearly states that only after PDA has been signed and the JIC has been formed, the evacuation could begin. At this evacuation stage, it is highly likely that the individuals in the affected areas might have been exposed to the radiative plume (Kyne 2014).

Fourth, another issue associated with radiological emergency planning is lack of sound preparedness, which deserves to be included in the list of priorities. Planning, training, and existing written plans, and their relatedness to one another contribute to the effectiveness of any emergency plan (Perry and Lindell 2003). The ultimate goal of emergency management is to minimize the potential negative impacts and save lives of individuals in the impacted areas. Before any evacuation activities could be initiated, there are certain detailed and technically oriented steps and procedure to follow. These steps include projecting the plume dispersion pathway, estimated radiation dose level, generating PAR, and generating the PAD, which precede the official notification. Carefully and exactly following the steps, and obtaining the desired outcomes from radioactive plume projection, in addition to correctly estimating population size could take a certain amount of time; it is highly likely that the radioactive plume might have dispersed and traveled more than 10-miles from the PVNGS when the notification has been completed by the JIC. It is vital to initiate evacuation activities as soon as the incident has begun at the NPP because any delay might have unsurmountable negative impacts (Kyne 2014).

Fifth, citizen participation is another priority in formulating an emergency plan. The potentially effected population size is enormous. It could be as big as more than three million of people depending on the weather conditions including wind speed and wind direction on that particular day. When it is time to evacuate, citizens must evacuate themselves. It is impossible for the responders go to every residence and evacuate the individuals who are in the affected areas. To do so, it is necessary that the emergency plan emphasizes citizen participation in the emergency planning and exercises. The plume exposure pathway EPZ, when drills are exercised every 2 years, the individuals living in those areas should participate in the exercise. As suggested, the citizen participation should also be expended into radiation ingestion zone: EPZ within a 50-mile radius.

In short, emergency planning clearly lacks defined priorities that are vital to achieve the ultimate goal of saving lives and minimizing impacts. The priorities must be clearly related to the ultimate goal, and they deserve attention by key stakeholders who are concerned with the radiological emergency planning.

5.4 Issues Associated with the Stakeholders' Participation

The key stakeholders involved in the radiological emergency management could be identified as follows:

1. Nuclear power plants,
2. State, local, and tribal organizations,
3. Federal agencies including NRC, FEMA, EPA, DOE,
4. Nongovernmental organizations, and
5. Citizen groups.

It is evident that the above-mentioned key stakeholders are not involved in the emergency planning. For example, the emergency drill exercises are required by the NRC's regulation. Every 2 years, a joint emergency drill exercise must be carried out by nuclear power plant personnel, and off-site responders including emergency responders. On November 1, 2011, PVNGS conducted a medical emergency response drill under FEMA's Radiological Emergency Response Plans (RERP). The drill included key stakeholders: ARRA, PVNGS, and Buckeye Valley Fire Department (BVFD). In contrast, in case of radiology emergency, more than seven stakeholders including other agencies and departments are anticipated to carry out response activities together. The objective drill was to evaluate the preparedness and readiness level of the key stakeholders, with a goal to carry out two hypothetical patients from the plant site, facing radiological injuries and contamination. There are two reasons that such drills do not represent a real scenario. In the event of a radiological emergency, citizens must evacuate on their own. In a real event, there are no sufficient responders for individuals living in areas within a 10-mile radius from an NPP. The drill also lacks other activities such as the actual practices of plume path projection, generating PAR, making decision to reach a PAD, and disseminating information by a JIC. It is essential to design the drill that resembles actual disaster events and key stakeholders (Kyne 2015).

Another evidence for ineffective preparedness was found in the evaluation report, as the emergency responders do not have permanent dosimeters, but borrowed ones; the dosimeters are required to read the radiation levels before entering into the incident areas, according to the EPA's PAGs (US FEMA 2011). This kind of issue with poor and ineffective preparedness related to radiological emergency planning is shared by 26 of the 31 states that host an NPP (Watkins et al. 2011). In particular, the emergency planning lacks effective preparedness in other activities including exposure assessment, environmental sampling, human specimen collection and analysis, and human health assessments, which were poorly prepared to meet their responsibilities as laid out in the written response plans (Kyne 2014).

In short, getting a PAD done in a timely manner is very important in any emergency planning process, in order to begin evacuation activities. It is entirely dependent on how quickly PVNGS and ARRA's work is performed in assessing radiation doses and producing PARs. It is evident that the previous training and drills did not embrace the other elements, focusing on risk assessment, field data collection, radiation dose estimation, and on compiling a PAR. Most important, the missing element of risk-communication could hinder other efforts in preparedness. Trainings are meant to facilitate communication between at-risk populations and other federal, state, and local agencies with no active role to play in emergency planning (Perry and Lindell 2003). The radiological emergency plan should include a training focusing on the important element of risk-communication with the individuals living the impacted areas near NPPs; such populations should be provided with KI tablets and clear and detailed instructions on when and how to properly use them to protect any potential exposure to radiation in case of a nuclear power emergency (Perry and Lindell 2003; Kyne 2014).

5.5 Issues Associated with the Evacuation

According to FEMA (US FEMA 2016), the minimum acceptable design objectives for coverage by the system are:

(1) "the capability to provide both an alert signal and an informational or instructional message to the population on an area wide basis throughout the 10-mile emergency planning zone (EPZ), within 15 min;
(2) that the initial notification system will assure direct coverage of essentially 100% of the population within 5 miles of the site;
(3) notification methods will be established to assure coverage within 45 min of essentially 100% of the population within the entire plume exposure EPZ who may not have received the initial notification. The basis for any special requirement exceptions (e.g., large water areas with transient boats or remote hiking trails) must be documented; and
(4) utility operators identify and develop, in conjunction with state, local, and tribal governmental officials, the administrative and physical means for a backup public alert and notification system capable of covering essentially 100% of the population within the entire plume exposure EPZ in the event that the primary method is unavailable. The backup means that alert and notification will be conducted within a reasonable time, with a recommended goal of 45 min (US FEMA 2016)."

There are several issues in evacuation: First, the findings of this study revealed that there are about 1.39–19.85 million population that could possibly be exposed to the radioactive plume pathway in case of a nuclear core-meltdown accident at the IPNG. This is a very significant finding as this would drastically impact a very large population and would have long-term negative consequences for many years after the initial incident. The projection conducted in this study has the following two limitations: First, the findings related to the radiation dose and plume path dispersion are based on an assumption that they are included in one of the two reactors as a core-meltdown accident. However, the Indian Point Nuclear Generating hosts two nuclear power reactors, namely Indian Point Unit 2 and Indian Point Unit 3 that are currently located and are in operation at the plant site. If both are to experience the core-meltdown accident scenario, then the impacts and radioactivity from that extreme event could be magnified in multiples of two, given the issue at hand and that would be twice as big as the numbers. Second, the computer code provides a maximum of 100-mile radius for projected plume pathway. In reality, the radioactive plume could travel more than 100 miles from the nuclear power plant. For example, the plume traveled more than 160 miles when the Fukushima Daiichi Nuclear Power Plant encountered the core-meltdown accident. In addition, there are estimated 253,977 individuals and 15,118,181 individuals live in areas within a 10-mile radius surrounding the plant and areas within a 50-mile radius surrounding the plant, respectively, which are officially designated as emergency planning zones (EPZs) by the NRC. In counting those individuals, this constitutes higher percentage of minority racial and ethnic groups than their counterparts living in areas outside of the EPZs. Thus, minority groups are again significantly more impacted and face greater consequences in such events. Having said this, the potential extreme event stemming from the IPNG's core-meltdown accident and the estimated negative impacts, the key stakeholders who are involved in the emergency management process could face tremendous issues in carrying out activities in the process. Among the issues they could potentially face, I would like to highlight the following emergency event-related issues below:

When the evacuation has been announced, the population living in the EPZ zone within a 10-mile radius around the IPNG will have been completely exposed to the radioactive plume. The well-documented emergency planning and preparedness including the EPA's *Manual of protective action guides and protective action for nuclear incidents* (US EPA 1992) and FEMA's action guide (US EPA 1992; US FEMA 2013) do not reflect the time required for actions in practice.

Another anticipated issue is shadow evacuation which is defined as "evacuations by persons outside of any officially declared evacuation zone" (Dotson and Jones 2008). The current emergency plan focuses on areas within a 10-mile radius and show evacuations beyond a 10-mile distance, which could become a serious problem in case of a nuclear power emergency. When individuals living beyond the 10-mile distance engage in shadow evacuations, they could have a negative impact on traffic flows especially in some large-scale evacuations. The risk awareness, personalization of that risks, and evaluation of alternative actions of those individuals would have impacts on the behavior of evacuation (Sorenson 1987). It is a concern that individuals living outside a 10-mile radius did not receive regular information on emergency planning (US GAO 2012) and their risk awareness, risks personalization, evaluation of alternative actions could be different from those of their counterpart living in the areas within the 10-mile radius zone. Therefore, to overcome issues associated with the shadow evacuation, the NRC must incorporate this important element of shadow evacuation in the emergency plan, drills, and evacuation exercises (Kyne 2016).

Another issue associated with evacuation is the important element of coordination among key stakeholders. In a nuclear emergency, to effectively carry out evacuation activities, the coordination among various federal, state, tribal, and local agencies is critical. The evacuation tasks become more complex when the emergency takes place in a mega city such as New York. Many years ago, a small number of organizations predominantly participated in the evacuation efforts in response to 911 attacks in New York (Schweinberger et al. 2014). It was evident that there was a direct communication among US NRC, FEMA and state authorities, but they lacked direct communication with county officials who are responsible to develop a radiological emergency plan (US GAO 2001). The importance of coordination among federal agencies was highlighted in case of weapon of mass destruction (WMD) response in mega cities, including New York (Bell and Dallas 2007). Thus, the element of coordination is essential among various key stakeholders at all levels (Kyne 2016).

The other issue is that the evacuation plan lacks the element of educating the public. In the context of mega cities, there are many high-rise buildings and evacuation from individuals living in those buildings needs special procedures. There is a continued need for educating the public from evacuating from such high-rises buildings after the 911 attacks (Zmud 2008). Emergency responders could face great challenges when evacuating the individuals living in high-rise buildings who lack education or training on special instruction and procedures required for evacuation. Thus educating the public living in the host communities is a vital element to include in emergency planning for a successful and effective evacuation (Kyne 2016).

5.6 Issues Associated with the Recovery

There are anticipated issues related to recovery activities in nuclear emergency planning. First, after nuclear core-meltdown accidents, nearby lands became irrecoverable and uninhabitable for humans. After the Fukushima accident, some areas were left contaminated and some of them become uninhabitable lands and were declared as permanent "exclusion" zones. It was estimated that there were about 11,580 squares miles of land which were contaminated with radioactive materials; about 4,500 squares miles (similar to the size of Connecticut) have radiation level of 1 mSV (millisievert) per year, which is higher than the US exposure limit; and a total of 310 square miles located northwest of the plant with the 12-mile radius from the plant were declared as permanent "exclusion" zones (Starr 2016). There are nearly 160,000 Japanese who are still displaced because they cannot return to their lands which were contaminated with radioactive materials or declared as permanent "exclusion" zones (Glinsky 2011). Therefore, it is essential to incorporate factors of radioactive contamination on lands nearby NPPs and situate displaced individuals into an effective recovery plan (Kyne 2016).

Another issue is associated with the loss of land due to contamination with radioactive materials after a nuclear power core-meltdown accident. Quantifying the economic and monetary values of land, which poses great challenges in carrying out recovery activities. Radioactive contamination is very complex to manage and sometime contamination leaves land permanently uninhabitable. For example, it was roughly estimated that the values of contaminated lands and property within the 310 square miles after the Fukushima disaster were between $250 and $500 billion US. In addition, there was a cost for small compensation provided to 160,000 people with no place to live (Starr 2016). The costs for radioactive contamination in some places in the U.S. would be even greater than those of Fukushima. If the IPNG NPP encounters a core-meltdown accident, the radioactive contamination of land and water could cause $1.702 trillion for taxable property in the New York City area (based on the estimated total market value of taxable property in New York City as of July 1, 2016) (Barbanel 2016). In addition to the estimated values, residents could face similar problems to Japanese victims who were forced to pay their mortgage for the houses they could no longer live in, located in the "exclusion" zones (Kyne 2016).

Third, the other issue associated with recovery is the economic loss of businesses which were located in the radioactive contaminated areas. For example, Wall Street, a world international financial hub is located only 25 miles away from the IPNG NNP in New York. In addition, the world's biggest economy is located in the U.S.; 2013 US gross domestic product (GDP) of $16.80 trillion, comprising 22.4% of global economic output (Picardo 2014). The business firms located on Wall Street control trillions of dollars in financial assets. Furthermore, the New York Foreign Exchange Market is ranked as the second-largest trading center in the world, with a daily volume exceeding $5 trillion (Picardo 2014). It should be noted that economic loss could vary with different geographical areas where the NPP is located. However, if the core-meltdown accident takes place in an area like New York, it is hard to deny the potential tremendous impacts on the nation's economy as well as the world's economy, in addition to impacts to the local community with regard to contamination and radiation exposure (Kyne 2016).

5.7 Conclusion

This chapter provides discussions on issues associated with the nuclear emergency management process. First, this chapter identifies issues associated with the four response activities in nuclear emergency process, namely projection of radioactive plume path dispersion, protective action recommendations, protective action decision, and evacuation. The government has developed an emergency plan containing detailed guidelines. This chapter investigates the emergency plan policies and guidelines. The issues identified are potential issues in relation to the nuclear emergency management process and policy, the redundancy, and a lengthy process of nuclear power emergency management. The ultimate goal in the nuclear power emergency management process is to provide a lead time to mitigate the impacts from the nuclear power accidents. To do so, the public must be notified of any alert as soon as the accident is initiated. A delay in evacuation activities would have undesired consequences of increasing mortality rates caused by the nuclear power accident. To elaborate the consequences of delayed evacuation, the study demonstrates the two simulation exercises of nuclear power emergency events at PVNG and INPG. The projection results show the demographic characteristics of populations that are vulnerable to any potential nuclear power accidents (as discussed details in Chap. 4). In the worst-case scenarios, people living within the area underneath the plume path would be close to 3.5 million in numbers (Table 4.6, Chap. 4), under typifying weather conditions in quarter 3 at PVNG and 19.86 million people (Table 4.11, Chap. 4) are likely to be exposed to high levels of radiation doses. Above all, the issues also places 87.5 million people living within a 50-mile radius surrounding the NPPs (including 45 million women, 8 million elderly who are 65 years or older, and one million kids who are younger than 5 years old) (Table 2.1, Chap. 2) at risks of exposure to radiation and radioactive nuclear materials. This chapter also identifies issues related to priority setting in the emergency management process. Findings show that the emergency planning clearly lacks defined priorities that are vital to achieve the ultimate goal of saving lives and minimizing impacts. This chapter also calls attention from key stakeholders, namely nuclear power plants; state, local, and tribal organizations; federal agencies including NRC, FEMA, EPA, DOE; nongovernmental organizations; and citizen groups to actively participate and collaborate in carrying out activities in managing nuclear power emergency process. Issues associated with evacuation are also identified and addressed within this chapter. As discussed in Chap. 1 under topic 1.3, the previous three historical nuclear power accidents demonstrated that there were delays in evacuation. For example, in the Three Mile Island accident, only women and children were advised to evacuate within a 5-mile radius, which was a day after the radioactive materials were released into the air. Men, women, seniors, and children were not included in the evacuation plans immediately. There are about eight million seniors who are 65 years or older live within a 50-mile radius from NPPs. Evacuating elders need to be a program with special and specific details relating to limitations, including possible mobility issues. There were nearly 10% of evacuations that were recorded as evacuation-related deaths in the Fukushima accident (Chap. 1). In addition, the places in New York and its vicinity areas pose challenges for evacuation because of the high-rise buildings which require special skills for evacuation. Finally, the recovery poses challenges including removing melted fuels and containing the disaster side.

References

ARRA (2013) Palo Verde Nuclear Generating Station. Retrieved from Arizona Radiation Regulatory Agency http://www.azrra.gov/er/pvngs.html

Barbanel J (2016) New York city property values surge; brooklyn leads way in largest increase in values since '08; taxes are set to rise. Retrieved from Wall Street Journal (Online) http://login.ezproxy1.lib.asu.edu/login?url=http://search.proquest.com/docview/1757132464?accountid=4485

Bell WC, Dallas CE (2007) Vulnerability of populations and the urban health care systems to nuclear weapon attack-examples from four American cities. Int J Health Geogr 6. Retrieved from WOS:000258119500001

Choi SO (2008) Emergency management: implications from a strategic management perspective. J Homeland Secur Emerg Manag 5(1). Retrieved from WOS:000254565600001

Dotson LJ, Jones J (2008) Identification and analysis of factors affecting emergency evacuations. U.S. Nuclear Regulatory Commission; Division of Preparedness and Response; Sandia National Laboratories

Glinsky V (2011) Indian point: the next Fukushima? Retrieved from http://www.nytimes.com/2011/12/17/opinion/is-indian-point-the-next-fukushima.html?_r=0

Kyne D (2014) *Environmental justice issues in communities hosting US nuclear power plants* (Order No. 3617964). Available from ProQuest Dissertations and Theses Global. (1528556250). Retrieved from http://ezhost.utrgv.edu:2048/login?url=http://search.proquest.com.ezhost.utrgv.edu:2048/docview/1528556250?accountid=7119

Kyne D (2015) Managing nuclear power plant induced disasters. J Emerg Manag (Weston, Mass) 13(5):417–429. doi:10.5055/jem.2015.0252. Retrieved from Prime National Pub. Corp

Kyne D (2016) A Fukushima-like disaster in New York city and possible issues associated with the extreme event management system. J Extreme Events 3(2). doi:10.1142/S234573761650007X

Lochbaum D (2012) The NRC and nuclear power plant safety in 2012: tolerating the intolerable. Union of Concerned Scientists: Citizens and Scientists for Environmental Solutions. Retrieved from: http://www.ucsusa.org/assets/documents/nuclear_power/NRC-nuclear-safety-2012-report.pdf

Maricopa County Emergency Management (2013) Nuclear plant to sound emergency sirens: annual test part of federal requirements. Retrieved from Maricopa County Emergency Management http://www.maricopa.gov/pr_detail.aspx?releaseID=2445

Morris DZ (2016) At Indian point, a small leak turns into a big political battle over nuclear energy. Retrieved from Fortune http://fortune.com/2016/02/28/indian-point-nuclear/

Picardo E (2014) Why Wall Street is a key player in the world's economy. Retrieved from Investopedia. http://www.investopedia.com/articles/investing/100814/wall-streetsenduring-impact-economy.asp

Perry RW, Lindell MK (2003) Preparedness for emergency response: guidelines for the emergency planning process. Disasters 27(4):336–350. Retrieved from WOS:000187255200005

Randazzo R (2013) Palo Verde nuclear plant leak could cost $10 M to $15 M to fix. Retrieved from AZCentral.com http://www.azcentral.com/business/consumer/articles/20131015palo-verde-nuclear-plant-leak-could-cost-m-m-fix.html

RSICC (2013) RASCAL 4.3: radiological assessment systems for consequence analysis Reactor and Nuclear Systems Division (RNSD). Oak Ridge National Laboratory, The Radiation Safety Information Computational Center (RSICC). Retrieved from https://rsicc.ornl.gov/PackageDetail.aspx?p=RASCAL%204.3&id=C00783&cpu=PCX86&v=02&t=Radiological%20Assessment%20Systems%20for%20Consequence%20AnaLysis

Schweinberger M, Petrescu-Prahova M, Vu DQ (2014). Disaster response on September 11, 2001 through the lens of statistical network analysis. Social Networks, 37:42–55. Retrieved from WOS:000334001400004

Sorenson JH (1987) Evacuations due to off-site releases from chemical accidents experience from 1980 to 1984. J Hazard Mater 14

Starr S (2016) Costs and consequences of the Fukushima Daiichi disaster. In: PSR's Environmental Health Policy Institute Retrieved from http://www.psr.org/environment-and-health/environmental-health-policy-institute/responses/costs-and-consequences-of-fukushima.html

The Associated Press (2013) Palo Verde unit cuts power for valve problem. Retrieved from Contra Costa Times http://www.contracostatimes.com/ci_24477243/palo-verde-unit-cuts-power-valve-problem

US EPA (1992) Manual of protective action guides and protective action for nuclear incidents. Office of Radiations Program, United States Environmental Protection Agency, Washington DC 20460. Retrieved from: http://www.epa.gov/radiation/docs/er/400-r-92-001.pdf

US FEMA (2007) National preparedness guidelines. US Department of Homeland Security. Retrieved from: http://www.fema.gov/pdf/emergency/nrf/National_Preparedness_Guidelines.pdf

US FEMA (2011) Palo Verde nuclear generating station after action report/ improvement plan. Radiological Emergency Preparedness (REP) Program, US FEMA. Retrieved from: http://pbadupws.nrc.gov/docs/ML1217/ML12174A197.pdf

US FEMA (2013) Communicating during and after a nuclear power plant incident. US Department of Homeland Security. Retrieved from: http://www.fema.gov/media-library-data/20130726-1919-25045-1433/communicating_during_and_after__npp_incident___june_2013__secure_.pdf

US FEMA (2016) REP program manual January 2016 (FEMA P-1028). US FEMA. Retrieved from: http://www.fema.gov/media-library-data/1452711021573-a9b920f4f7ac34ea9f32738f51982afe/DHS_FEMA_REP_Program_Manual_Jan2016_Secure.pdf

US GAO (2001) Nuclear regulations: progress made in emergency preparedness at Indian point 2, but additional improvements needed (GAO-01-605). United States Government Accountable Office (US GAO). Retrieved from: http://www.gao.gov/assets/240/231903.pdf

US GAO (2012) Emergency preparedness: NRC needs to better understand likely public response to radiological incidents at nuclear power plants (GAO-13-243). United States Government Accountable Office (US GAO). Retrieved from: https://www.nirs.org/reactorwatch/emergency/gaoshadowevacreport413.pdf

US NRC (2011) Nuclear Regulartory Commission. Retrieved from US Nuclear Regulatory Commission www.nrc.gov

US NRC (2014) Protective action recommendations. Retrieved from US Nuclear Regulatory Commission http://www.nrc.gov/about-nrc/emerg-preparedness/about-emerg-preparedness/protective-action-recommendations.html

Watkins SM, Perrotta DM, Stanbury M, Heumann M, Anderson H, Simms E et al (2011) State-level emergency preparedness and response capabilities. Disaster Med Public Health Preparedness 5:S134-S142. Retrieved from WOS:000288748000014

Zmud M (2008) Public perceptions of high-rise building emergency evacuation preparedness. Fire Technol 44(4):329–336. Retrieved from WOS:000259908600003

Chapter 6
Conclusions

6.1 Dealing with the Inevitable Risks of NPPs

The previous chapters discussed the risks associated with the US commercial nuclear power plants. The catastrophic nature of nuclear power plants is evident in the three historic nuclear power disasters, namely the Three Mile Island (TMI) in USA, Chernobyl in Ukraine, and Fukushima Daiichi in Japan. This powerful and destructive nature of power plants is regarded as inevitable. The inevitability of the risks means that as long as nuclear power plants exist, individuals living in close proximity to these facilities must burden the risks and must embrace the catastrophic nature of a possible powerful destruction. As long as our country places nuclear power plants in operation, it has to find ways to deal with the risks and pitfalls that come with nuclear facilities. There are many possible ways to better deal with the nature of powerful destruction of nuclear power plants. This chapter addresses three such possible ways to address these potential disasters:

(1) Complete elimination of inevitable risks
(2) Transforming the inevitable risks into evitable risks
(3) Minimizing the inevitable risks (Fig. 6.1).

One possible way is to completely eliminate inevitable risks of nuclear power plants by getting rid of the source of the risks. In other words, a complete shutdown of the fleet of nuclear power plants could be one possible way to completely eliminate the inevitable risks. This also requires consensus agreement of key stakeholders.

Second, another possible way to deal with the inevitable risks of nuclear power plants is to transform the inevitable risks into evitable risks. This requires a close monitoring system and making great efforts in transforming the risks.

Finally, another possible way to deal with inevitable risks is to minimize them, so that the inevitable risks become minimal, which could lead to reduced impacts.

D. Kyne, *Nuclear Power Plant Emergencies in the USA*,
DOI 10.1007/978-3-319-50343-1_6

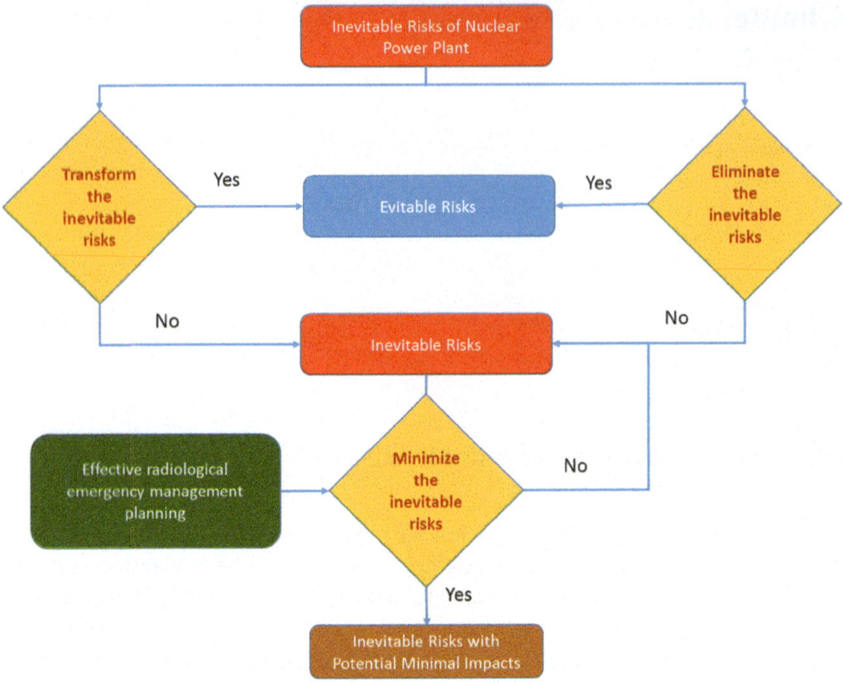

Fig. 6.1 Dealing with the inevitable risks of nuclear power plants

Because the risks are inevitable, a possible intervention to minimize these risks is to create an effective radiological emergency plan, which could possibly lead to minimizing the negative impacts.

6.2 Eliminating the Inevitable Risks

To completely eliminate the inevitable risks requires the shutting down permanently of all existing US commercial nuclear power plants. To do so, all stakeholders must have a common consensus to discontinue nuclear power as an energy source. The feasibility of the discontinuation of nuclear energy depends on key stakeholders and their perspectives on the issue. This section enhances the understanding of key stakeholders' perspectives toward the future of nuclear energy. The government, regulators, NPP owners, nuclear safety advocates, pro-nuclear power advocates, scientists, and the public are identified as key stakeholders in such projects (Kyne 2014).

Dr. Gregory B. Jaczko, who was a regulator and former chairman of the NRC (2009–2012), saw some weaknesses and limitations of nuclear technology. He asserted that the current NPPs in operation were not designed to withstand severe accidents (Jaczko 2013; Kyne 2014). His suggestion was that the U.S. take a pause for now and spend some time to collect more information about the current technology and their limitations; then find solutions to cover such limitations. In his view, ideal technology should not have such limitations to withstand accidents, to safely operate, and to dispose spent fuel. He provided a question to the key stakeholders: "Why do we continue to move forward with the technology that presents all of these weaknesses and pursue it without stopping?" (Jaczko 2013; Kyne 2014). In fact, Dr. Jaczko might not expect the answers from the stakeholders; his intention might have been to encourage them to reflect the current limitations with the existing nuclear technology (Kyne 2014).

Another key stakeholder, namely Arnie Gundersen, who is a nuclear safety advocate with 40 years of experience in nuclear power engineering asked all key stakeholders including federal- and state-level governments, business interest groups and the nuclear power industry at large to watch the video footage of the at Japan's Fukushima NPP (Gundersen 2013; Kyne 2014). His intention was to increase the awareness of the key stakeholders, especially policy makers that such a nuclear power accident could occur in the U.S. and they should discontinue their beliefs that such accident could occur only outside of the U.S. (Kyne 2014).

There are climate-change scientists who think that nuclear power energy is the only viable source of energy to counter climate change issues from other forms of energy. Among them, James Hansen, Ken Caldeira, Kerry Emanuel and Tom Wigley are a group of four climate-change scientists who sent an open letter in which they called world leaders to support nuclear power energy (CNN 2013; Kyne 2014). They raised concerns with the current energy supply and consumption rates exceeding the planet's ability to reverse the effects of carbon dioxide pollution; nuclear power energy is seen as solution to the pollution problem (Kyne 2014). However, in my view, when taking into account of potential harms from nuclear power core-meltdown accidents, radioactive pollution from day-to-day operations, irreversible effects of radioactive contamination, risks associated with transport and storage of radioactive waste could easily far outweigh solving the carbon dioxide pollution (Kyne 2014).

Like the climate-change scientists, there are some key stakeholders who have thoughts on nuclear renaissance. However, one important key stakeholder, namely Peter Bradford, a former member of the NRC and former chair of the Maine and New York Utility Regulatory Commission has negative views on the possibility of "nuclear renaissance" taking place (Kyne 2014). He argued that "there is no coming nuclear renaissance—instead, the nuclear renaissance bubble will soon burst (Bradford 2013; Kyne 2014)." His notion was that the requirement of immerse financial commitments caused nuclear power energy to be a less acceptable and competent form of energy and his prediction was that the NPPs in current operation will shut down of their own accord, due to economic considerations (Kyne 2014).

Another key stakeholder is Naoto Kan, the former prime minister of Japan and leader of the administration who managed the nuclear emergency during the Fukushima nuclear catastrophe. He mentioned that nuclear power core-meltdown accidents are the only events that could negatively impact 50 million people, short of a global war (Kan 2013; Kyne 2014). After the aftermath of the Fukushima nuclear accident, Mr. Kan admitted that his belief of safe nuclear power energy has gone a 180° change and now he believes that the risks associated with nuclear power plants are not inevitable; his suggestion in eliminating the risks is to get rid of nuclear power itself (Kyne 2014).

Another key stakeholder, President Barack Obama presented his vision on nuclear energy. He asserted that nuclear energy could become part of the "global energy mix" in his "Sustainable Energy for All Initiative" (Gerhardt 2013; Kyne 2014). He stated the following:

> The United States will continue to promote the safe and secure use of nuclear power worldwide, through a variety of bilateral and multilateral engagements. For example, the U. S. Nuclear Regulatory Commission advises international partners on safety and regulatory best practices, and the Department of Energy works with international partners on research and development, nuclear waste and storage, training, regulations, quality control, and comprehensive fuel leasing options. Going forward, we will expand these efforts to promote nuclear energy generation consistent with maximizing safety and nonproliferation goals (Barack Obama, as cited in (Gerhardt 2013).

My notion is that the President's remarks share the vision of President Eisenhower, who launched the Atoms for Peace program some 60 years prior (Kyne 2014):

> The more important responsibility of this Atomic Energy Agency would be to devise methods whereby this fissionable material would be allocated to serve the peaceful pursuits of mankind. Experts would be mobilized to apply atomic energy to the needs of agriculture, medicine, and other peaceful activities. A special purpose would be to provide abundant electrical energy in the power-starved areas of the world. Thus the contributing powers would be dedicating some of their strength to serve the needs rather than the fears of mankind (Eisenhower 1953).

Other key stakeholders include the U.S. general public. There was a poll conducted by CBS, the American broadcast-television station, which measure U.S. public opinion of nuclear power just after the event of the nuclear disaster in Japan. The survey results showed that 50% of Americans were opposed to building new nuclear plants, an increase from 34% in 2008; and 62% of Americans objected to the construction of an NPP anywhere near their communities (Madrigal 2011; Kyne 2014). In my view, the public does not benefit from the current nuclear power plants. One reason is that they have to bare the costs in the event of a core-meltdown accident with their tax-payer money under the Price-Anderson Act of 1957 (Huffington Post 2013; Kyne 2014). In addition, nuclear power poses potential safety issues to the public as unnecessary, uneconomic, uninsurable, and unsafe (Kyne 2014). Moreover, the public appears concerned with the ageing of the existing fleet of U.S.-based commercial NPPs and they are even called "zombie nukes" (Gerhardt 2013; Kyne 2014).

In short, one notion is that there is an on-going tension between two groups of the key stakeholders with pro- and anti-nuclear views. It is unlikely that the two groups could have a common consensus to discontinue the use of nuclear power energy in the near future. As a result, it is very likely that the use of nuclear power energy will go on, given the fact that political and economic factors outweigh the risks individuals living around the nuclear power plants are faced with, including the potential for powerful destruction scenarios.

6.3 Transforming the Inevitable Risks

The NRC has initiated reactor oversight framework which emphasizes on three performance areas, namely reactor, radiation, and safeguards. These three areas consist of seven cornerstones that reflect the essential safety aspects of facility operation: (1) initiating events, (2) mitigating systems, (3) barrier integrity, (4) emergency preparedness, (5) public radiation safety, (6) occupational radiation safety, and (7) physical protection (Fig. 6.2) (US NRC 2015).

The NRC claimed that it utilizes the well-defined reactor oversight process (ROP) in order to monitor the performance of reactors and their adherence to the NRC's rules, regulations, and license requirements (US NRC 2014). With this ROP in place, the NRC expects to transform the inevitable risks into evitable risks. In fact, no matter how much ROP benefited from the past 30 years of experience that NRC has, it is unlikely that the regulatory agency could realize its overall objective for two reasons: The powerful destructive nature of nuclear fuels could be changed as long as the nuclear fuel rods are used in the nuclear core reactor, and the nuclear power plants consist of very complex and tightly coupled systems and accidents in the system are normal (Perrow 1999).

The reasons for focusing the three areas reveal that each goal in each of the three areas share a common goal which is to "avoid" accidents or risks associated with the nuclear power plants (US NRC 2014). The focus on reactor safety area calls for an avoidance of accidents and reducing their impacts. Radiation safety aims to avoid unnecessary radiation exposure for plant workers and the public. It also aims to avoid risks of sabotage and other security threats (US NRC 2014).

Another notion in carrying out the reactor oversight framework is that the seven cornerstones are impacted by human interactions with the technology. Human interactions could be classified into three areas which are named as "cross-cutting" by the NRC. They are human performance, problem identification and resolution, and safety-conscious work environment. Human errors in human interactions with the complex and tightly coupled system could not be completely avoided.

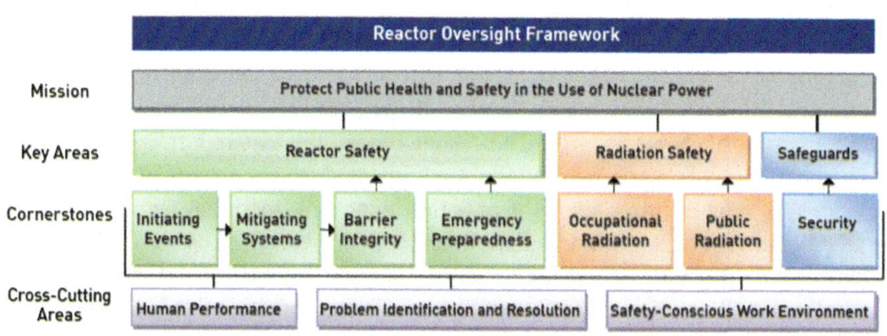

Fig. 6.2 Reactor oversight framework, *Source* (US NRC 2014)

Above all, the transformation of the inevitable risks associated with nuclear power plants could not be feasible for the reason provided by Perrow (1999), who explained that nuclear power plants are complex and tightly coupled systems and accidents are normal and unavoidable. Therefore, the option of transforming the inevitable risks associated with nuclear power plants is less likely to be a feasible option.

6.4 Minimizing the Inevitable Risks

There are two options with regard to minimizing of inevitable risks: to completely eliminate the inevitable risks associated with nuclear power plants, and to transform the inevitable risks associated with such facilities, which are discussed in the previous sections, and they are highly unlikely to be the feasible options. This section focuses on the option of minimizing the inevitable risks.

Avoiding the inevitable risks associated with a nuclear power plant is not a vital element in this option to minimize the risks. Rather this option embraces the avoidance of the risks associated with the nuclear power plants, and unlike the previous two options to deal with the inevitable risks, this minimizing option does not attempt to avoid the inevitable risks by eliminating or transforming them. It rather provides preparedness options to accept the inevitable risks and attempts to minimize the inevitable risks and their negative impacts.

Embracing the inevitability of the risks can mean several key points that require further explanation. First, when this option embraces the inevitable risks, it assumes that there is no way to transform the risks into evitable or avoidable. Second, it also assumes that the inevitable risks associated with the nuclear power plants are catastrophic and powerful in destruction. What are the immediate risks from nuclear power plants after an incident? The immediate risks are ironizing radiation which normally comes from meltdown of nuclear fuels, from the reactor, and other cooling system structures. The US EPA has, for the past 30 years, provided guidelines to protect from radiation (Fig. 6.3) (US EPA 2013). There are various thresholds of radiation does for different scenarios.

Are the EPA's protective action guidelines acceptable? There were two studies conducted by two organizations that invalidate the radiation dose levels described in the US EPA protective action manual. The first organization is the US National Academy of Sciences (US NRC 2008; Royal 2008) which publishes reports on Biological Effects of Ionizing Radiation (BEIR) VII. This is the most recent publication from the NAS's series. The recent report, the last report BEIR V, being published 16 years ago, reveals the findings that show statistically significant relationship between any levels of radiation exposure, including exposure to naturally occurring background radiation and the increased risks of cancer it can create. Similarly, the second organization, namely International Commission on Radiological Protection (ICRP 2007), concluded in its study of The 2007 Recommendations of the International Commission on Radiological Protection Publication 103 that any low-level radiation including naturally occurring

Phase	Protective Action Recommendation	Protective Action Guide or Planning Guide
Early	Sheltering-in-place or evacuation of the public[a]	1 to 5 rem (10 mSv to 50 mSv) projected dose/4 days[b]
	Administration of prophylactic drugs KI[c]	5 rem (50 mSv) projected child thyroid dose[d] from radioactive iodine
	Limit emergency worker exposure	5 rem (50 mSv)/year (or greater under exceptional circumstances)[e]
Intermediate	Relocation of the public	2 rem (20 mSv) projected dose first year[b] Subsequent years, 0.5 rem (5 mSv)/year projected dose
	Food interdiction[f]	0.5 rem (5 mSv)/year projected dose, or 5 rem (50 mSv)/year to any individual organ or tissue, whichever is limiting
	Limit emergency worker exposure	5 rem (50 mSv)/year[b]
	Reentry	Operational Guidelines[g] (Stay times and concentrations) for specific activities (see Section 3.7)
Late	Cleanup	Brief description of planning process
	Waste Disposal	Brief description of planning process

[a] Should begin at 1 rem (10 mSv); take whichever action (or combination of actions) that results in the lowest exposure for the majority of the population. Sheltering may begin at lower levels if advantageous.
[b] Projected dose- the sum of the effective dose from external radiation exposure (i.e., groundshine and cloudshine) and the committed effective dose from inhaled radioactive material.
[c] Provides thyroid protection from internal exposure to radioactive iodines only. For other information on radiological prophylactics and treatment, refer to http://www.fda.gov/Drugs/EmergencyPreparedness/BioterrorismandDrugPreparedness/ucm063807.htm http://www.bt.cdc.gov/radiation and www.orau.gov/reacts
[d] Thyroid equivalent dose. For more information, refer to http://www.fda.gov/Drugs/EmergencyPreparedness/BioterrorismandDrugPreparedness/ucm072265.htm.
[e] When radiation control options are not available, or, due to the magnitude of the incident, are not sufficient, doses to emergency workers above 5 rem (50 mSv) may be unavoidable and are generally approved by competent authority. For further discussion see Chapter 2, Section 2.6.
[f] For more information on food and animal feeds guidance, the complete FDA guidance may be found at http://www.fda.gov/downloads/MedicalDevices/DeviceRegulationandGuidance/GuidanceDocuments/UCM094513.pdf
[g] For extensive technical and practical implementation information please see "Preliminary Report on Operational Guidelines Developed for Use in Emergency Preparedness and Response to a Radiological Dispersal Device Incident" (DOE 2009)

Fig. 6.3 Planning guidance and protective action guides for radiological incidents, *Source* (US EPA 2013)

background radiation is not safe and it causes cancer. Specifically, the LNT model assumes that, in the low-dose range, radiation doses greater than zero will increase the risk of excess cancer and/or heritable disease in a simple proportionate manner. While further research is needed to undoubtedly prove this as true, current findings support this model, and it has been adopted in standards and regulations worldwide. Regardless of whether or not the source of radiation is human-made, as in the case of NPPs, or natural, and as exposure increases so does risk.

When any low-level radiation dose is not safe for public to be exposed to, the high-level radiation dose from a radiological emergency could certainly be unsafe. To minimize the impacts from the inevitable risks, the best way is to mitigate from the high-level radiation. This kind of mitigation must be incorporated into the radiological emergency planning. The most important element in the radiological emergency planning is to ensure that there is a lead time to mitigate from the ironizing radiation.

6.5 Revisiting Radiological Emergency Planning

The two options proposed above, elimination and transformation of the inevitable risks associated with NPPs, are not feasible. It would not be possible to immediately shut down all 61 US commercial NPPs. Similarly, the US NRC efforts to transform the inevitable risks into manageable risks are not practical because the risks associated with the NPPs are inevitable in nature. As a result, the only practical and feasible option is to minimize the inevitable risks.

As discussed earlier, minimizing the inevitable risks requires an effective radiological emergency plan, with a priority on lead time to mitigate before the public is exposed to the high-level radiation during a nuclear power emergency event. This study identified issues associated with the nuclear power emergency management process. To transform the current emergency plan into an effective one, this section revisits the current radiological emergency planning.

Transforming the current emergency plan into an effective one is vital for at least two primary reasons. There are more than 87.5 million people who are living within a 50-mile radius surrounding the 62 US commercial nuclear power plants (Table 2.1, Chap. 2). In addition, the population living within the 50-mile radius includes 45 million women, 8 million elders who are 65 years old or older, and 1 million children who are younger than five years (Table 2.1, Chap. 2). All of the mentioned population groups are placed in danger of exposure to high levels of radioactive materials from the NPPs when a core-meltdown accident takes place. Another reason is that the effective plan could save the lives of 87.5 million people. The vulnerable population groups mentioned above could be saved from exposure to radiation, and they could be prevented from evacuation-related deaths.

The issues identified (in Chap. 5) in the current radiological emergency management plan are (Fig. 6.4) as follows:

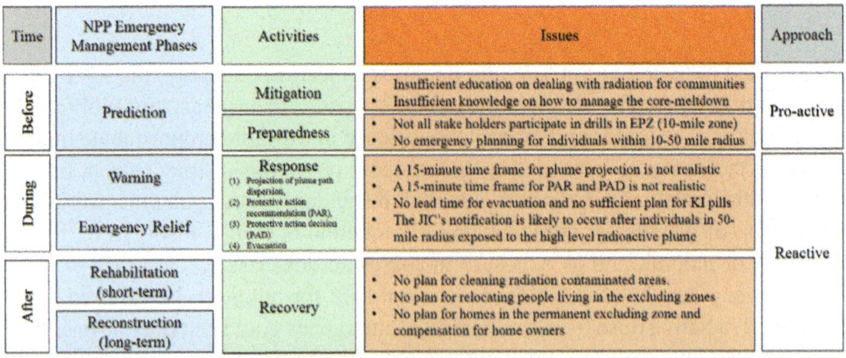

Fig. 6.4 Issues associated with the radiological emergency management process

- Insufficient education on dealing with radiation for communities;
- Insufficient knowledge on how to manage the core meltdown;
- Not all stake holders participate in drills in EPZ (10-mile zone);
- No emergency planning for individuals within 10–50 mile radius;
- A 15-min time frame for plume projection is not realistic;
- A 15-min time frame for PAR and PAD is not realistic;
- There is no lead time for evacuation;
- The JIC's notification is likely to occur after individuals in 50-mile radius exposed to the high-level radioactive plume;
- No plan for cleaning radiation-contaminated areas;
- No plan for relocating people living in the excluding zones; and
- No plan for homes in the permanent excluding zone and compensation for home owners.

One notion is that responsible authorities do not provide sufficient information and education on handling and responding to radiation exposure. Risk communication is very important for the radiological emergency management process. The public must be educated with information on how to deal with radiation when a nuclear core-meltdown accident occurs at NPPs. In addition, the public should be provided with necessary resources to prevent themselves from possible radiation impacts resulting from nuclear power accident. For example, potassium pills (KI) should be distributed in advance, and public education must be provided on how to administer the medication prior to radiation exposure.

There is not sufficient knowledge on how to manage the core-meltdown accident within the existing literature and knowledge accumulation on core-meltdown accidents. When a nuclear power plant encounters a nuclear core-meltdown accident, there is no manual on how to manage such an event. In the previous three historic nuclear power disasters, Three Mile Island, Chernobyl, and Fukushima Daiichi nuclear power plants, the procedures for cooling down the fuel in the core reactor were all different. In the most recent nuclear power plant core-meltdown accident at the Fukushima Daiichi power plant, the seawater was injected into the nuclear core in order to cool down the temperature inside the reactor core. Even though the injection of seawater was ordered to be ceased by the Office of the Prime Minister, the plant management did not follow those directions. Finally, the temperature in the reactor core was high, and the nuclear fuel submerged from water. The fuel rods had zirconium cladding with steam after the water level dropped. The zirconium cladding created the formation of hydrogen bubbles, and later the increasing pressure inside the reactor core exploded. More scientific studies should be conducted to understand how to manage nuclear core-meltdown accident at NPPs.

An additional issue to consider is to encourage all key stakeholders to participate in annual drills in the EPZs (10-mile zone). For example, when the PVNG had a biannual drill, only firemen from the closest town to the nuclear power plant participated. In case of a nuclear power accident event taking place, the radioactive plume could be dispersed to a longer distance than the 10-mile radius. Therefore, it is recommended that all fire and emergency departments within the 10-mile radius should be participating in drills. Their participation will increase the key stakeholders' awareness of nuclear power accidents. In addition, they will become aware of the need to use radiation dosimeters to measure the radiation while they evacuate the victims at the nuclear power plant site as well as effected areas.

It is evident that in the event of a nuclear power accident, the radioactive materials could be carried away by the winds and its direction. The previous three nuclear power accidents show that wind could carry radioactive materials more than a 50-mile distance. As a result, the individuals living in areas within 10–50 mile radius from the nuclear power plants are exposed to potential nuclear power plants. The emergency plan was developed for the 10-mile radius zone which was designated as an EPZ zone. However, there is no emergency planning for the people who live between 10- and 50-mile radii. Because they are also equally exposed to the radiation risk as those living within 10-mile radius from NPPs, a nuclear emergency plan should be developed for the areas within 10–50 mile radius from the NPPs.

It is not practical or feasible to conduct a projection of radioactive plume path dispersion within a 15-min time frame. In fact, the projection requires use of RASCAL computer code and special training. Turning on a computer and opening the RASCAL computer code alone might need at least 15 min. A required time frame should be revised to reflect sufficient amount of time to carry out the projection. Under the estimation of the required time for projection of plume dispersion could contribute to under estimation of actual required time frame for entire nuclear power emergency plan. Similarly, allocated time frames for PAR and PAD—a 15-min time frame for each of PAR and PAD—are not realistic.

There is no lead time for mitigation from the nuclear power accident before the residents expose to the radioactive materials. The critical issue is that there is no lead time to evaluate the people beneath the high-level ironizing plume in the response stage. As soon after an incident has been initiated at the plant facility, the authorities could not notify the public about the incident, before completing the four tasks, namely projection of plume path dispersion, protective action recommendation (PAR), protective action decision (PAD), and evacuation. To minimize the impacts, it is necessary to find a lead time to evacuate the public before they are being exposed to the radioactive plume path. As soon as the incident is initiated, the public should be notified to evacuate. Spending longer than 45 min to find out estimated radiation doses is not worthwhile and keeping individuals and families waiting could be lethal or dramatically increase their exposure to radiation. As noted previously, any dose level is not safe and the public should be notified as soon as the incident is initiated.

The JIC formation could be seen as the last step before the evacuation activities could be carried out. In fact, as discussed earlier in the Chaps. 3–5, the JIC's notification is likely to occur after individuals in 50-mile radius zones are exposed to the high-level radioactive plume. In that case, the nuclear power plant accident took place on a day with a wind speed of 10 miles, then radioactive plume might have carried out by winds to about 9-mile distance. Therefore, in any case, the evacuation activities are likely to take place after the individuals who lived in the 50-mile radius have exposed to the radiation. FEMA has set guidelines to complete the tasks within a recommended 45-min time frame. In fact, the four tasks require great efforts and technological skills. In practice, it is unlikely to complete the tasks within the recommended 45 min. The purpose of carrying out the tasks is to know the radiation dose level and plume pathway. While the concerned key stakeholders are carrying out the tasks to know the radiation dose and plume pathway, the individuals living around the nuclear power plants would be exposed to the radiation in the meantime. Upon the completion of the procedure and obtaining the results of estimated dose and plume pathway with PAD, the JIC notified the public for evacuation. When the public receives notification, they would have been exposed to radiation. It is too late to administer the KI pills after exposing radiation exposure.

Another challenge in the nuclear power emergency management is cleaning radiation contaminated areas. In the Chernobyl nuclear power accident, there were more than 5000 tons of sand, nuclear fuels, and other radioactive dusts at the plant site, which could not be easily removed. After eight months of the explosion taking place (December 14), a "sarcophagus," a concrete roof was constructed over the troubled reactor (Chap. 1). That sarcophagus lasted only 30 years, and a new containment structure which costs $ 1.7 billion was recently replaced to entomb the entire plant site. This structure is anticipated to last for 100 years. Therefore, a detailed plan on how to clean the radiation-contaminated materials should be developed for each plant. In addition, an emergency plan should include how to cool down the melted fuel rods and types of materials such as sand to be used. For example, PVNG is not located next to water sources such as river or sea. In case of nuclear core-meltdown accident, it is necessary to develop an emergency plan that describes the types of materials to be used to cool down the temperature of melted fuel rods.

Emergency plans should include relocating people in the excluding zones. Looking at the previous nuclear core-meltdown accidents at Chernobyl and Fukushima Daiichi, the governments determined the excluding zones which were areas contaminated with high level of radioactive materials and it was not safe for people to live. In case of nuclear core-meltdown accident at a NPP, there could expect an excluding zone. However, it is difficult to estimate how large the excluding zones would be and how many people live in the excluding zones before the event takes place. With the limitation of estimation, an emergency planning should include a plan for relocating people living in the excluding zones.

Nuclear emergency plans lack specific preparations and recommendations for relocating the individuals living in the permanent excluding zones and how to compensate the home owners. In case of a nuclear core-meltdown accident at IPNG in New York, the relocation of the residents could be very costly because the property values in the New York area are very high, relative to other areas in the country that face similar risks of exposure. In addition, the relocation could be very complex and difficult to handle. If the accident required to designate permanent excluding zones, then the costs will be much higher because of the high property values in the areas.

In short, revisiting the radiological emergency plan is vital for minimizing the inevitable risks associated with NPPs. The above-mentioned issues associated with the plan must be well addressed as soon as possible. There are 23 Mark I reactors, similar to the reactors used at Fukushima Daiichi plant site in current operation. In addition, the nuclear reactors in the USA are called zombie reactors because of their aging properties. The only option available to save lives of more than 87 million people living in the areas within 50-mile radius from the NPPs is to transform current radiological emergency plan into an effective one. The clock is ticking, and any delay in actions against transformation of the plan cannot be afforded and will have significant impacts on lives, in addition to the increased costs of managing the aftermath of such events.

References

Bradford P (2013) Panel discussion. In: Panel discussion. The Fukushima nuclear accident: ongoing lessons for New York, New York

CNN (2013) Climate change warriors: it's time to go nuclear. Retrieved from CNN News http://www.cnn.com/2013/11/03/world/nuclear-energy-climate-change-scientists/index.html?hpt=hp_t2

Eisenhower DD (1953) Atoms for peace. Retrieved from Voices of Democracy http://voicesofdemocracy.umd.edu/eisenhower-atoms-for-peace-speech-text/

Gerhardt T (2013) Obama's climate action plan: nuclear energy? Retrieved from http://www.huffingtonpost.com/tina-gerhardt/obamas-climate-action-plan4_b_3546999.html

Gundersen A (2013) Panel discussion. In: Panel discussion. The Fukushima nuclear accident: ongoing lessons for New York, New York

Huffington Post (2013) Atomic energy: unnecessary, uneconomic, uninsurable, unevacuable and unsafe. Retrieved from Huffington Post http://www.huffingtonpost.com/ralph-nader/atomic-energy_b_4086042.html?view=print&comm_ref=false

ICRP (2007) The 2007 recommendations of the International Commission on radiological protection publication 103 (Annual ICRP 37). International Commission on Radiological Protection (ICRP). Retrieved from: http://www.icrp.org/docs/ICRP_Publication_103-Annals_of_the_ICRP_37%282-4%29-Free_extract.pdf

Jaczko GB (2013) Panel discussion. In: Panel discussion. The Fukushima nuclear accident: ongoing lessons for New York, New York

Kan N (2013) Panel discussion. In: Panel discussion. The Fukushima nuclear accident: ongoing lessons for New York, New York

Kyne D (2014) *Environmental justice issues in communities hosting US nuclear power plants* (Order No. 3617964). Available from ProQuest Dissertations and Theses Global. (1528556250). Retrieved from http://ezhost.utrgv.edu:2048/login?url=http://search.proquest.com.ezhost.utrgv.edu:2048/docview/1528556250?accountid=7119

Madrigal AC (2011) In Fukushima aftermath, half of Americans now oppose building new nuclear plants. Retrieved from http://www.theatlantic.com/technology/print/2011/03/in-fukushima-aftermath-half-of-americans-now-oppose-building-new-nuclear-plants/72911/

Perrow C (1999) Normal accidents living with high-risk technologies: with a new afterword and a postscript on the Y2 K problem. Retrieved from Princeton University Press
Royal DH (2008) Effects of low level radiation—what's new? Semin Nucl Med 392–402. doi:10.1053/j.semnuclmed.2008.05.006
US EPA (2013) Manual of protective action guides and protective action for nuclear incidents. Office of Radiations Program, United States Environmental Protection Agency, Washington DC 20460. http://www.epa.gov/sites/production/files/2015-06/documents/pag-manual-interim-public-comment-4-2-2013.pdf
US NRC (2008) Review of ICRP publication 103—the 2007 recommendations of the International Commission on Radiological Protection (ICRP). US Nuclear Regulatory Commission (NRC). http://pbadupws.nrc.gov/docs/ML0807/ML080790301.pdf
US NRC (2014) Reactor oversight process (NUREG-0654/FEMA-REP-1, Revision 1, Supplement 3). U.S. Nuclear Regulatory Commission. http://pbadupws.nrc.gov/docs/ML1405/ML14052A306.pdf
US NRC (2015. Reactor oversight process (ROP), U.S. Nuclear Regulatory Commission. http://www.nrc.gov/NRR/OVERSIGHT/ASSESS/

Appendix A

See Table A.1.

Table A.1 Construction of study variables. *Source* (Kyne 2014)

Category	1990 Census (SF3)	2000 Census (SF3)	2010 CAS 5-year estimate
Race/Ethnicity			
Total population	Total population (P0010001)	Total population (P006001)	Total population (B01001_2)
% White	White (P0080001)/ Total population * 100	White (P006002)/ Total population * 100	White (B02001_2)/Total population * 100
% Black	Black (P0080002)/ Total population * 100	Black (P006003)/ Total population * 100	Black (B02001_3)/Total population * 100
% Asian	Asia-Pacific (P0080004)/Total population * 100	Asia-Pacific (P006005-6)/Total population * 100	Asia-Pacific (B02001_5, B02001_6)/Total population * 100
% Native American	Native American (P0080003)/Total population * 100	Native American (P006004)/Total population * 100	Native American (B02001_4)/Total population * 100
% Others	Other race (P0080005)/Total population * 100	Other race (P006007-8)/Total population * 100	Other race (B02001_7, B02001_8)/Total population * 100
% Hispanic	Total persons of Hispanic origin (P0100001)/Total population * 100	Total persons of Hispanic origin (P007010)/Total population * 100	Total persons of Hispanic origin (B03002_12)/ Total population * 100
% Color	Total population (P0010001)-Not Hispanic origin, White (P0120001)/Total population * 100	Total population (P006001)-Not Hispanic origin, White (P007003)/ Total population * 100	Total population (B01001_2)-Not Hispanic origin, White (B03002_3)/Total population * 100
Households			
Mean household income ($)	Aggregate household income (P0810001-2)/ Total household (P0050001) * 100	Aggregate household income (P054001)/Total household (P014001) * 100	Aggregate household income (B19025_1)/ Total household (B11001_1) * 100

(continued)

D. Kyne, *Nuclear Power Plant Emergencies in the USA*,
DOI 10.1007/978-3-319-50343-1

Table A.1 (continued)

Category	1990 Census (SF3)	2000 Census (SF3)	2010 CAS 5-year estimate
Poverty			
Poverty rate	Individuals below poverty threshold (P1170013-24)/Total population poverty determined (P1170001-12) * 100	Individuals below poverty threshold (P08002)/Total population poverty determined (P087001) * 100	Individuals below poverty threshold (B17001_2)/Total population poverty determined (B17001_1) * 100
Housing			
Occupancy rate	Occupied housing units (H0040001)/ Total housing units (H0010001) * 100	Occupied housing units (H006002)/ Total housing units (H001001) * 100	Occupied housing units (B25002_2)/Total housing units (B25001_1) * 100
Vacant rate	Vacant housing units (H0040002)/Total housing units (H0010001) * 100	Vacant housing units (H006003)/Total housing units (H001001) * 100	Vacant housing units (B25002_3)/Total housing units (B25001_1) * 100
% Owner occupied housing units	Owner occupied housing units (H0080001) /Total housing units (H0010001) * 100	Owner occupied housing units (H007002) /Total housing units (H001001) * 100	Owner occupied housing units (B25003_2)/ Total housing units (B25001_1) * 100
% Renter occupied housing units	Renter occupied housing units (H0080002)/Total housing units (H0010001) * 100	Renter occupied housing units (H007003)/Total housing units (H001001) * 100	Renter occupied housing units (B25003_3)/ Total housing units (B25001_1) * 100
Unemployment			
% Unemployment rate	Unemployed population (P0700003 + P0700007)/Civilian employed population (P0700002-3, P0700006-7) * 100	Unemployed population (P043007 + P043014)/Civilian employed population (P043005, P043012) * 100	Unemployed population (B23001_8,15,22,29,36,43,50,57,64,71,76, 81,86,94,101,108,115,122,129,136,143,150, 157,162,167,172)/Civilian employed population (unemployed, B23001_7,14,21, 28,35,42,49,56,63,70,75,80,85,93,100,107, 114,121,128,135,142,149,156,161,166,171)

See Table A.2.

Table A.2 Study variables of US demographic characteristics in 1990, 2000, and 2010. *Source* (Kyne 2014)

	US_1990	US_2000	US_2010
Tract	73,057	73,057	73,057
Tract area (sq. mile)	3,796,743	3,796,743	3,796,743
Total population	248,709,360	281,421,920	303,965,280
White	199,826,608	211,353,728	224,895,696
Black	29,930,428	34,361,740	37,978,752
Asian	7,226,882	10,550,602	14,677,166
Native American	2,015,044	2,447,989	2,480,465
Others	9,710,097	22,707,850	23,933,188
Hispanic	21,900,100	35,238,480	47,727,532

(continued)

Table A.2 (continued)

	US_1990	US_2000	US_2010
Color	60,284,992	86,907,768	107,392,496
White (%)	80.35	75.10	73.99
Black (%)	12.03	12.21	12.49
Asian (%)	2.91	3.75	4.83
Native American (%)	0.81	0.87	0.82
Others (%)	3.90	8.07	7.87
Hispanic (%)	8.81	12.52	15.70
Color (%)	24.24	30.88	35.33
Female (%)	51.28	50.99	50.85
Old (65+ years) (%)	12.54	12.43	12.75
Kid (<5 years) (%)	8.83	8.18	6.62
Native-born (%)	92.05	88.95	87.28
Renter housing units (%)	32.19	30.77	29.33
College degree or higher (%)	20.34	24.40	27.90
Unemployment (%)	6.31	5.77	7.92
Poverty (%)	13.12	12.38	13.82
Mean household income ($)	63,119	70,570	70,881

Appendix B
Descriptive Analyses

B.1 Browns Ferry Nuclear Plant, Alabama

See Fig. B.1.
See Table B.1.

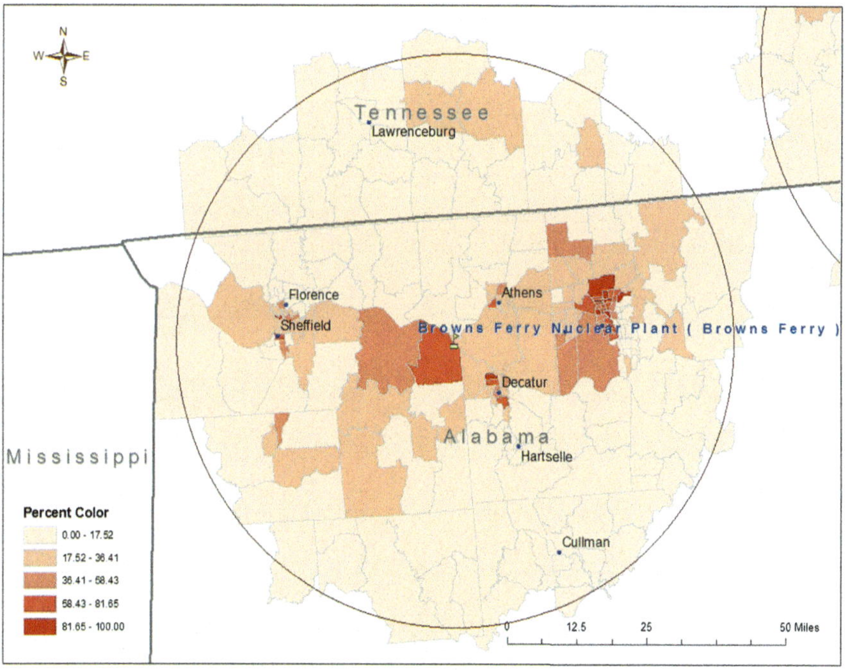

Fig. B.1 Spatial distribution of percent color among the populations surrounding Browns Ferry Nuclear Plant in Alabama in 2010

D. Kyne, *Nuclear Power Plant Emergencies in the USA*,
DOI 10.1007/978-3-319-50343-1

Table B.1 Demographic composition of population, as sorted by distance from Browns Ferry Nuclear Plant, Alabama, in 1990, 2000, and 2010 *Source* (Kyne 2014)

	0–10	11–20	21–30	31–40	41–50	0–50	Outside
Year 1990							
Tract	10	35	71	64	27	207	2,471
Tract area (sq. mile)	302	964	1,607	1,877	1,603	6,353	88,211
Total population	36,259	118,087	240,080	219,938	101,947	716,311	8,201,456
White	24,454	105,244	186,051	200,891	97,148	613,788	6,409,774
Black	11,481	10,323	49,527	15,824	4,342	91,497	1,705,299
Asian	92	545	2,328	2,038	181	5,184	47,166
Native American	194	1,856	1,640	945	247	4,882	25,891
Others	38	115	544	243	27	967	13,340
Hispanic	119	686	2,090	1,285	343	4,523	50,135
Color	11,881	13,371	55,147	20,054	5,094	105,547	1,824,165
White (%)	67.44	89.12	77.50	91.34	95.29	85.69	78.15
Black (%)	31.66	8.74	20.63	7.19	4.26	12.77	20.79
Asian (%)	0.25	0.46	0.97	0.93	0.18	0.72	0.58
Native American (%)	0.54	1.57	0.68	0.43	0.24	0.68	0.32
Others (%)	0.10	0.10	0.23	0.11	0.03	0.13	0.16
Hispanic (%)	0.33	0.58	0.87	0.58	0.34	0.63	0.61
Color (%)	32.77	11.32	22.97	9.12	5.00	14.73	22.24
Female (%)	51.78	50.77	50.89	51.75	51.60	51.28	52.03
Old (65+ years) (%)	13.05	10.88	10.26	13.07	14.78	12.01	12.86
Kid (<5 years) (%)	8.92	8.28	8.38	8.01	8.43	8.28	8.29
Native-born (%)	99.70	99.07	97.86	98.52	99.64	98.61	98.87
Renter housing units (%)	26.48	25.52	32.30	22.05	23.74	26.52	28.20
Education (%)	12.45	16.74	20.59	21.38	7.96	18.00	15.64
Unemployment (%)	7.55	5.46	5.96	5.12	6.59	5.78	6.69
Poverty (%)	17.47	12.53	13.71	11.72	17.50	13.63	17.19
Mean household income ($)	49,881	56,000	54,512	59,278	42,778	54,307	51,020
Year 2000							
Total population	38,640	143,349	263,503	251,145	116,327	812,964	9,323,419
White	25,495	120,917	189,607	223,740	109,114	668,873	7,055,252
Black	11,681	15,402	61,373	18,372	4,733	111,561	1,971,347
Asian	113	1,261	2,976	2,145	328	6,823	80,563
Native American	329	1,829	2,334	1,403	399	6,294	32,144
Others	1,022	3,930	7,221	5,495	1,743	19,411	184,115
Hispanic	1,283	3,916	4,416	5,369	1,919	16,903	175,149
Color	13,776	24,724	75,857	30,103	8,107	152,567	2,348,154
White (%)	65.98	84.35	71.96	89.09	93.80	82.28	75.67
Black (%)	30.23	10.74	23.29	7.32	4.07	13.72	21.14
Asian (%)	0.29	0.88	1.13	0.85	0.28	0.84	0.86

(continued)

Table B.1 (continued)

	0–10	11–20	21–30	31–40	41–50	0–50	Outside
Native American (%)	0.85	1.28	0.89	0.56	0.34	0.77	0.34
Others (%)	2.64	2.74	2.74	2.19	1.50	2.39	1.97
Hispanic (%)	3.32	2.73	1.68	2.14	1.65	2.08	1.88
Color (%)	35.65	17.25	28.79	11.99	6.97	18.77	25.19
Female (%)	51.16	50.27	51.29	51.53	51.30	51.18	51.57
Old (65+ years) (%)	12.31	11.26	11.46	14.28	14.82	12.82	12.64
Kid (<5 years) (%)	7.88	8.20	7.97	7.46	7.62	7.80	7.95
Native-born (%)	97.88	97.37	97.04	97.44	98.61	97.49	97.57
Renter housing units (%)	24.97	22.96	29.04	20.19	21.63	23.98	26.30
College degree or higher (%)	14.80	21.21	23.36	24.83	10.97	21.26	19.16
Unemployment (%)	5.97	4.90	6.83	4.00	5.87	5.45	5.81
Poverty (%)	18.08	11.60	12.97	11.12	15.11	12.70	14.80
Mean household income ($)	55,722	62,486	59,569	65,931	51,161	60,667	59,018
Year 2010							
Total population	40,292	164,818	294,974	268,473	116,817	885,374	10,062,245
White	27,807	133,604	206,691	235,486	109,217	712,805	7,491,857
Black	9,843	20,368	71,901	21,230	3,892	127,234	2,142,467
Asian	138	2,687	4,721	3,402	442	11,390	131,842
Native American	610	2,097	2,102	1,528	498	6,835	34,216
Others	1,894	6,062	9,559	6,827	2,768	27,110	261,863
Hispanic	3,196	9,223	11,974	8,942	3,608	36,943	386,876
Color	14,133	36,909	96,997	38,088	9,945	196,072	2,812,238
White (%)	69.01	81.06	70.07	87.71	93.49	80.51	74.46
Black (%)	24.43	12.36	24.38	7.91	3.33	14.37	21.29
Asian (%)	0.34	1.63	1.60	1.27	0.38	1.29	1.31
Native American (%)	1.51	1.27	0.71	0.57	0.43	0.77	0.34
Others (%)	4.70	3.68	3.24	2.54	2.37	3.06	2.60
Hispanic (%)	7.93	5.60	4.06	3.33	3.09	4.17	3.84
Color (%)	35.08	22.39	32.88	14.19	8.51	22.15	27.95
Female (%)	51.91	50.17	51.06	51.15	51.15	50.97	51.39
Old (65+ years) (%)	13.91	12.29	12.17	15.65	16.71	13.93	13.18
Kid (<5 years) (%)	6.57	6.19	6.38	6.20	5.18	6.14	6.48
Native-born (%)	95.02	94.81	95.57	96.66	97.94	96.05	96.06
Renter housing units (%)	24.24	23.86	28.39	20.58	21.25	24.00	25.98
College degree or higher (%)	17.45	24.33	26.78	27.70	12.66	24.27	22.10

(continued)

Table B.1 (continued)

	0–10	11–20	21–30	31–40	41–50	0–50	Outside
Unemployment (%)	12.09	7.17	9.30	6.77	8.43	8.15	8.69
Poverty (%)	17.70	12.49	14.94	13.88	17.47	14.62	16.95
Mean household income ($)	54,044	63,724	61,702	64,257	49,219	60,858	58,340
Index	1	1	1	1	1	1	1

B.2 Joseph M. Farley Nuclear Plant, Alabama

See Fig. B.2.
See Table B.2.

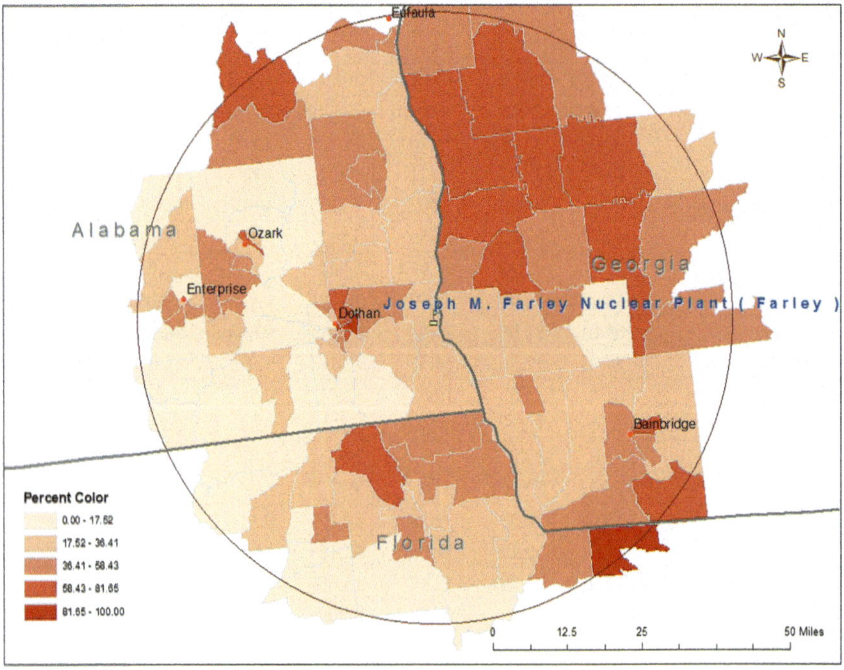

Fig. B.2 Spatial distribution of percent color among the populations surrounding Joseph M. Farley Nuclear Plant in Alabama in 2010

Table B.2 Demographic composition of population, as sorted by distance from Joseph M. Farley Nuclear Plant, Alabama, in 1990, 2000, and 2010 *Source* (Kyne 2014)

	0–10	11–20	21–30	31–40	41–50	0–50	Outside
Year 1990							
Tract	5	22	21	28	18	94	7,301
Tract area (sq. mile)	358	885	1,745	2,026	1,629	6,642	170,960
Total population	15,104	76,597	60,730	101,613	62,945	316,989	23,139,704
White	11,933	53,836	44,653	72,411	43,971	226,804	18,107,526
Black	3,096	22,242	15,579	27,185	18,101	86,203	4,434,358
Asian	2	317	154	898	325	1,696	244,015
Native American	62	157	239	534	335	1,327	74,867
Others	12	44	106	583	218	963	278,930
Hispanic	57	373	1,336	1,796	791	4,353	1,675,669
Color	3,212	22,963	17,149	30,155	19,461	92,940	6,365,508
White (%)	79.01	70.28	73.53	71.26	69.86	71.55	78.25
Black (%)	20.50	29.04	25.65	26.75	28.76	27.19	19.16
Asian (%)	0.01	0.41	0.25	0.88	0.52	0.54	1.05
Native American (%)	0.41	0.20	0.39	0.53	0.53	0.42	0.32
Others (%)	0.08	0.06	0.17	0.57	0.35	0.30	1.21
Hispanic (%)	0.38	0.49	2.20	1.77	1.26	1.37	7.24
Color (%)	21.27	29.98	28.24	29.68	30.92	29.32	27.51
Female (%)	51.60	52.97	51.31	50.38	52.70	51.70	51.69
Old (65+ years) (%)	14.04	13.41	14.09	12.41	14.17	13.40	15.13
Kid (<5 years) (%)	8.58	8.68	8.11	9.16	7.89	8.56	8.28
Native-born (%)	99.69	99.32	97.58	98.05	98.38	98.41	91.90
Renter housing units (%)	19.63	31.89	18.76	28.94	26.46	26.73	28.42
Education (%)	6.81	15.19	8.88	12.05	11.78	11.87	18.20
Unemployment (%)	4.75	5.21	5.29	6.20	6.29	5.70	5.95
Poverty (%)	18.22	19.50	20.77	19.73	22.73	20.38	14.12
Mean household income ($)	43,205	50,080	44,969	42,653	41,130	44,687	58,591
Year 2000							
Total population	15,882	80,324	70,181	107,605	69,667	343,659	28,272,272
White	12,428	52,701	50,919	73,826	45,531	235,405	20,716,744
Black	3,283	25,867	17,510	29,538	21,878	98,076	5,709,183
Asian	20	383	261	737	437	1,838	475,775
Native American	32	361	246	612	284	1,535	99,478
Others	120	1,011	1,248	2,887	1,552	6,818	1,271,080
Hispanic	151	921	1,342	2,810	1,944	7,168	3,175,749
Color	3,516	28,030	19,912	34,956	25,075	111,489	9,791,218
White (%)	78.25	65.61	72.55	68.61	65.36	68.50	73.28
Black (%)	20.67	32.20	24.95	27.45	31.40	28.54	20.19

(continued)

Table B.2 (continued)

	0–10	11–20	21–30	31–40	41–50	0–50	Outside
Asian (%)	0.13	0.48	0.37	0.68	0.63	0.53	1.68
Native American (%)	0.20	0.45	0.35	0.57	0.41	0.45	0.35
Others (%)	0.76	1.26	1.78	2.68	2.23	1.98	4.50
Hispanic (%)	0.95	1.15	1.91	2.61	2.79	2.09	11.23
Color (%)	22.14	34.90	28.37	32.49	35.99	32.44	34.63
Female (%)	52.49	51.75	50.75	50.47	50.25	50.87	51.24
Old (65+ years) (%)	14.14	14.27	14.57	13.19	14.48	14.03	14.59
Kid (<5 years) (%)	7.85	8.08	7.51	8.38	7.78	7.99	7.69
Native-born (%)	99.07	98.52	98.95	97.64	97.56	98.16	88.22
Renter housing units (%)	16.05	29.76	17.23	25.65	25.99	24.45	26.74
College degree or higher (%)	10.45	17.31	13.21	12.86	13.36	13.97	22.47
Unemployment (%)	4.88	6.09	4.94	7.31	5.92	6.11	5.63
Poverty (%)	15.64	17.81	16.32	19.38	20.57	18.43	13.14
Mean household income ($)	51,195	53,088	54,466	46,642	48,381	50,377	66,419
Year 2010							
Total population	17,021	84,626	77,175	108,538	72,992	360,352	32,332,734
White	12,261	53,343	57,249	72,860	48,375	244,088	22,953,312
Black	4,218	28,075	17,356	31,014	21,146	101,809	6,903,997
Asian	47	654	621	827	698	2,847	807,358
Native American	119	301	178	403	435	1,436	102,182
Others	376	2,253	1,771	3,434	2,338	10,172	1,565,885
Hispanic	376	2,209	1,850	5,110	3,115	12,660	4,929,747
Color	4,982	32,178	21,139	38,629	26,626	123,554	13,066,692
White (%)	72.03	63.03	74.18	67.13	66.27	67.74	70.99
Black (%)	24.78	33.18	22.49	28.57	28.97	28.25	21.35
Asian (%)	0.28	0.77	0.80	0.76	0.96	0.79	2.50
Native American (%)	0.70	0.36	0.23	0.37	0.60	0.40	0.32
Others (%)	2.21	2.66	2.29	3.16	3.20	2.82	4.84
Hispanic (%)	2.21	2.61	2.40	4.71	4.27	3.51	15.25
Color (%)	29.27	38.02	27.39	35.59	36.48	34.29	40.41
Female (%)	52.69	51.25	50.64	48.80	49.71	50.14	51.18
Old (65+ years) (%)	15.64	13.79	15.44	14.36	15.31	14.71	14.50
Kid (<5 years) (%)	4.97	7.14	5.99	6.53	6.38	6.46	6.34
Native-born (%)	98.32	97.80	97.49	97.17	96.11	97.23	85.75
Renter housing units (%)	18.65	30.07	18.34	27.80	26.48	25.58	25.47
College degree or higher (%)	12.53	17.62	14.57	13.74	15.20	15.05	25.79

(continued)

Table B.2 (continued)

	0–10	11–20	21–30	31–40	41–50	0–50	Outside
Unemployment (%)	6.11	7.71	8.34	7.78	9.55	8.14	8.85
Poverty (%)	15.77	20.05	17.96	18.53	21.44	19.21	14.79
Mean household income ($)	48,218	51,861	55,221	47,416	47,774	50,227	65,299
Index	2	2	2	2	2	2	2

B.3 Palo Verde Nuclear Generating Station, Arizona

See Fig. B.3.
See Table B.3.

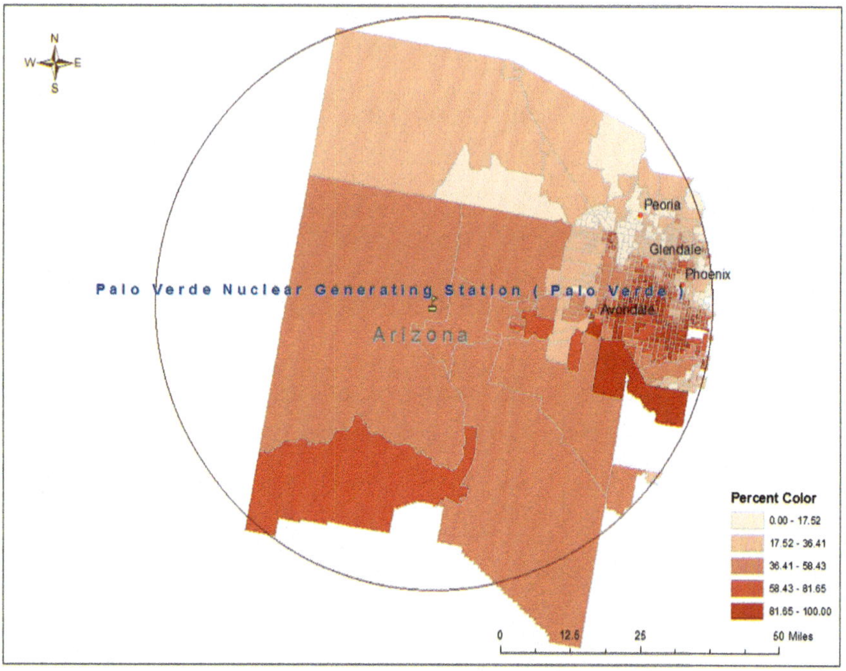

Fig. B.3 Spatial distribution of percent color among the populations surrounding Palo Verde Nuclear Generating Station, Arizona, in 2010

Table B.3 Demographic composition of population, as sorted by distance from Palo Verde Nuclear Generating Station, Arizona, in 1990, 2000, and 2010 *Source* (Kyne 2014)

	0–10	11–20	21–30	31–40	41–50	0–50	Outside
Year 1990							
Tract	1	8	39	192	294	534	992
Tract area (sq. mile)	120	1,658	672	2,661	1,831	6,942	107,049
Total population	1,365	10,128	38,556	379,343	866,410	1,295,802	2,369,419
White	1,090	8,112	28,273	299,286	714,771	1,051,532	1,916,149
Black	0	328	2,073	15,153	42,671	60,225	49,833
Asian	0	44	563	5,254	14,728	20,589	33,539
Native American	72	209	496	6,985	18,992	26,754	177,831
Others	203	1,440	7,144	52,651	75,260	136,698	192,071
Hispanic	203	2,580	11,418	85,971	162,063	262,235	418,401
Color	275	3,137	14,272	112,206	234,538	364,428	668,883
White (%)	79.85	80.09	73.33	78.90	82.50	81.15	80.87
Black (%)	0.00	3.24	5.38	3.99	4.93	4.65	2.10
Asian (%)	0.00	0.43	1.46	1.39	1.70	1.59	1.42
Native American (%)	5.27	2.06	1.29	1.84	2.19	2.06	7.51
Others (%)	14.87	14.22	18.53	13.88	8.69	10.55	8.11
Hispanic (%)	14.87	25.47	29.61	22.66	18.71	20.24	17.66
Color (%)	20.15	30.97	37.02	29.58	27.07	28.12	28.23
Female (%)	48.21	49.81	47.30	51.82	50.43	50.74	50.64
Old (65+ years) (%)	9.23	8.18	7.12	19.38	10.08	12.70	13.19
Kid (<5 years) (%)	6.23	11.33	11.99	9.84	9.80	9.89	9.30
Native-born (%)	92.89	90.32	91.61	92.68	91.30	91.71	92.80
Renter housing units (%)	20.21	22.27	34.78	23.84	37.08	32.95	27.74
Education (%)	14.12	8.44	12.64	13.97	20.77	18.45	21.31
Unemployment (%)	12.74	10.17	8.92	7.56	6.49	6.85	7.36
Poverty (%)	26.30	22.84	20.52	13.05	14.66	14.42	16.47
Mean household income ($)	38,481	48,753	54,608	54,603	60,623	58,632	57,882
Year 2000							
Total population	2,112	15,826	62,558	570,281	1,149,152	1,799,929	3,330,703
White	1,615	11,767	45,916	398,763	841,429	1,299,490	2,572,225
Black	15	583	2,661	26,280	55,123	84,662	69,654
Asian	10	99	922	8,189	28,148	37,368	60,021
Native American	49	309	703	9,584	25,715	36,360	217,182
Others	423	3,073	12,357	127,445	198,728	342,026	411,644
Hispanic	712	5,260	20,013	208,105	347,341	581,431	713,886
Color	779	6,310	24,968	258,156	469,773	759,986	1,098,581

(continued)

Table B.3 (continued)

	0–10	11–20	21–30	31–40	41–50	0–50	Outside
White (%)	76.47	74.35	73.40	69.92	73.22	72.20	77.23
Black (%)	0.71	3.68	4.25	4.61	4.80	4.70	2.09
Asian (%)	0.47	0.63	1.47	1.44	2.45	2.08	1.80
Native American (%)	2.32	1.95	1.12	1.68	2.24	2.02	6.52
Others (%)	20.03	19.42	19.75	22.35	17.29	19.00	12.36
Hispanic (%)	33.71	33.24	31.99	36.49	30.23	32.30	21.43
Color (%)	36.88	39.87	39.91	45.27	40.88	42.22	32.98
Female (%)	48.48	43.41	48.87	51.28	49.04	49.69	50.43
Old (65+ years) (%)	10.18	7.92	8.98	17.07	8.55	11.26	13.96
Kid (<5 years) (%)	9.14	8.72	9.78	10.24	9.79	9.92	8.34
Native-born (%)	81.53	86.71	87.52	83.72	81.68	82.57	89.72
Renter housing units (%)	23.38	26.40	24.29	23.10	38.25	32.89	25.37
College degree or higher (%)	7.02	8.83	17.09	15.40	23.71	20.68	25.00
Unemployment (%)	6.13	5.84	4.55	5.80	5.62	5.64	5.63
Poverty (%)	21.02	17.93	12.74	13.00	15.61	14.70	13.48
Mean household income ($)	53,204	59,309	71,249	60,305	68,888	66,197	67,691
Year 2010							
Total population	3,090	41,543	164,444	805,703	1,193,109	2,207,889	4,038,927
White	2,446	32,437	132,810	612,665	929,449	1,709,807	3,173,799
Black	224	2,368	9,639	49,776	70,875	132,882	107,773
Asian	10	852	4,670	19,844	41,677	67,053	111,648
Native American	0	761	1,769	13,084	26,470	42,084	236,252
Others	410	5,125	15,556	110,334	124,638	256,063	409,455
Hispanic	1,066	16,096	49,966	357,283	402,723	827,134	987,540
Color	1,319	20,479	67,230	446,487	555,346	1,090,861	1,488,924
White (%)	79.16	78.08	80.76	76.04	77.90	77.44	78.58
Black (%)	7.25	5.70	5.86	6.18	5.94	6.02	2.67
Asian (%)	0.32	2.05	2.84	2.46	3.49	3.04	2.76
Native American (%)	0.00	1.83	1.08	1.62	2.22	1.91	5.85
Others (%)	13.27	12.34	9.46	13.69	10.45	11.60	10.14
Hispanic (%)	34.50	38.75	30.38	44.34	33.75	37.46	24.45
Color (%)	42.69	49.30	40.88	55.42	46.55	49.41	36.86
Female (%)	49.94	47.17	51.24	51.15	49.38	50.13	50.31
Old (65+ years) (%)	9.00	5.97	8.00	14.46	8.65	10.67	14.80
Kid (<5 years) (%)	6.89	11.41	9.96	9.04	7.76	8.46	6.84

(continued)

Table B.3 (continued)

	0–10	11–20	21–30	31–40	41–50	0–50	Outside
Native-born (%)	87.57	88.74	88.10	79.95	80.63	81.10	88.43
Renter housing units (%)	21.78	22.18	20.49	27.12	36.01	31.61	25.23
College degree or higher (%)	7.27	15.58	24.73	18.38	26.31	23.13	27.96
Unemployment (%)	10.90	7.32	7.31	8.89	7.15	7.75	7.60
Poverty (%)	21.23	12.79	9.81	16.41	18.84	17.18	14.21
Mean household income ($)	66,449	67,996	79,633	58,958	67,297	65,194	68,577

B.4 Arkansas Nuclear One, Arkansas

See Fig. B.4.
See Table B.4.

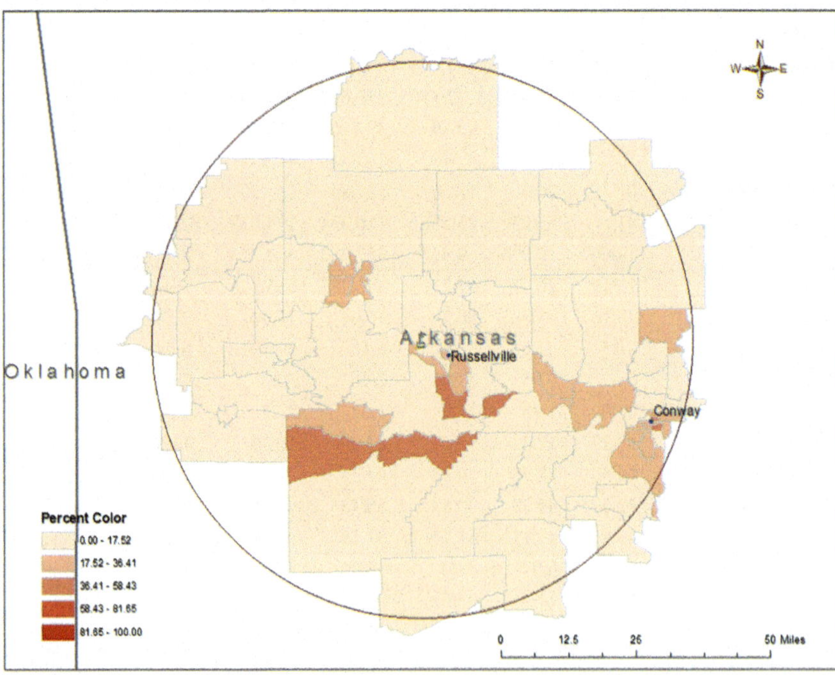

Fig. B.4 Spatial distribution of percent color among the populations surrounding Arkansas Nuclear One, Arkansas, in 2010

Table B.4 Demographic composition of population, as sorted by distance from Arkansas Nuclear One, Arkansas, in 1990, 2000, and 2010 *Source* (Kyne 2014)

	0–10	11–20	21–30	31–40	41–50	0–50	Outside
Year 1990							
Tract	8	10	12	11	4	45	641
Tract area (sq. mile)	196	851	1,943	2,258	799	6,046	47,133
Total population	32,038	34,024	38,439	38,745	11,044	154,290	2,196,442
White	30,524	32,899	36,164	36,706	10,898	147,191	1,797,202
Black	1,070	624	1,766	1,666	62	5,188	368,268
Asian	134	167	178	84	19	582	11,513
Native American	223	221	245	224	66	979	13,339
Others	93	106	87	63	0	349	6,109
Hispanic	292	315	273	279	53	1,212	18,372
Color	1,700	1,335	2,432	2,219	193	7,879	410,211
White (%)	95.27	96.69	94.08	94.74	98.68	95.40	81.82
Black (%)	3.34	1.83	4.59	4.30	0.56	3.36	16.77
Asian (%)	0.42	0.49	0.46	0.22	0.17	0.38	0.52
Native American (%)	0.70	0.65	0.64	0.58	0.60	0.63	0.61
Others (%)	0.29	0.31	0.23	0.16	0.00	0.23	0.28
Hispanic (%)	0.91	0.93	0.71	0.72	0.48	0.79	0.84
Color (%)	5.31	3.92	6.33	5.73	1.75	5.11	18.68
Female (%)	51.26	51.17	51.10	51.27	49.09	51.05	51.89
Old (65+ years) (%)	12.93	14.50	16.97	17.38	33.51	16.88	14.74
Kid (<5 years) (%)	8.62	8.25	8.09	8.11	4.62	7.99	8.50
Native-born (%)	99.37	98.86	99.20	99.48	99.19	99.23	98.92
Renter housing units (%)	32.26	20.41	20.44	18.02	13.20	21.51	27.49
Education (%)	17.43	10.09	8.78	7.39	19.45	11.31	13.46
Unemployment (%)	6.33	6.03	7.60	7.77	5.48	6.88	6.75
Poverty (%)	14.10	17.28	18.65	21.98	13.54	17.89	19.15
Mean household income ($)	47,160	39,906	38,991	35,133	53,826	41,096	45,209
Year 2000							
Total population	37,734	41,883	42,616	44,654	15,982	182,869	2,490,531
White	34,594	39,037	38,594	41,468	15,478	169,171	1,967,995
Black	1,240	384	1,727	1,540	233	5,124	412,757
Asian	255	114	193	188	23	773	19,842
Native American	309	271	151	271	27	1,029	17,463
Others	1,341	2,072	1,951	1,186	225	6,775	72,471
Hispanic	1,167	2,331	2,090	692	184	6,464	79,112
Color	3,445	3,806	4,717	3,460	622	16,050	557,285

(continued)

Table B.4 (continued)

	0–10	11–20	21–30	31–40	41–50	0–50	Outside
White (%)	91.68	93.20	90.56	92.87	96.85	92.51	79.02
Black (%)	3.29	0.92	4.05	3.45	1.46	2.80	16.57
Asian (%)	0.68	0.27	0.45	0.42	0.14	0.42	0.80
Native American (%)	0.82	0.65	0.35	0.61	0.17	0.56	0.70
Others (%)	3.55	4.95	4.58	2.66	1.41	3.70	2.91
Hispanic (%)	3.09	5.57	4.90	1.55	1.15	3.53	3.18
Color (%)	9.13	9.09	11.07	7.75	3.89	8.78	22.38
Female (%)	51.50	50.47	50.62	50.34	51.03	50.73	51.29
Old (65+ years) (%)	13.49	13.75	15.60	15.78	39.99	16.92	13.80
Kid (<5 years) (%)	8.08	7.54	7.73	7.69	3.44	7.37	8.21
Native-born (%)	97.43	96.00	96.16	98.78	98.26	97.21	97.25
Renter housing units (%)	31.61	21.31	19.94	18.44	9.06	20.98	27.68
College degree or higher (%)	22.69	13.20	10.58	9.89	27.36	15.14	16.78
Unemployment (%)	6.17	6.62	5.56	5.64	6.42	6.02	6.10
Poverty (%)	14.44	15.36	16.39	17.20	7.09	15.13	15.89
Mean household income ($)	52,502	51,909	47,867	45,765	61,536	50,583	53,505
Year 2010							
Total population	43,212	44,933	44,141	45,623	17,054	194,963	2,677,721
White	38,960	42,063	39,805	42,668	16,321	179,817	2,074,058
Black	1,737	730	1,786	1,339	48	5,640	439,557
Asian	471	319	242	287	52	1,371	36,089
Native American	374	246	281	464	66	1,431	17,761
Others	1,670	1,575	2,027	865	567	6,704	110,256
Hispanic	3,402	4,384	3,533	972	317	12,608	155,819
Color	6,576	6,500	6,521	3,647	821	24,065	689,814
White (%)	90.16	93.61	90.18	93.52	95.70	92.23	77.46
Black (%)	4.02	1.62	4.05	2.93	0.28	2.89	16.42
Asian (%)	1.09	0.71	0.55	0.63	0.30	0.70	1.35
Native American (%)	0.87	0.55	0.64	1.02	0.39	0.73	0.66
Others (%)	3.86	3.51	4.59	1.90	3.32	3.44	4.12
Hispanic (%)	7.87	9.76	8.00	2.13	1.86	6.47	5.82
Color (%)	15.22	14.47	14.77	7.99	4.81	12.34	25.76
Female (%)	50.42	49.82	50.76	50.92	50.85	50.51	50.99
Old (65+ years) (%)	12.40	14.41	15.94	17.48	44.97	17.70	13.87
Kid (<5 years) (%)	6.60	7.30	6.31	6.37	1.96	6.24	6.85

(continued)

Table B.4 (continued)

	0–10	11–20	21–30	31–40	41–50	0–50	Outside
Native-born (%)	95.08	94.21	94.99	98.83	97.82	95.98	95.72
Renter housing units (%)	32.84	23.43	19.56	17.42	9.98	21.39	28.21
College degree or higher (%)	22.33	13.80	12.47	11.63	27.09	16.22	19.35
Unemployment (%)	6.28	7.74	7.24	6.72	8.87	7.09	7.90
Poverty (%)	17.63	17.86	16.99	20.15	10.03	17.46	18.05
Mean household income ($)	52,363	44,991	45,846	45,433	56,786	48,185	53,615
Index	4	4	4	4	4	4	4

B.5 Diablo Canyon Nuclear Power Plant, California

See Fig. B.5.
See Table B.5.

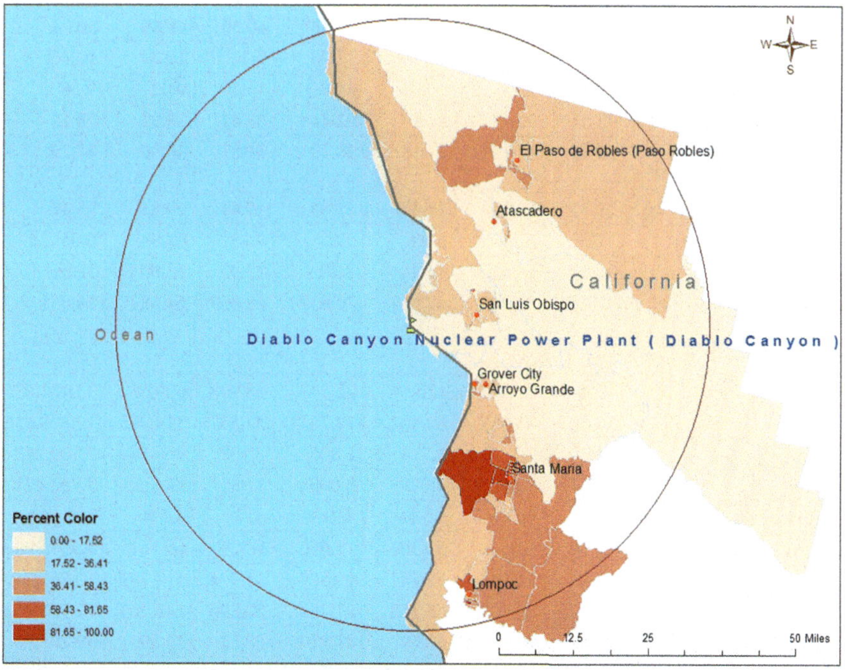

Fig. B.5 Spatial distribution of percent color among the populations surrounding Diablo Canyon Nuclear Power Plant, California, in 2010

Table B.5 Demographic composition of population, as sorted by distance from Diablo Canyon Nuclear Power Plant, California, in 1990, 2000, and 2010 *Source* (Kyne 2014)

	0–10	11–20	21–30	31–40	41–50	0–50	Outside
Year 1990							
Tract	7	23	27	22	11	90	7,967
Tract area (sq. mile)	358	181	1,051	1,352	106	3,048	160,647
Total population	39,157	89,483	107,194	80,430	45,168	361,432	29,398,594
White	36,325	77,343	77,687	68,790	33,571	293,716	20,261,924
Black	398	3,568	1,570	2,624	3,016	11,176	2,187,592
Asian	1,374	3,678	4,372	2,466	2,244	14,134	2,833,691
Native American	328	860	1,308	833	712	4,041	244,866
Others	768	4,011	22,240	5,714	5,618	38,351	3,870,512
Hispanic	3,136	11,985	36,995	12,530	9,663	74,309	7,483,224
Color	5,001	19,405	43,312	17,898	15,241	100,857	12,565,203
White (%)	92.77	86.43	72.47	85.53	74.32	81.26	68.92
Black (%)	1.02	3.99	1.46	3.26	6.68	3.09	7.44
Asian (%)	3.51	4.11	4.08	3.07	4.97	3.91	9.64
Native American (%)	0.84	0.96	1.22	1.04	1.58	1.12	0.83
Others (%)	1.96	4.48	20.75	7.10	12.44	10.61	13.17
Hispanic (%)	8.01	13.39	34.51	15.58	21.39	20.56	25.45
Color (%)	12.77	21.69	40.41	22.25	33.74	27.90	42.74
Female (%)	51.30	45.82	49.80	49.58	49.50	48.89	50.01
Old (65+ years) (%)	17.81	13.99	11.49	12.98	9.21	12.84	10.46
Kid (<5 years) (%)	7.23	5.73	10.92	9.66	11.44	9.02	9.56
Native-born (%)	92.81	91.18	83.17	92.44	88.41	88.92	78.17
Renter housing units (%)	32.46	44.19	37.18	28.39	41.51	36.90	41.25
Education (%)	31.11	22.73	14.37	18.84	17.49	19.81	23.40
Unemployment (%)	3.76	6.14	6.93	5.65	6.95	6.09	6.65
Poverty (%)	9.40	18.65	14.44	7.20	13.02	13.00	12.50
Mean household income ($)	72,435	58,966	62,857	70,037	62,859	64,682	76,051
Year 2000							
Total population	40,444	97,522	130,641	92,047	48,908	409,562	33,462,086
White	35,637	81,088	89,862	74,754	33,562	314,903	19,808,056
Black	376	2,814	1,774	1,807	3,177	9,948	2,209,242
Asian	1,417	3,865	4,099	3,000	1,981	14,362	3,782,471
Native American	247	693	1,584	830	817	4,171	308,044
Others	2,810	9,033	33,303	11,656	9,359	66,161	7,354,290
Hispanic	4,324	16,140	57,347	19,507	15,974	113,292	10,855,840
Color	7,169	25,603	66,358	27,243	22,890	149,263	17,951,222
White (%)	88.11	83.15	68.79	81.21	68.62	76.89	59.20

(continued)

Table B.5 (continued)

	0–10	11–20	21–30	31–40	41–50	0–50	Outside
Black (%)	0.93	2.89	1.36	1.96	6.50	2.43	6.60
Asian (%)	3.50	3.96	3.14	3.26	4.05	3.51	11.30
Native American (%)	0.61	0.71	1.21	0.90	1.67	1.02	0.92
Others (%)	6.95	9.26	25.49	12.66	19.14	16.15	21.98
Hispanic (%)	10.69	16.55	43.90	21.19	32.66	27.66	32.44
Color (%)	17.73	26.25	50.79	29.60	46.80	36.44	53.65
Female (%)	51.37	46.66	49.13	50.67	47.92	48.96	50.29
Old (65+ years) (%)	19.20	13.93	11.20	15.02	11.21	13.50	10.55
Kid (<5 years) (%)	5.69	5.00	9.34	7.70	8.79	7.51	8.84
Native-born (%)	91.40	91.12	77.10	89.17	83.63	85.34	73.69
Renter housing units (%)	32.04	43.83	35.79	27.02	41.32	36.00	40.64
College degree or higher (%)	37.33	27.66	15.47	21.37	17.39	22.35	26.67
Unemployment (%)	4.18	7.79	6.94	5.78	8.50	6.76	7.02
Poverty (%)	11.02	17.80	15.49	9.44	13.34	13.88	14.22
Mean household income ($)	75,723	62,642	64,956	72,633	64,607	67,384	81,943
Year 2010							
Total population	27,776	108,423	157,599	104,511	49,983	448,292	36,189,000
White	23,841	89,252	130,401	84,814	33,275	361,583	22,031,130
Black	90	3,469	2,187	2,303	2,744	10,793	2,235,518
Asian	1,538	4,780	5,083	3,957	1,708	17,066	4,870,615
Native American	148	866	1,375	1,111	761	4,261	279,367
Others	2,159	10,056	18,553	12,326	11,495	54,589	6,772,368
Hispanic	4,023	19,014	78,580	31,345	22,195	155,157	13,301,000
Color	6,652	30,258	87,482	40,018	28,213	192,623	21,337,624
White (%)	85.83	82.32	82.74	81.15	66.57	80.66	60.88
Black (%)	0.32	3.20	1.39	2.20	5.49	2.41	6.18
Asian (%)	5.54	4.41	3.23	3.79	3.42	3.81	13.46
Native American (%)	0.53	0.80	0.87	1.06	1.52	0.95	0.77
Others (%)	7.77	9.27	11.77	11.79	23.00	12.18	18.71
Hispanic (%)	14.48	17.54	49.86	29.99	44.41	34.61	36.75
Color (%)	23.95	27.91	55.51	38.29	56.45	42.97	58.96
Female (%)	50.04	46.95	49.56	51.10	47.76	49.12	50.27
Old (65+ years) (%)	16.26	14.70	11.11	14.49	10.58	13.02	11.06
Kid (<5 years) (%)	5.32	3.68	8.10	7.12	7.48	6.56	6.95
Native-born (%)	89.02	90.48	75.29	86.82	76.47	82.63	72.69

(continued)

Table B.5 (continued)

	0–10	11–20	21–30	31–40	41–50	0–50	Outside
Renter housing units (%)	32.40	43.05	36.24	27.95	43.16	36.55	39.00
College degree or higher (%)	38.33	32.63	19.80	24.50	17.11	24.92	30.12
Unemployment (%)	5.77	7.80	8.21	7.29	8.86	7.79	9.00
Poverty (%)	14.25	17.98	14.37	8.29	15.62	13.82	13.71
Mean household income ($)	78,370	67,640	73,873	76,408	65,405	72,270	83,626
Index	5	5	5	5	5	5	5

B.6 Millstone Power Station, Connecticut

See Fig. B.6.
See Table B.6.

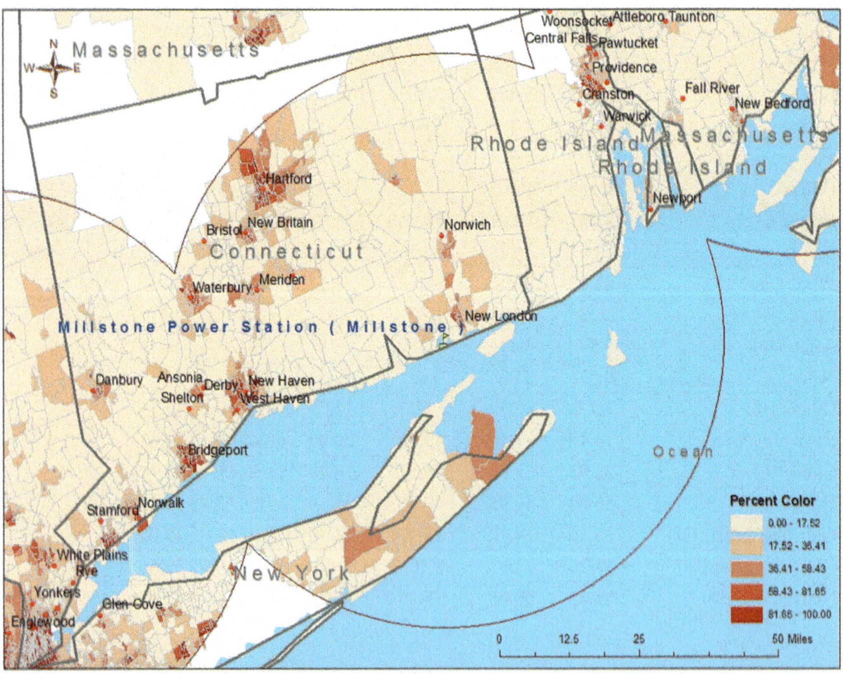

Fig. B.6 Spatial distribution of percent color among the populations surrounding Millstone Power Station, Connecticut, in 2010

Table B.6 Demographic composition of population, as sorted by distance from Millstone Power Station, Connecticut, in 1990, 2000, and 2010 *Source* (Kyne 2014)

	0–10	11–20	21–30	31–40	41–50	0–50	Outside
Year 1990							
Tract	32	47	56	161	99	395	5,601
Tract area (sq. mile)	266	1,698	1,112	1,375	651	5,103	56,540
Total population	118,811	177,020	230,951	629,978	365,327	1,522,087	20,758,812
White	104,298	169,316	221,012	588,085	269,809	1,352,520	15,827,065
Black	8,990	4,061	5,988	22,550	56,558	98,147	3,073,983
Asian	2,059	1,908	1,434	8,264	6,058	19,723	736,121
Native American	604	752	813	2,186	1,062	5,417	64,730
Others	2,863	978	1,715	8,896	31,835	46,287	1,056,919
Hispanic	5,706	3,328	4,360	23,402	52,940	89,736	2,309,453
Color	17,009	10,018	12,513	56,031	113,106	208,677	5,922,545
White (%)	87.78	95.65	95.70	93.35	73.85	88.86	76.24
Black (%)	7.57	2.29	2.59	3.58	15.48	6.45	14.81
Asian (%)	1.73	1.08	0.62	1.31	1.66	1.30	3.55
Native American (%)	0.51	0.42	0.35	0.35	0.29	0.36	0.31
Others (%)	2.41	0.55	0.74	1.41	8.71	3.04	5.09
Hispanic (%)	4.80	1.88	1.89	3.71	14.49	5.90	11.13
Color (%)	14.32	5.66	5.42	8.89	30.96	13.71	28.53
Female (%)	48.26	50.85	50.79	51.88	51.88	51.31	52.08
Old (65+ years) (%)	12.72	13.90	12.49	14.80	12.15	13.55	13.26
Kid (<5 years) (%)	8.79	8.37	8.39	7.61	9.00	8.24	8.29
Native-born (%)	95.35	95.72	95.83	92.94	89.25	93.00	84.97
Renter housing units (%)	35.27	24.61	17.99	28.65	45.72	30.88	42.70
Education (%)	23.61	23.75	27.76	24.41	20.54	23.89	23.64
Unemployment (%)	6.05	5.42	4.95	4.56	6.79	5.35	6.73
Poverty (%)	6.33	5.37	5.19	5.34	14.03	7.50	12.28
Mean household income ($)	70,684	77,408	86,104	77,147	64,401	74,916	74,244
Year 2000							
Total population	115,203	188,366	257,325	666,051	362,705	1,589,650	21,840,692
White	95,049	172,513	239,689	580,046	244,041	1,331,338	15,228,340
Black	8,948	5,042	5,792	34,218	56,329	110,329	3,227,051
Asian	3,031	2,475	2,820	14,919	9,431	32,676	1,127,550
Native American	740	1,619	836	2,520	1,042	6,757	87,100
Others	7,456	6,702	8,199	34,364	51,848	108,569	2,170,631
Hispanic	8,583	6,851	10,193	50,324	65,674	141,625	3,132,790
Color	23,862	19,475	23,021	109,403	137,811	313,572	7,858,900
White (%)	82.51	91.58	93.15	87.09	67.28	83.75	69.72

(continued)

Table B.6 (continued)

	0–10	11–20	21–30	31–40	41–50	0–50	Outside
Black (%)	7.77	2.68	2.25	5.14	15.53	6.94	14.78
Asian (%)	2.63	1.31	1.10	2.24	2.60	2.06	5.16
Native American (%)	0.64	0.86	0.32	0.38	0.29	0.43	0.40
Others (%)	6.47	3.56	3.19	5.16	14.29	6.83	9.94
Hispanic (%)	7.45	3.64	3.96	7.56	18.11	8.91	14.34
Color (%)	20.71	10.34	8.95	16.43	38.00	19.73	35.98
Female (%)	50.99	50.87	50.98	51.84	51.82	51.52	51.85
Old (65+ years) (%)	14.29	14.28	13.18	14.75	12.18	13.82	13.06
Kid (<5 years) (%)	7.83	7.39	7.55	7.21	8.26	7.57	7.86
Native-born (%)	93.52	94.49	94.35	91.43	87.42	91.50	80.67
Renter housing units (%)	33.36	23.66	16.94	28.78	43.93	29.76	42.07
College degree or higher (%)	28.60	29.06	33.64	28.51	23.68	28.37	27.85
Unemployment (%)	4.59	4.08	4.33	4.50	7.43	5.07	6.87
Poverty (%)	7.22	5.86	5.09	7.12	14.99	8.45	13.86
Mean household income ($)	78,503	81,782	93,395	80,057	66,283	79,102	78,923
Year 2010							
Total population	118,025	201,538	259,552	708,022	380,788	1,667,925	22,164,052
White	94,588	174,106	240,142	591,583	251,480	1,351,899	15,075,107
Black	8,032	7,211	6,350	38,874	61,244	121,711	3,273,726
Asian	5,210	6,954	3,637	25,273	14,822	55,896	1,504,402
Native American	457	1,401	841	1,958	1,292	5,949	73,407
Others	9,738	11,866	8,582	50,334	51,950	132,470	2,237,411
Hispanic	12,836	13,482	14,476	82,304	77,811	200,909	3,659,514
Color	28,679	33,599	28,663	157,284	159,913	408,138	8,663,811
White (%)	80.14	86.39	92.52	83.55	66.04	81.05	68.02
Black (%)	6.81	3.58	2.45	5.49	16.08	7.30	14.77
Asian (%)	4.41	3.45	1.40	3.57	3.89	3.35	6.79
Native American (%)	0.39	0.70	0.32	0.28	0.34	0.36	0.33
Others (%)	8.25	5.89	3.31	7.11	13.64	7.94	10.09
Hispanic (%)	10.88	6.69	5.58	11.62	20.43	12.05	16.51
Color (%)	24.30	16.67	11.04	22.21	42.00	24.47	39.09
Female (%)	50.76	50.49	50.47	51.45	51.25	51.09	51.63
Old (65+ years) (%)	15.09	15.77	15.10	14.66	12.10	14.31	13.35
Kid (<5 years) (%)	5.47	5.47	5.05	5.13	6.41	5.47	6.01
Native-born (%)	90.86	90.75	93.38	89.22	84.96	89.20	79.23

(continued)

Table B.6 (continued)

	0–10	11–20	21–30	31–40	41–50	0–50	Outside
Renter housing units (%)	30.91	21.27	13.56	25.61	40.87	26.78	38.99
College degree or higher (%)	33.29	33.20	37.91	32.62	27.32	32.43	32.51
Unemployment (%)	6.86	5.83	5.87	7.30	10.07	7.49	7.56
Poverty (%)	7.63	6.90	4.96	8.91	16.13	9.61	13.61
Mean household income ($)	80,662	88,699	101,587	82,663	67,734	82,791	82,038
Index	6	6	6	6	6	6	6

B.7 Saint Lucie, Florida

See Fig. B.7.
See Table B.7.

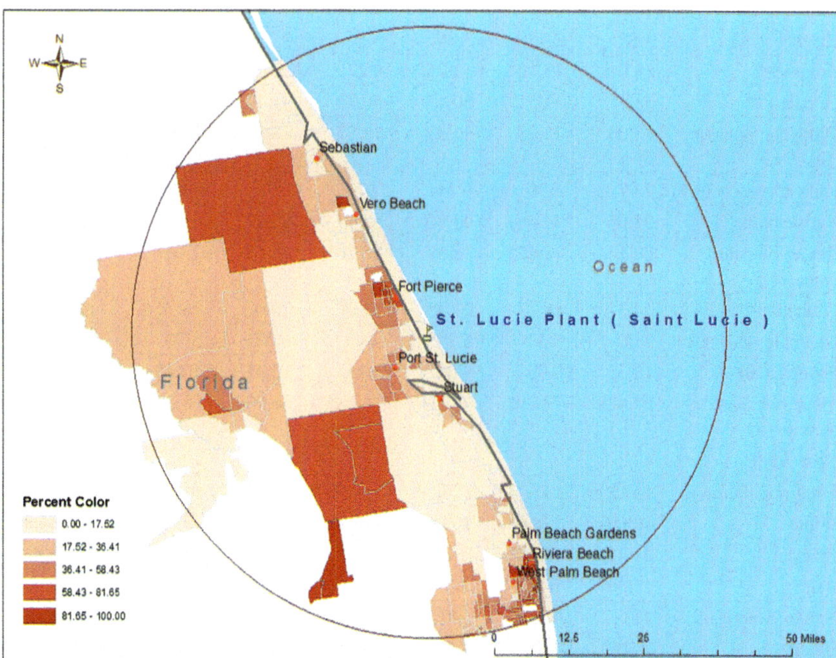

Fig. B.7 Spatial distribution of percent color among the populations surrounding Saint Lucie, Florida, in 2010

Table B.7 Demographic composition of population, as sorted by distance from Saint Lucie, Florida, in 1990, 2000, and 2010 *Source* (Kyne 2014)

	0–10	11–20	21–30	31–40	41–50	0–50	Outside
Year 1990							
Tract	35	41	34	60	96	266	3,979
Tract area (sq. mile)	234	623	759	1,130	548	3,296	62,462
Total population	132,150	108,530	102,822	126,571	262,246	732,319	12,205,582
White	111,476	96,285	90,925	118,866	195,351	612,903	10,142,813
Black	18,060	10,615	9,469	3,328	58,913	100,385	1,655,553
Asian	778	592	609	1,026	2,977	5,982	144,219
Native American	341	187	199	375	570	1,672	40,941
Others	1,489	863	1,611	2,969	4,449	11,381	222,066
Hispanic	4,855	2,983	3,965	6,965	22,856	41,624	1,513,424
Color	23,803	14,237	14,140	11,732	83,969	147,881	3,301,312
White (%)	84.36	88.72	88.43	93.91	74.49	83.69	83.10
Black (%)	13.67	9.78	9.21	2.63	22.46	13.71	13.56
Asian (%)	0.59	0.55	0.59	0.81	1.14	0.82	1.18
Native American (%)	0.26	0.17	0.19	0.30	0.22	0.23	0.34
Others (%)	1.13	0.80	1.57	2.35	1.70	1.55	1.82
Hispanic (%)	3.67	2.75	3.86	5.50	8.72	5.68	12.40
Color (%)	18.01	13.12	13.75	9.27	32.02	20.19	27.05
Female (%)	51.67	51.51	50.47	50.73	51.49	51.25	51.68
Old (65+ years) (%)	23.04	25.28	23.59	18.57	18.34	20.99	18.15
Kid (<5 years) (%)	7.97	6.90	6.74	7.70	8.78	7.88	7.81
Native-born (%)	93.06	94.66	93.80	93.37	87.41	91.43	86.89
Renter housing units (%)	22.10	18.62	18.89	20.20	28.65	23.07	27.86
Education (%)	14.27	18.77	21.85	21.99	18.07	18.71	18.25
Unemployment (%)	7.20	5.33	5.39	4.08	6.04	5.70	5.79
Poverty (%)	12.32	8.99	9.02	9.14	12.41	10.85	12.80
Mean household income ($)	57,422	73,203	76,239	73,685	66,000	68,322	59,432
Year 2000							
Total population	157,349	147,427	124,539	179,907	321,611	930,833	15,051,545
White	126,970	128,081	109,638	160,956	213,372	739,017	11,724,285
Black	21,221	13,421	8,984	6,928	80,707	131,261	2,180,844
Asian	1,436	1,157	921	2,703	5,504	11,721	259,468
Native American	413	451	330	366	1,040	2,600	51,828
Others	7,302	4,331	4,653	8,921	21,014	46,221	835,133
Hispanic	13,508	8,159	8,239	18,018	47,076	95,000	2,585,314

(continued)

Table B.7 (continued)

	0–10	11–20	21–30	31–40	41–50	0–50	Outside
Color	38,521	24,559	19,342	29,831	139,481	251,734	5,274,186
White (%)	80.69	86.88	88.04	89.47	66.34	79.39	77.89
Black (%)	13.49	9.10	7.21	3.85	25.09	14.10	14.49
Asian (%)	0.91	0.78	0.74	1.50	1.71	1.26	1.72
Native American (%)	0.26	0.31	0.26	0.20	0.32	0.28	0.34
Others (%)	4.64	2.94	3.74	4.96	6.53	4.97	5.55
Hispanic (%)	8.58	5.53	6.62	10.02	14.64	10.21	17.18
Color (%)	24.48	16.66	15.53	16.58	43.37	27.04	35.04
Female (%)	51.60	51.43	50.26	50.28	51.34	51.05	51.29
Old (65+ years) (%)	23.15	27.36	25.16	19.70	16.24	21.03	17.34
Kid (<5 years) (%)	6.64	5.49	5.97	6.46	7.47	6.62	7.14
Native-born (%)	89.03	92.79	91.97	89.39	82.09	87.69	83.02
Renter housing units (%)	20.50	15.67	17.77	16.61	28.75	21.27	26.27
College degree or higher (%)	16.22	22.95	27.15	26.87	21.50	22.69	22.31
Unemployment (%)	5.39	3.97	4.32	3.63	5.56	4.74	5.62
Poverty (%)	14.10	8.69	8.10	7.93	14.10	11.25	12.59
Mean household income ($)	58,800	77,356	84,862	88,879	71,357	75,463	66,100
Year 2010							
Total population	206,961	186,605	144,462	232,232	380,764	1,151,024	17,360,596
White	157,045	151,285	127,353	204,978	240,086	880,747	13,234,481
Black	34,488	23,619	10,053	11,994	104,682	184,836	2,715,576
Asian	3,400	2,543	1,409	4,697	8,452	20,501	434,880
Native American	1,268	736	383	608	1,078	4,073	50,497
Others	10,760	8,422	5,264	9,955	26,466	60,867	925,162
Hispanic	31,454	22,129	15,235	32,462	78,316	179,596	3,815,728
Color	71,721	51,118	28,928	51,951	196,294	400,012	7,179,352
White (%)	75.88	81.07	88.16	88.26	63.05	76.52	76.23
Black (%)	16.66	12.66	6.96	5.16	27.49	16.06	15.64
Asian (%)	1.64	1.36	0.98	2.02	2.22	1.78	2.50
Native American (%)	0.61	0.39	0.27	0.26	0.28	0.35	0.29
Others (%)	5.20	4.51	3.64	4.29	6.95	5.29	5.33
Hispanic (%)	15.20	11.86	10.55	13.98	20.57	15.60	21.98
Color (%)	34.65	27.39	20.02	22.37	51.55	34.75	41.35
Female (%)	51.10	50.48	52.08	50.58	51.37	51.11	51.11
Old (65+ years) (%)	21.43	22.60	24.35	21.65	15.33	20.01	16.73

(continued)

Table B.7 (continued)

	0–10	11–20	21–30	31–40	41–50	0–50	Outside
Kid (<5 years) (%)	6.02	5.11	4.83	5.08	6.07	5.55	5.86
Native-born (%)	84.74	87.78	90.01	86.13	76.52	83.46	80.65
Renter housing units (%)	20.81	14.73	16.85	16.92	25.91	20.05	24.75
College degree or higher (%)	18.16	26.19	29.63	33.01	24.66	26.14	25.89
Unemployment (%)	11.90	10.29	9.08	7.97	9.69	9.74	8.83
Poverty (%)	14.10	10.29	11.94	9.36	15.67	12.77	13.89
Mean household income ($)	56,660	76,319	81,158	95,132	69,089	75,053	65,733
Index	7	7	7	7	7	7	7

B.8 Turkey Point Nuclear Generating, Florida

See Fig. B.8.
See Table B.8.

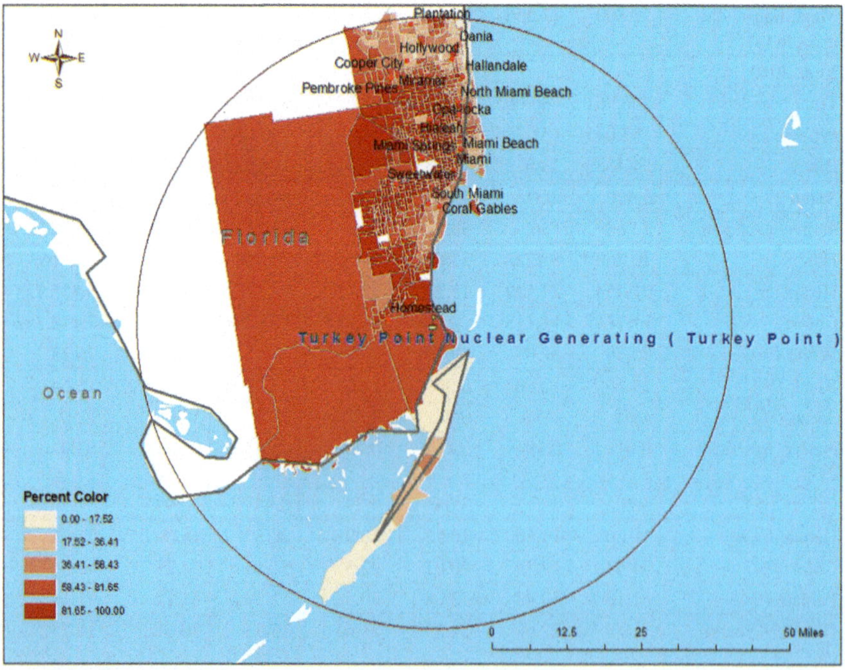

Fig. B.8 Spatial distribution of percent color among the populations surrounding Turkey Point Nuclear Generating, Florida, in 2010

Table B.8 Demographic composition of population, as sorted by distance from Turkey Point Nuclear Generating, Florida, in 1990, 2000, and 2010 *Source* (Kyne 2014)

	0–10	11–20	21–30	31–40	41–50	0–50	Outside
Year 1990							
Tract	26	97	236	193	139	691	3,554
Tract area (sq. mile)	208	599	588	2,794	242	4,431	61,326
Total population	89,348	362,555	847,153	721,636	468,445	2,489,137	10,448,764
White	58,339	294,607	658,329	475,188	389,620	1,876,083	8,879,633
Black	22,707	46,829	129,598	208,634	64,701	472,469	1,283,469
Asian	1,251	9,012	6,571	9,146	6,461	32,441	117,760
Native American	355	552	976	1,185	1,397	4,465	38,148
Others	6,704	11,544	51,705	27,477	6,252	103,682	129,765
Hispanic	27,217	127,347	557,806	249,329	48,145	1,009,844	545,204
Color	50,859	181,670	681,481	457,442	118,755	1,490,207	1,958,986
White (%)	65.29	81.26	77.71	65.85	83.17	75.37	84.98
Black (%)	25.41	12.92	15.30	28.91	13.81	18.98	12.28
Asian (%)	1.40	2.49	0.78	1.27	1.38	1.30	1.13
Native American (%)	0.40	0.15	0.12	0.16	0.30	0.18	0.37
Others (%)	7.50	3.18	6.10	3.81	1.33	4.17	1.24
Hispanic (%)	30.46	35.12	65.84	34.55	10.28	40.57	5.22
Color (%)	56.92	50.11	80.44	63.39	25.35	59.87	18.75
Female (%)	50.02	51.65	52.05	52.91	51.65	52.09	51.55
Old (65+ years) (%)	9.31	8.66	16.14	15.00	16.87	14.61	19.19
Kid (<5 years) (%)	11.87	9.21	7.83	8.71	7.59	8.38	7.68
Native-born (%)	81.77	69.11	42.53	62.01	84.47	61.35	93.29
Renter housing units (%)	39.97	29.04	49.39	34.49	27.42	37.28	25.58
Education (%)	10.72	31.65	16.61	15.57	18.96	18.70	18.18
Unemployment (%)	7.53	4.86	8.65	7.70	5.49	7.14	5.43
Poverty (%)	21.95	8.97	22.76	15.35	10.76	16.29	11.83
Mean household income ($)	49,888	88,232	54,126	60,719	66,146	63,151	59,246
Year 2000							
Total population	99,529	461,953	920,354	913,887	628,598	3,024,321	12,958,057
White	55,125	356,258	722,966	520,497	476,049	2,130,895	10,332,407
Black	30,816	55,493	110,717	297,104	96,481	590,611	1,721,494
Asian	1,187	11,351	8,006	14,613	15,638	50,795	220,394
Native American	137	1,389	1,634	1,984	1,976	7,120	47,308
Others	12,273	37,454	77,017	79,689	38,437	244,870	636,484
Hispanic	45,524	240,853	651,320	391,989	128,148	1,457,834	1,222,480

(continued)

Table B.8 (continued)

	0–10	11–20	21–30	31–40	41–50	0–50	Outside
Color	78,747	313,670	767,526	715,076	253,323	2,128,342	3,397,578
White (%)	55.39	77.12	78.55	56.95	75.73	70.46	79.74
Black (%)	30.96	12.01	12.03	32.51	15.35	19.53	13.29
Asian (%)	1.19	2.46	0.87	1.60	2.49	1.68	1.70
Native American (%)	0.14	0.30	0.18	0.22	0.31	0.24	0.37
Others (%)	12.33	8.11	8.37	8.72	6.11	8.10	4.91
Hispanic (%)	45.74	52.14	70.77	42.89	20.39	48.20	9.43
Color (%)	79.12	67.90	83.39	78.25	40.30	70.37	26.22
Female (%)	51.22	52.03	51.17	52.50	51.25	51.72	51.17
Old (65+ years) (%)	7.75	9.56	16.27	13.10	12.56	13.24	18.57
Kid (<5 years) (%)	11.47	8.43	6.78	8.26	7.69	7.82	6.94
Native-born (%)	69.49	58.29	40.41	53.52	75.88	55.43	89.79
Renter housing units (%)	44.55	26.43	46.13	31.96	27.29	34.79	24.25
College degree or higher (%)	11.72	31.26	21.02	18.61	27.13	22.86	22.22
Unemployment (%)	11.41	6.04	9.08	9.18	4.78	7.70	5.07
Poverty (%)	27.84	10.58	20.36	16.83	10.52	15.98	11.70
Mean household income ($)	47,347	88,385	60,468	62,458	76,392	68,310	66,321
Year 2010							
Total population	147,705	491,195	1,006,372	987,234	686,042	3,318,548	15,193,072
White	97,921	395,950	828,717	568,334	487,871	2,378,793	11,736,435
Black	37,054	52,053	117,434	329,001	123,671	659,213	2,241,199
Asian	2,897	12,580	13,448	19,416	23,470	71,811	383,570
Native American	143	647	1,228	1,819	1,648	5,485	49,085
Others	9,690	29,965	45,545	68,664	49,382	203,246	782,783
Hispanic	82,831	307,830	755,234	484,952	199,051	1,829,898	2,165,426
Color	122,889	375,537	873,067	826,040	356,808	2,554,341	5,025,023
White (%)	66.29	80.61	82.35	57.57	71.11	71.68	77.25
Black (%)	25.09	10.60	11.67	33.33	18.03	19.86	14.75
Asian (%)	1.96	2.56	1.34	1.97	3.42	2.16	2.52
Native American (%)	0.10	0.13	0.12	0.18	0.24	0.17	0.32
Others (%)	6.56	6.10	4.53	6.96	7.20	6.12	5.15
Hispanic (%)	56.08	62.67	75.05	49.12	29.01	55.14	14.25
Color (%)	83.20	76.45	86.75	83.67	52.01	76.97	33.07
Female (%)	52.12	52.53	50.21	52.89	50.95	51.59	51.01
Old (65+ years) (%)	7.30	11.99	15.89	13.55	12.28	13.49	17.68

(continued)

Table B.8 (continued)

	0–10	11–20	21–30	31–40	41–50	0–50	Outside
Kid (<5 years) (%)	10.02	6.00	5.59	6.48	5.87	6.17	5.76
Native-born (%)	65.01	56.35	41.90	52.40	69.62	53.92	86.70
Renter housing units (%)	40.03	25.16	41.58	29.91	25.89	32.38	22.99
College degree or higher (%)	17.99	34.25	26.39	23.91	32.65	27.77	25.51
Unemployment (%)	9.68	7.72	7.88	9.43	8.03	8.42	8.99
Poverty (%)	24.41	11.06	19.12	15.67	11.01	15.43	13.47
Mean household income ($)	52,294	86,037	60,694	61,531	79,165	68,233	65,956
Index	8	8	8	8	8	8	8

B.9 Edwin I. Hatch Nuclear Plant, Georgia

See Fig. B.9.
See Table B.9.

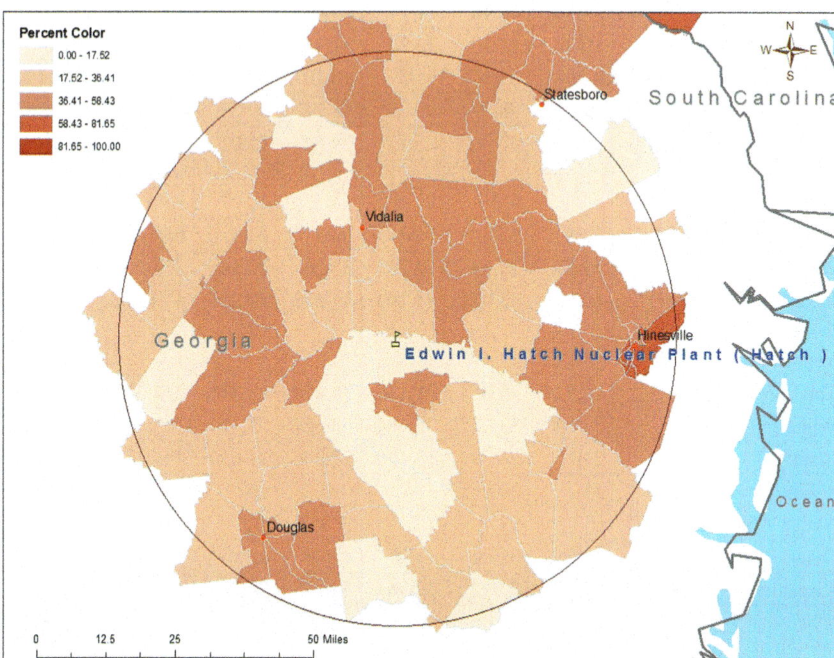

Fig. B.9 Spatial distribution of percent color among the populations surrounding Edwin I. Hatch Nuclear Plant, Georgia, in 2010

Table B.9 Demographic composition of population, as sorted by distance from Edwin I. Hatch Nuclear Plant, Georgia, in 1990, 2000, and 2010 *Source* (Kyne 2014)

	0–10	11–20	21–30	31–40	41–50	0–50	Outside
Year 1990							
Tract	3	10	16	23	20	72	1,897
Tract area (sq. mile)	333	842	1,516	2,060	1,835	6,586	52,839
Total population	9,459	39,605	45,207	82,460	67,051	243,782	6,234,427
White	6,608	31,273	33,461	57,591	50,374	179,307	4,424,063
Black	2,595	7,488	11,279	23,146	16,157	60,665	1,684,207
Asian	1	64	76	534	269	944	72,812
Native American	22	118	39	186	104	469	14,813
Others	237	663	344	1,000	158	2,402	38,495
Hispanic	264	861	592	1,839	400	3,956	97,435
Color	2,878	8,518	11,952	25,591	16,887	65,826	1,862,637
White (%)	69.86	78.96	74.02	69.84	75.13	73.55	70.96
Black (%)	27.43	18.91	24.95	28.07	24.10	24.88	27.01
Asian (%)	0.01	0.16	0.17	0.65	0.40	0.39	1.17
Native American (%)	0.23	0.30	0.09	0.23	0.16	0.19	0.24
Others (%)	2.51	1.67	0.76	1.21	0.24	0.99	0.62
Hispanic (%)	2.79	2.17	1.31	2.23	0.60	1.62	1.56
Color (%)	30.43	21.51	26.44	31.03	25.19	27.00	29.88
Female (%)	50.66	49.44	52.17	49.57	52.49	50.88	51.53
Old (65+ years) (%)	10.97	11.30	14.20	11.14	12.04	11.97	10.00
Kid (<5 years) (%)	8.30	8.55	8.45	10.10	9.61	9.34	9.14
Native-born (%)	99.03	98.79	99.09	98.74	99.33	98.99	97.26
Renter housing units (%)	18.48	27.00	25.12	29.62	25.02	26.56	31.63
Education (%)	5.79	9.11	9.54	9.41	10.07	9.43	19.69
Unemployment (%)	6.68	5.76	5.61	6.23	6.29	6.06	5.73
Poverty (%)	22.77	19.83	23.72	23.56	20.18	22.01	14.37
Mean household income ($)	40,800	43,952	44,102	41,438	44,229	43,121	61,069
Year 2000							
Total population	10,524	45,652	51,136	96,476	80,828	284,616	7,901,837
White	7,454	33,027	35,664	63,657	55,714	195,516	5,131,659
Black	2,465	9,230	13,406	27,346	21,630	74,077	2,268,033
Asian	11	309	184	699	562	1,765	173,564
Native American	0	93	187	304	140	724	22,964
Others	597	3,002	1,676	4,475	2,795	12,545	305,606
Hispanic	663	3,992	1,886	5,908	2,838	15,287	414,689

(continued)

Table B.9 (continued)

	0–10	11–20	21–30	31–40	41–50	0–50	Outside
Color	3,180	13,770	15,976	34,963	25,865	93,754	2,962,972
White (%)	70.83	72.35	69.74	65.98	68.93	68.69	64.94
Black (%)	23.42	20.22	26.22	28.34	26.76	26.03	28.70
Asian (%)	0.10	0.68	0.36	0.72	0.70	0.62	2.20
Native American (%)	0.00	0.20	0.37	0.32	0.17	0.25	0.29
Others (%)	5.67	6.58	3.28	4.64	3.46	4.41	3.87
Hispanic (%)	6.30	8.74	3.69	6.12	3.51	5.37	5.25
Color (%)	30.22	30.16	31.24	36.24	32.00	32.94	37.50
Female (%)	51.16	47.95	49.12	48.13	51.01	49.21	50.93
Old (65+ years) (%)	11.16	10.79	12.87	10.12	11.62	11.19	9.57
Kid (<5 years) (%)	9.13	8.62	8.41	9.21	8.88	8.87	8.67
Native-born (%)	95.43	94.26	96.90	95.88	97.12	96.14	92.83
Renter housing units (%)	14.63	24.77	22.43	26.20	23.22	23.93	29.99
College degree or higher (%)	6.42	9.63	11.15	9.58	12.31	10.54	24.78
Unemployment (%)	6.58	5.31	5.34	6.70	5.61	5.89	5.48
Poverty (%)	21.69	21.75	21.48	20.52	19.46	20.61	12.72
Mean household income ($)	46,407	46,599	48,493	47,793	49,401	48,147	71,311
Year 2010							
Total population	10,422	47,478	57,197	98,638	90,482	304,217	9,164,598
White	7,395	33,734	39,614	66,518	59,717	206,978	5,581,277
Black	2,449	8,986	16,033	26,473	25,290	79,231	2,793,838
Asian	0	367	412	241	718	1,738	300,098
Native American	0	248	50	233	137	668	22,566
Others	578	4,143	1,088	5,173	4,620	15,602	466,819
Hispanic	640	6,562	2,146	8,851	5,475	23,674	761,038
Color	3,200	16,518	18,791	36,632	32,836	107,977	3,979,068
White (%)	70.96	71.05	69.26	67.44	66.00	68.04	60.90
Black (%)	23.50	18.93	28.03	26.84	27.95	26.04	30.49
Asian (%)	0.00	0.77	0.72	0.24	0.79	0.57	3.27
Native American (%)	0.00	0.52	0.09	0.24	0.15	0.22	0.25
Others (%)	5.55	8.73	1.90	5.24	5.11	5.13	5.09
Hispanic (%)	6.14	13.82	3.75	8.97	6.05	7.78	8.30
Color (%)	30.70	34.79	32.85	37.14	36.29	35.49	43.42
Female (%)	53.03	47.99	47.84	48.12	49.98	48.77	51.21
Old (65+ years) (%)	13.83	11.57	13.40	11.54	12.48	12.25	10.19

(continued)

Table B.9 (continued)

	0–10	11–20	21–30	31–40	41–50	0–50	Outside
Kid (<5 years) (%)	6.77	7.90	6.41	7.29	6.84	7.07	7.28
Native-born (%)	96.78	91.37	96.89	93.65	96.39	94.83	90.25
Renter housing units (%)	19.33	28.34	24.96	28.84	25.42	26.63	28.33
College degree or higher (%)	8.15	10.66	13.69	10.42	12.00	11.48	27.74
Unemployment (%)	6.08	6.01	5.88	8.09	7.58	7.12	8.90
Poverty (%)	24.15	23.90	20.96	21.13	19.39	21.12	15.53
Mean household income ($)	42,325	44,850	48,634	43,926	49,169	46,491	67,268
Index	9	9	9	9	9	9	9

B.10 Vogtle Electric Generating Plant, Georgia

See Fig. B.10.
See Table B.10.

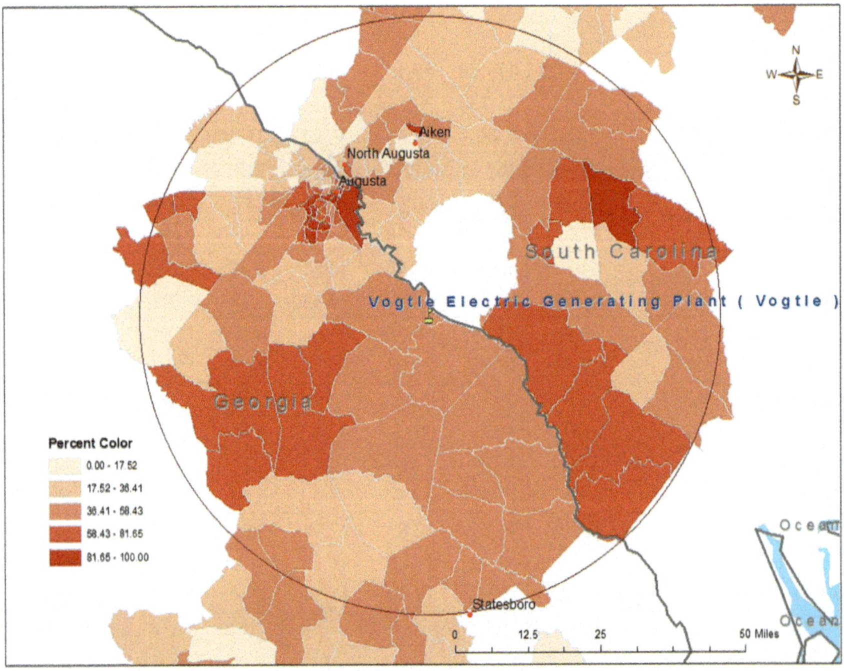

Fig. B.10 Spatial distribution of percent color among the populations surrounding Vogtle Electric Generating Plant, Georgia, in 2010

Table B.10 Demographic composition of population, as sorted by distance from Vogtle Electric Generating Plant, Georgia, in 1990, 2000, and 2010 *Source* (Kyne 2014)

	0–10	11–20	21–30	31–40	41–50	0–50	Outside
Year 1990							
Tract	2	11	62	50	14	139	2,933
Tract area (sq. mile)	325	1,146	1,365	1,978	1,434	6,248	85,198
Total population	6,629	36,043	245,643	180,329	49,337	517,981	9,446,914
White	3,583	21,485	141,389	129,742	25,072	321,271	6,689,779
Black	3,009	14,313	98,548	47,408	24,121	187,399	2,597,474
Asian	0	67	3,180	2,481	31	5,759	89,294
Native American	6	102	679	311	64	1,162	23,053
Others	31	78	1,846	378	49	2,382	47,291
Hispanic	38	189	3,876	1,738	160	6,001	123,720
Color	3,055	14,671	105,948	51,775	24,350	199,799	2,823,437
White (%)	54.05	59.61	57.56	71.95	50.82	62.02	70.81
Black (%)	45.39	39.71	40.12	26.29	48.89	36.18	27.50
Asian (%)	0.00	0.19	1.29	1.38	0.06	1.11	0.95
Native American (%)	0.09	0.28	0.28	0.17	0.13	0.22	0.24
Others (%)	0.47	0.22	0.75	0.21	0.10	0.46	0.50
Hispanic (%)	0.57	0.52	1.58	0.96	0.32	1.16	1.31
Color (%)	46.09	40.70	43.13	28.71	49.35	38.57	29.89
Female (%)	50.20	52.51	51.69	51.40	52.07	51.66	51.55
Old (65+ years) (%)	9.49	10.47	10.83	9.45	12.38	10.46	10.53
Kid (<5 years) (%)	10.42	10.18	9.57	9.50	9.90	9.63	8.99
Native-born (%)	99.70	99.31	97.74	97.82	99.70	98.09	97.74
Renter housing units (%)	16.76	22.45	34.44	24.41	23.61	28.97	29.80
Education (%)	5.62	10.53	15.24	20.52	9.46	16.11	18.50
Unemployment (%)	7.39	7.50	6.87	5.24	6.36	6.27	5.65
Poverty (%)	28.92	20.85	18.86	13.99	25.33	18.06	14.73
Mean household income ($)	40,703	48,701	49,739	62,449	41,184	53,117	58,054
Year 2000							
Total population	7,929	43,024	264,222	213,861	58,281	587,317	11,611,148
White	4,311	24,830	134,664	148,505	30,561	342,871	7,679,982
Black	3,560	17,122	118,375	56,432	26,662	222,151	3,302,686
Asian	6	181	3,122	4,016	187	7,512	205,706
Native American	0	158	801	731	208	1,898	36,478
Others	52	739	7,256	4,171	660	12,878	386,303
Hispanic	110	981	7,174	3,824	1,062	13,151	509,653
Color	3,672	18,751	132,525	67,414	28,316	250,678	4,163,659

(continued)

Table B.10 (continued)

	0–10	11–20	21–30	31–40	41–50	0–50	Outside
White (%)	54.37	57.71	50.97	69.44	52.44	58.38	66.14
Black (%)	44.90	39.80	44.80	26.39	45.75	37.82	28.44
Asian (%)	0.08	0.42	1.18	1.88	0.32	1.28	1.77
Native American (%)	0.00	0.37	0.30	0.34	0.36	0.32	0.31
Others (%)	0.66	1.72	2.75	1.95	1.13	2.19	3.33
Hispanic (%)	1.39	2.28	2.72	1.79	1.82	2.24	4.39
Color (%)	46.31	43.58	50.16	31.52	48.59	42.68	35.86
Female (%)	50.36	51.92	51.68	51.93	50.09	51.61	51.03
Old (65+ years) (%)	9.04	10.19	11.73	10.90	11.39	11.25	10.40
Kid (<5 years) (%)	9.71	8.58	8.48	8.21	8.17	8.38	8.42
Native-born (%)	98.85	98.34	97.26	96.63	98.79	97.28	94.17
Renter housing units (%)	12.79	20.15	32.67	22.13	20.30	26.51	27.94
College degree or higher (%)	8.41	11.69	17.57	24.76	11.72	19.12	23.20
Unemployment (%)	7.42	7.49	8.58	5.76	7.12	7.26	5.53
Poverty (%)	21.22	20.93	19.47	12.25	20.27	17.03	13.17
Mean household income ($)	45,999	51,797	54,547	66,926	49,348	58,262	67,495
Year 2010							
Total population	8,333	41,084	272,766	243,472	62,698	628,353	13,351,890
White	4,383	23,614	130,725	166,074	34,830	359,626	8,462,629
Black	3,870	16,067	128,319	64,543	25,438	238,237	3,904,316
Asian	0	154	4,007	5,517	677	10,355	347,822
Native American	0	184	656	246	193	1,279	36,050
Others	80	1,065	9,059	7,092	1,560	18,856	601,073
Hispanic	117	589	11,525	9,143	2,062	23,436	970,030
Color	4,067	17,758	147,781	82,823	28,865	281,294	5,409,504
White (%)	52.60	57.48	47.93	68.21	55.55	57.23	63.38
Black (%)	46.44	39.11	47.04	26.51	40.57	37.91	29.24
Asian (%)	0.00	0.37	1.47	2.27	1.08	1.65	2.61
Native American (%)	0.00	0.45	0.24	0.10	0.31	0.20	0.27
Others (%)	0.96	2.59	3.32	2.91	2.49	3.00	4.50
Hispanic (%)	1.40	1.43	4.23	3.76	3.29	3.73	7.27
Color (%)	48.81	43.22	54.18	34.02	46.04	44.77	40.51
Female (%)	52.78	51.17	51.63	51.57	48.99	51.33	51.19
Old (65+ years) (%)	10.15	10.66	12.48	12.50	11.11	12.20	11.16
Kid (<5 years) (%)	8.29	6.67	6.84	6.97	6.64	6.88	7.05

(continued)

Table B.10 (continued)

	0–10	11–20	21–30	31–40	41–50	0–50	Outside
Native-born (%)	99.11	98.49	96.42	95.56	96.51	96.27	91.78
Renter housing units (%)	15.25	23.31	32.10	22.09	21.49	26.46	27.23
College degree or higher (%)	6.42	14.91	20.08	26.87	15.81	21.84	26.34
Unemployment (%)	11.84	9.94	10.55	7.65	10.52	9.35	8.96
Poverty (%)	28.36	22.54	21.82	14.06	21.54	18.90	15.79
Mean household income ($)	44,522	50,799	51,582	66,663	49,153	57,078	64,378
Index	10	10	10	10	10	10	10

B.11 Braidwood Station, Illinois

See Fig. B.11.
See Table B.11.

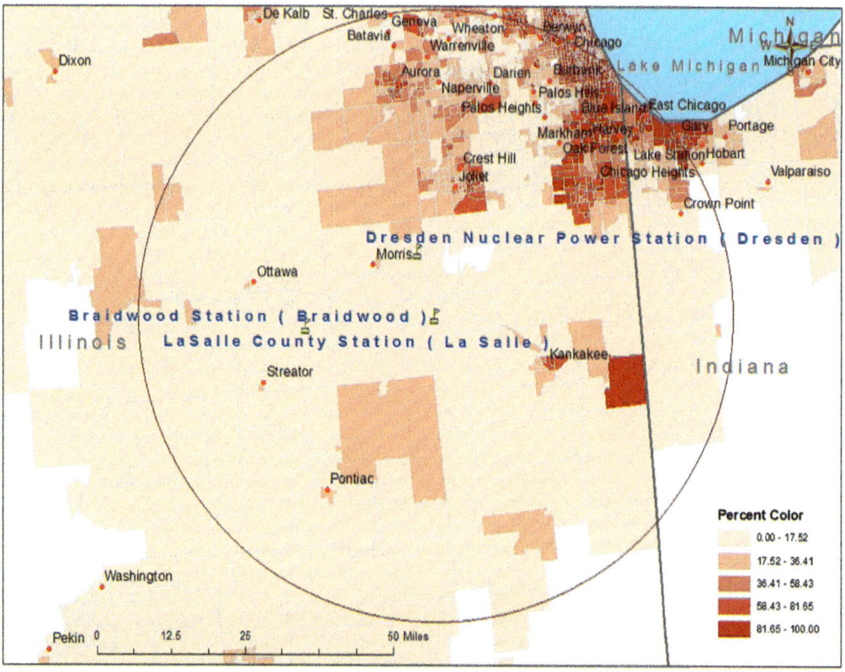

Fig. B.11 Spatial distribution of percent color among the populations surrounding Braidwood Station, Illinois, in 2010

Table B.11 Demographic composition of population, as sorted by distance from Braidwood Station, Illinois, in 1990, 2000, and 2010 *Source* (Kyne 2014)

	0–10	11–20	21–30	31–40	41–50	0–50	Outside
Year 1990							
Tract	10	35	115	248	252	660	2,463
Tract area (sq. mile)	327	1,083	1,322	2,181	893	5,806	52,108
Total population	28,159	103,904	298,603	927,921	1,008,609	2,367,196	9,063,387
White	27,795	97,315	246,012	756,921	917,354	2,045,397	6,912,516
Black	68	4,658	41,091	122,487	50,272	218,576	1,472,281
Asian	80	650	2,059	20,754	24,501	48,044	236,884
Native American	124	198	631	1,498	1,558	4,009	20,059
Others	92	1,077	8,820	26,250	14,923	51,162	421,643
Hispanic	323	2,281	16,004	53,399	42,049	114,056	764,632
Color	580	7,725	59,335	196,653	117,630	381,923	2,477,934
White (%)	98.71	93.66	82.39	81.57	90.95	86.41	76.27
Black (%)	0.24	4.48	13.76	13.20	4.98	9.23	16.24
Asian (%)	0.28	0.63	0.69	2.24	2.43	2.03	2.61
Native American (%)	0.44	0.19	0.21	0.16	0.15	0.17	0.22
Others (%)	0.33	1.04	2.95	2.83	1.48	2.16	4.65
Hispanic (%)	1.15	2.20	5.36	5.75	4.17	4.82	8.44
Color (%)	2.06	7.43	19.87	21.19	11.66	16.13	27.34
Female (%)	49.91	51.32	50.25	51.13	51.51	51.18	51.55
Old (65+ years) (%)	11.27	12.20	10.90	9.67	13.35	11.52	12.82
Kid (<5 years) (%)	9.28	8.87	9.44	9.77	8.89	9.31	8.77
Native-born (%)	98.77	98.15	96.39	94.15	92.73	94.06	91.05
Renter housing units (%)	20.53	27.60	25.39	23.40	24.13	24.12	35.58
Education (%)	7.54	14.85	14.39	23.12	27.07	23.29	20.46
Unemployment (%)	7.49	4.77	6.26	5.22	3.92	4.78	7.14
Poverty (%)	7.89	6.07	8.91	6.79	4.12	5.89	13.49
Mean household income ($)	60,876	65,366	65,609	78,615	83,354	78,358	64,316
Year 2000							
Total population	32,863	119,309	391,018	1,085,281	1,089,926	2,718,397	9,700,896
White	31,993	108,074	313,445	814,590	902,339	2,170,441	6,953,123
Black	162	6,652	50,610	164,410	89,077	310,911	1,553,708
Asian	20	767	4,769	39,221	37,641	82,418	344,833
Native American	63	224	862	2,094	2,005	5,248	25,159
Others	623	3,596	21,342	64,954	58,861	149,376	824,076
Hispanic	948	4,866	35,026	106,439	100,098	247,377	1,281,764
Color	1,436	13,623	94,633	324,334	242,765	676,791	3,319,094

(continued)

Table B.11 (continued)

	0–10	11–20	21–30	31–40	41–50	0–50	Outside
White (%)	97.35	90.58	80.16	75.06	82.79	79.84	71.68
Black (%)	0.49	5.58	12.94	15.15	8.17	11.44	16.02
Asian (%)	0.06	0.64	1.22	3.61	3.45	3.03	3.55
Native American (%)	0.19	0.19	0.22	0.19	0.18	0.19	0.26
Others (%)	1.90	3.01	5.46	5.98	5.40	5.50	8.49
Hispanic (%)	2.88	4.08	8.96	9.81	9.18	9.10	13.21
Color (%)	4.37	11.42	24.20	29.88	22.27	24.90	34.21
Female (%)	50.34	50.92	49.89	51.24	51.33	51.05	51.10
Old (65+ years) (%)	11.02	11.59	10.31	10.35	13.04	11.48	12.23
Kid (<5 years) (%)	8.27	7.89	9.68	9.12	8.47	8.88	8.39
Native-born (%)	98.75	97.30	94.03	90.05	88.14	90.28	86.96
Renter housing units (%)	17.68	24.15	19.96	20.33	22.74	21.40	33.20
College degree or higher (%)	10.37	18.73	19.58	29.02	32.83	28.59	25.35
Unemployment (%)	5.04	5.15	4.85	4.90	4.22	4.63	6.45
Poverty (%)	5.24	6.03	6.98	6.41	5.15	5.95	12.02
Mean household income ($)	71,905	75,332	78,226	88,157	93,188	88,091	73,612
Year 2010							
Total population	35,921	143,022	510,817	1,212,228	1,116,411	3,018,399	9,726,960
White	34,793	128,578	400,286	833,626	876,852	2,274,135	6,862,545
Black	221	9,007	62,644	203,715	106,057	381,644	1,478,661
Asian	133	1,502	9,591	61,370	48,741	121,337	453,490
Native American	67	256	492	2,249	2,031	5,095	19,634
Others	707	3,679	37,804	111,268	82,730	236,188	912,630
Hispanic	1,270	9,617	77,223	191,369	171,079	450,558	1,489,370
Color	2,096	21,427	157,319	473,862	338,307	993,011	3,549,698
White (%)	96.86	89.90	78.36	68.77	78.54	75.34	70.55
Black (%)	0.62	6.30	12.26	16.81	9.50	12.64	15.20
Asian (%)	0.37	1.05	1.88	5.06	4.37	4.02	4.66
Native American (%)	0.19	0.18	0.10	0.19	0.18	0.17	0.20
Others (%)	1.97	2.57	7.40	9.18	7.41	7.82	9.38
Hispanic (%)	3.54	6.72	15.12	15.79	15.32	14.93	15.31
Color (%)	5.84	14.98	30.80	39.09	30.30	32.90	36.49
Female (%)	49.39	50.20	50.62	51.11	51.23	51.01	50.95
Old (65+ years) (%)	12.54	11.31	10.36	11.08	12.62	11.56	12.54
Kid (<5 years) (%)	6.36	6.86	7.57	6.82	6.58	6.85	6.56

(continued)

Table B.11 (continued)

	0–10	11–20	21–30	31–40	41–50	0–50	Outside
Native-born (%)	98.54	96.46	90.89	86.60	85.33	87.46	86.04
Renter housing units (%)	16.45	20.16	17.21	19.41	21.69	19.93	30.13
College degree or higher (%)	11.65	22.01	24.50	32.24	37.15	32.08	29.72
Unemployment (%)	11.90	7.26	8.40	8.85	8.01	8.42	8.64
Poverty (%)	8.47	7.09	8.69	8.87	7.01	8.06	14.07
Mean household income ($)	67,059	76,135	77,469	84,061	91,234	85,146	72,227
Index	11	11	11	11	11	11	11

B.12 Byron Station, Illinois

See Fig. B.12.
See Table B.12.

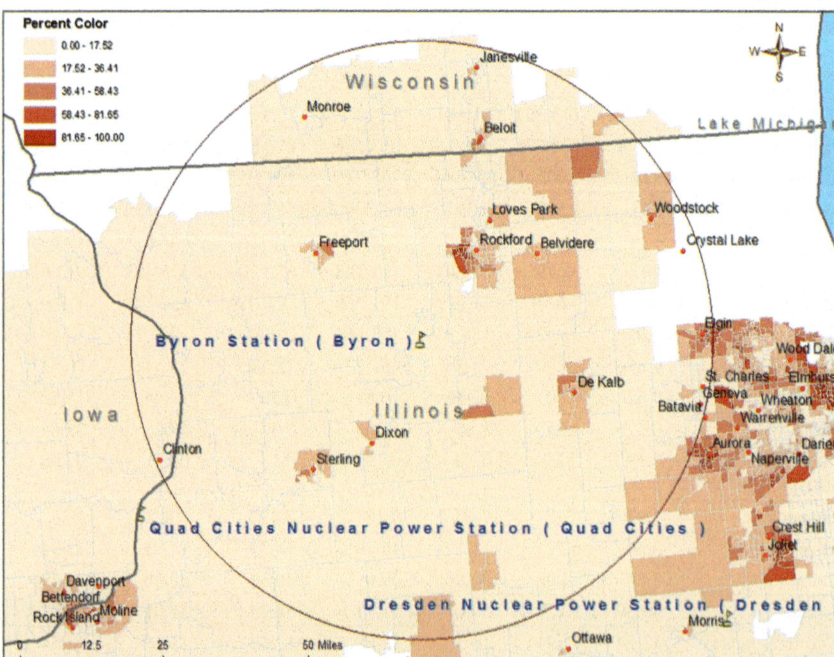

Fig. B.12 Spatial distribution of percent color among the populations surrounding Byron Station, Illinois, in 2010

Table B.12 Demographic composition of population, as sorted by distance from Byron Station, Illinois, in 1990, 2000, and 2010 *Source* (Kyne 2014)

	0–10	11–20	21–30	31–40	41–50	0–50	Outside
Year 1990							
Tract	5	60	66	38	30	199	4,333
Tract area (sq. mile)	300	945	1,657	1,443	1,283	5,627	117,783
Total population	19,736	206,595	243,810	125,999	110,481	706,621	15,615,730
White	19,556	176,787	229,393	116,345	108,690	650,771	12,821,452
Black	9	22,840	6,733	6,699	396	36,677	1,898,484
Asian	74	2,364	3,500	603	539	7,080	330,908
Native American	66	593	543	312	371	1,885	61,900
Others	29	4,006	3,647	2,044	484	10,210	502,957
Hispanic	161	7,220	7,513	3,985	1,495	20,374	945,920
Color	314	33,108	18,181	11,627	2,793	66,023	3,216,509
White (%)	99.09	85.57	94.09	92.34	98.38	92.10	82.11
Black (%)	0.05	11.06	2.76	5.32	0.36	5.19	12.16
Asian (%)	0.37	1.14	1.44	0.48	0.49	1.00	2.12
Native American (%)	0.33	0.29	0.22	0.25	0.34	0.27	0.40
Others (%)	0.15	1.94	1.50	1.62	0.44	1.44	3.22
Hispanic (%)	0.82	3.49	3.08	3.16	1.35	2.88	6.06
Color (%)	1.59	16.03	7.46	9.23	2.53	9.34	20.60
Female (%)	50.78	51.49	51.66	51.16	51.15	51.41	51.35
Old (65+ years) (%)	15.13	13.75	12.33	13.27	13.40	13.16	12.76
Kid (<5 years) (%)	8.72	9.24	8.19	9.13	9.15	8.83	8.91
Native-born (%)	98.97	96.32	96.75	97.96	98.43	97.17	93.25
Renter housing units (%)	24.78	34.11	29.13	28.33	29.60	30.46	32.23
Education (%)	15.36	14.27	18.65	11.75	13.05	15.13	20.28
Unemployment (%)	5.27	5.84	4.85	5.46	5.14	5.30	6.24
Poverty (%)	5.90	11.95	9.59	10.43	8.55	10.17	11.61
Mean household income ($)	59,673	54,360	62,984	56,810	58,473	58,488	64,540
Year 2000							
Total population	22,901	212,942	281,564	136,609	119,531	773,547	17,009,420
White	22,342	168,157	252,652	122,351	114,697	680,199	13,216,918
Black	74	27,743	11,306	6,562	879	46,564	2,118,410
Asian	50	2,824	4,622	984	591	9,071	502,834
Native American	37	816	620	493	447	2,413	77,655
Others	398	13,398	12,369	6,218	2,914	35,297	1,093,607
Hispanic	415	19,204	15,692	9,248	3,596	48,155	1,672,035
Color	724	53,299	35,945	18,634	6,525	115,127	4,556,784

(continued)

Table B.12 (continued)

	0–10	11–20	21–30	31–40	41–50	0–50	Outside
White (%)	97.56	78.97	89.73	89.56	95.96	87.93	77.70
Black (%)	0.32	13.03	4.02	4.80	0.74	6.02	12.45
Asian (%)	0.22	1.33	1.64	0.72	0.49	1.17	2.96
Native American (%)	0.16	0.38	0.22	0.36	0.37	0.31	0.46
Others (%)	1.74	6.29	4.39	4.55	2.44	4.56	6.43
Hispanic (%)	1.81	9.02	5.57	6.77	3.01	6.23	9.83
Color (%)	3.16	25.03	12.77	13.64	5.46	14.88	26.79
Female (%)	50.42	50.89	51.05	50.81	50.68	50.89	50.95
Old (65+ years) (%)	14.05	13.28	12.64	13.18	13.81	13.13	12.35
Kid (<5 years) (%)	7.19	8.55	7.84	8.12	8.11	8.11	8.27
Native-born (%)	98.23	92.99	95.40	95.87	97.85	95.28	90.09
Renter housing units (%)	21.66	31.55	27.17	24.32	26.80	27.69	30.08
College degree or higher (%)	19.42	15.86	21.22	14.62	15.94	17.68	25.29
Unemployment (%)	4.06	6.69	5.38	5.63	4.61	5.61	5.62
Poverty (%)	6.48	12.16	8.21	7.67	6.48	8.88	10.13
Mean household income ($)	70,621	59,496	70,338	65,104	66,589	65,820	74,050
Year 2010							
Total population	23,750	214,649	314,446	145,769	123,031	821,645	17,561,660
White	23,038	167,268	275,854	129,839	115,327	711,326	13,339,905
Black	81	31,547	17,387	6,820	1,975	57,810	2,151,157
Asian	142	3,037	7,093	1,525	982	12,779	686,043
Native American	0	673	832	230	256	1,991	71,927
Others	489	12,124	13,280	7,355	4,491	37,739	1,312,629
Hispanic	544	28,712	28,391	14,340	6,498	78,485	2,171,992
Color	1,030	67,676	56,879	24,883	11,158	161,626	5,283,605
White (%)	97.00	77.93	87.73	89.07	93.74	86.57	75.96
Black (%)	0.34	14.70	5.53	4.68	1.61	7.04	12.25
Asian (%)	0.60	1.41	2.26	1.05	0.80	1.56	3.91
Native American (%)	0.00	0.31	0.26	0.16	0.21	0.24	0.41
Others (%)	2.06	5.65	4.22	5.05	3.65	4.59	7.47
Hispanic (%)	2.29	13.38	9.03	9.84	5.28	9.55	12.37
Color (%)	4.34	31.53	18.09	17.07	9.07	19.67	30.09
Female (%)	52.07	50.66	50.53	51.25	50.42	50.72	50.79
Old (65+ years) (%)	17.44	13.24	13.33	13.29	14.36	13.57	12.59
Kid (<5 years) (%)	5.28	6.97	6.22	6.59	6.74	6.53	6.53
Native-born (%)	97.81	91.81	93.79	94.91	96.47	93.99	88.92

(continued)

Table B.12 (continued)

	0–10	11–20	21–30	31–40	41–50	0–50	Outside
Renter housing units (%)	20.74	30.58	25.64	20.99	23.89	25.77	27.60
College degree or higher (%)	20.27	17.16	23.91	17.78	18.18	20.06	29.30
Unemployment (%)	8.18	10.99	8.96	8.87	7.91	9.27	7.92
Poverty (%)	6.96	19.34	12.03	12.21	11.58	13.77	12.26
Mean household income ($)	68,972	53,238	66,216	62,255	60,945	61,359	72,448
Index	12	12	12	12	12	12	12

B.13 Clinton Power Station, Illinois

See Fig. B.13.
See Table B.13.

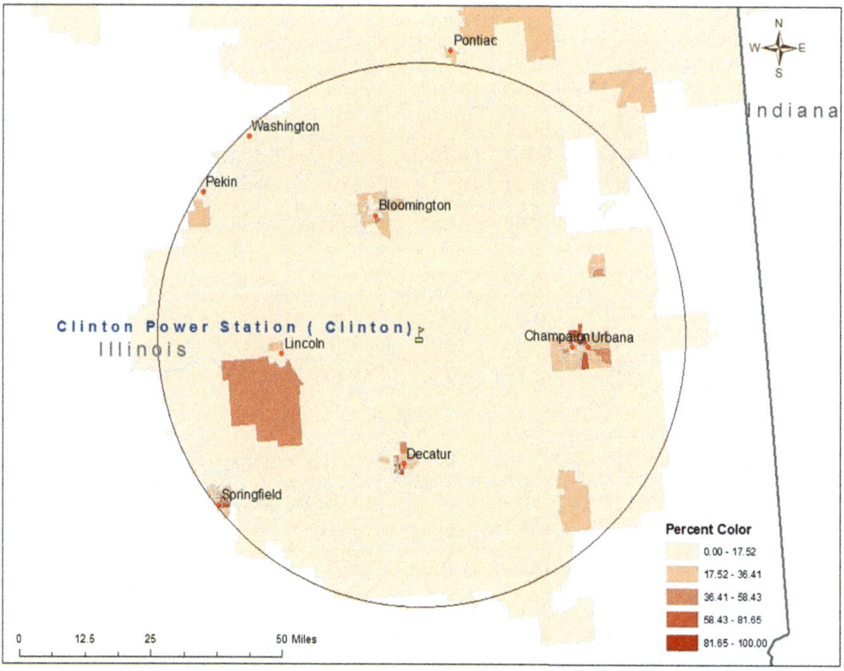

Fig. B.13 Spatial distribution of percent color among the populations surrounding Clinton Power Station, Illinois, in 2010

Table B.13 Demographic composition of population, as sorted by distance from Clinton Power Station, Illinois, in 1990, 2000, and 2010 *Source* (Kyne 2014)

	0–10	11–20	21–30	31–40	41–50	0–50	Outside
Year 1990							
Tract	4	10	90	37	13	154	2,969
Tract area (sq. mile)	254	823	1,713	1,783	1,278	5,851	52,063
Total population	13,021	35,172	303,709	147,650	48,624	548,176	10,882,407
White	12,902	34,568	272,592	128,355	48,287	496,704	8,461,209
Black	58	428	26,241	10,690	27	37,444	1,653,413
Asian	32	32	3,518	6,753	98	10,433	274,495
Native American	23	76	560	331	91	1,081	22,987
Others	6	65	800	1,522	121	2,514	470,291
Hispanic	74	157	2,410	3,242	505	6,388	872,300
Color	187	705	32,694	20,989	722	55,297	2,804,560
White (%)	99.09	98.28	89.75	86.93	99.31	90.61	77.75
Black (%)	0.45	1.22	8.64	7.24	0.06	6.83	15.19
Asian (%)	0.25	0.09	1.16	4.57	0.20	1.90	2.52
Native American (%)	0.18	0.22	0.18	0.22	0.19	0.20	0.21
Others (%)	0.05	0.18	0.26	1.03	0.25	0.46	4.32
Hispanic (%)	0.57	0.45	0.79	2.20	1.04	1.17	8.02
Color (%)	1.44	2.00	10.76	14.22	1.48	10.09	25.77
Female (%)	51.10	50.73	52.37	48.68	51.86	51.20	51.49
Old (65+ years) (%)	17.16	11.98	12.50	9.60	17.29	12.22	12.57
Kid (<5 years) (%)	8.47	8.58	8.22	7.51	8.48	8.08	8.92
Native-born (%)	99.52	99.42	98.21	94.04	99.03	97.27	91.39
Renter housing units (%)	29.21	18.49	32.93	42.39	24.29	33.57	33.35
Education (%)	11.75	16.89	24.90	25.90	11.40	22.95	20.95
Unemployment (%)	6.19	4.54	5.01	4.53	4.21	4.81	6.73
Poverty (%)	10.30	5.76	12.27	16.59	9.31	12.59	11.87
Mean household income ($)	52,091	64,651	59,178	51,492	54,360	57,000	67,629
Year 2000							
Total population	13,049	39,455	325,882	148,972	50,051	577,409	11,841,884
White	12,783	38,295	276,875	123,571	48,937	500,461	8,623,103
Black	38	597	33,812	12,609	130	47,186	1,817,433
Asian	42	151	7,211	8,633	140	16,177	411,074
Native American	16	82	523	337	53	1,011	29,396
Others	170	328	7,460	3,821	791	12,570	960,882
Hispanic	197	330	6,547	3,481	994	11,549	1,517,592
Color	426	1,363	52,396	27,178	1,621	82,984	3,912,901
White (%)	97.96	97.06	84.96	82.95	97.77	86.67	72.82

(continued)

Table B.13 (continued)

	0–10	11–20	21–30	31–40	41–50	0–50	Outside
Black (%)	0.29	1.51	10.38	8.46	0.26	8.17	15.35
Asian (%)	0.32	0.38	2.21	5.80	0.28	2.80	3.47
Native American (%)	0.12	0.21	0.16	0.23	0.11	0.18	0.25
Others (%)	1.30	0.83	2.29	2.56	1.58	2.18	8.11
Hispanic (%)	1.51	0.84	2.01	2.34	1.99	2.00	12.82
Color (%)	3.26	3.45	16.08	18.24	3.24	14.37	33.04
Female (%)	50.90	51.06	52.04	49.08	51.52	51.14	51.09
Old (65+ years) (%)	16.77	12.25	12.42	10.25	17.26	12.36	12.05
Kid (<5 years) (%)	7.69	7.98	7.53	6.73	7.51	7.36	8.55
Native-born (%)	98.66	99.00	96.48	93.46	98.60	96.10	87.28
Renter housing units (%)	25.45	14.79	32.01	39.58	20.98	31.67	30.71
College degree or higher (%)	13.97	23.13	30.04	28.10	15.49	27.30	26.00
Unemployment (%)	5.53	2.78	6.38	5.58	3.06	5.64	6.07
Poverty (%)	8.67	3.78	11.31	16.81	6.13	11.60	10.64
Mean household income ($)	60,631	76,044	65,546	57,685	63,521	64,024	77,324
Year 2010							
Total population	12,773	42,132	345,193	156,890	51,188	608,176	12,137,183
White	12,443	40,037	278,517	124,655	49,878	505,530	8,631,150
Black	75	590	40,759	15,948	136	57,508	1,802,797
Asian	17	415	13,732	11,405	232	25,801	549,026
Native American	0	68	433	469	36	1,006	23,723
Others	238	1,022	11,752	4,413	906	18,331	1,130,487
Hispanic	306	798	12,117	6,464	1,576	21,261	1,918,667
Color	460	2,508	73,374	36,393	2,267	115,002	4,427,707
White (%)	97.42	95.03	80.68	79.45	97.44	83.12	71.11
Black (%)	0.59	1.40	11.81	10.17	0.27	9.46	14.85
Asian (%)	0.13	0.98	3.98	7.27	0.45	4.24	4.52
Native American (%)	0.00	0.16	0.13	0.30	0.07	0.17	0.20
Others (%)	1.86	2.43	3.40	2.81	1.77	3.01	9.31
Hispanic (%)	2.40	1.89	3.51	4.12	3.08	3.50	15.81
Color (%)	3.60	5.95	21.26	23.20	4.43	18.91	36.48
Female (%)	51.00	50.25	51.67	49.06	51.48	50.87	50.96
Old (65+ years) (%)	17.11	14.28	12.43	10.23	16.46	12.43	12.30
Kid (<5 years) (%)	4.96	6.20	6.53	5.07	6.94	6.13	6.66
Native-born (%)	98.26	97.80	94.60	91.27	98.03	94.33	85.98

(continued)

Table B.13 (continued)

	0–10	11–20	21–30	31–40	41–50	0–50	Outside
Renter housing units (%)	24.75	13.74	30.59	37.37	19.42	30.09	27.78
College degree or higher (%)	13.51	27.62	34.75	30.53	17.01	31.14	30.24
Unemployment (%)	7.42	4.84	6.50	7.09	5.47	6.47	8.70
Poverty (%)	8.62	7.18	14.40	20.83	10.62	14.95	12.53
Mean household income ($)	54,808	77,212	65,965	56,181	60,541	63,666	75,738
Index	13	13	13	13	13	13	13

B.14 Dresden Nuclear Power Station, Illinois

See Fig. B.14.
See Table B.14.

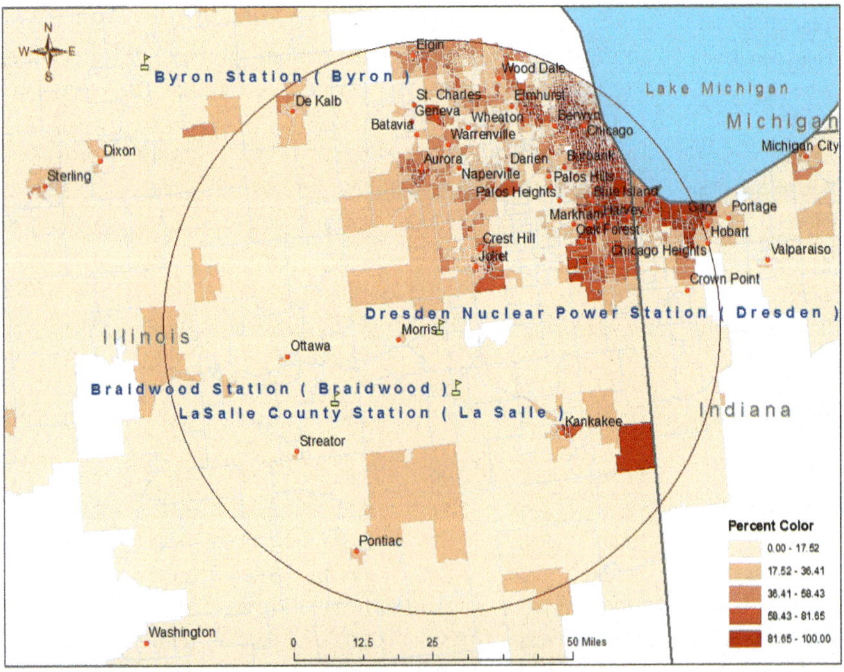

Fig. B.14 Spatial distribution of percent color among the populations surrounding Dresden Nuclear Power Station, Illinois, in 2010

Table B.14 Demographic composition of population, as sorted by distance from Dresden Nuclear Power Station, Illinois, in 1990, 2000, and 2010 *Source* (Kyne 2014)

	0–10	11–20	21–30	31–40	41–50	0–50	Outside
Year 1990							
Tract	17	85	212	413	310	1,037	2,086
Tract area (sq. mile)	251	952	1,729	1,820	1,085	5,838	52,075
Total population	44,647	200,027	726,503	1,677,784	1,252,112	3,901,073	7,529,510
White	43,455	165,920	629,535	1,293,755	882,651	3,015,316	5,942,597
Black	520	24,232	59,011	318,961	227,662	630,386	1,060,471
Asian	216	1,291	17,344	31,150	39,954	89,955	194,973
Native American	96	516	1,171	2,577	1,965	6,325	17,743
Others	363	8,076	19,417	31,348	99,874	159,078	313,727
Hispanic	957	13,800	40,275	74,537	173,573	303,142	575,546
Color	1,736	39,486	116,706	425,491	441,282	1,024,701	1,835,156
White (%)	97.33	82.95	86.65	77.11	70.49	77.29	78.92
Black (%)	1.16	12.11	8.12	19.01	18.18	16.16	14.08
Asian (%)	0.48	0.65	2.39	1.86	3.19	2.31	2.59
Native American (%)	0.22	0.26	0.16	0.15	0.16	0.16	0.24
Others (%)	0.81	4.04	2.67	1.87	7.98	4.08	4.17
Hispanic (%)	2.14	6.90	5.54	4.44	13.86	7.77	7.64
Color (%)	3.89	19.74	16.06	25.36	35.24	26.27	24.37
Female (%)	49.54	50.14	50.91	51.88	51.23	51.37	51.52
Old (65+ years) (%)	9.79	11.21	8.70	12.78	10.48	11.16	13.27
Kid (<5 years) (%)	9.45	9.21	10.10	8.88	9.52	9.34	8.65
Native-born (%)	98.27	95.86	94.17	93.58	86.33	91.53	91.74
Renter housing units (%)	23.06	26.18	23.93	24.58	34.79	27.80	35.93
Education (%)	14.00	14.24	24.78	22.52	18.44	21.12	21.00
Unemployment (%)	5.00	6.31	4.38	6.01	6.97	6.02	6.97
Poverty (%)	5.18	8.27	5.23	7.30	9.55	7.67	14.13
Mean household income ($)	70,896	65,575	80,845	76,836	68,798	74,364	63,637
Year 2000							
Total population	54,839	265,028	922,960	1,773,233	1,416,976	4,433,036	7,986,257
White	52,169	215,458	739,064	1,227,569	867,821	3,102,081	6,021,483
Black	957	28,127	94,872	387,663	255,674	767,293	1,097,326
Asian	222	3,269	35,311	47,345	66,231	152,378	274,873
Native American	64	691	1,861	3,553	4,621	10,790	19,617
Others	1,430	17,485	51,847	107,099	222,622	400,483	572,969
Hispanic	2,211	29,986	87,602	172,023	374,398	666,220	862,921

(continued)

Table B.14 (continued)

	0–10	11–20	21–30	31–40	41–50	0–50	Outside
Color	3,938	64,131	229,701	631,684	714,761	1,644,215	2,351,670
White (%)	95.13	81.30	80.08	69.23	61.24	69.98	75.40
Black (%)	1.75	10.61	10.28	21.86	18.04	17.31	13.74
Asian (%)	0.40	1.23	3.83	2.67	4.67	3.44	3.44
Native American (%)	0.12	0.26	0.20	0.20	0.33	0.24	0.25
Others (%)	2.61	6.60	5.62	6.04	15.71	9.03	7.17
Hispanic (%)	4.03	11.31	9.49	9.70	26.42	15.03	10.81
Color (%)	7.18	24.20	24.89	35.62	50.44	37.09	29.45
Female (%)	49.57	49.71	50.96	51.81	50.88	51.18	51.04
Old (65+ years) (%)	9.53	10.06	9.12	12.85	9.47	10.78	12.78
Kid (<5 years) (%)	8.44	9.99	9.40	8.58	9.79	9.22	8.09
Native-born (%)	97.63	93.12	90.10	89.40	77.47	86.06	88.59
Renter housing units (%)	19.83	21.05	19.47	23.43	32.69	25.27	33.43
College degree or higher (%)	18.28	18.79	31.09	27.35	22.06	25.86	26.17
Unemployment (%)	4.46	4.96	4.23	5.93	6.63	5.70	6.24
Poverty (%)	3.99	6.96	5.10	7.91	10.34	8.00	12.19
Mean household income ($)	82,461	76,446	93,541	84,662	77,382	83,743	73,119
Year 2010							
Total population	73,784	359,685	1,099,335	1,769,057	1,399,088	4,700,949	8,044,410
White	69,424	283,718	808,646	1,159,665	806,821	3,128,274	6,008,406
Black	2,059	37,128	127,911	402,725	230,682	800,505	1,059,800
Asian	436	7,531	61,015	55,901	87,939	212,822	362,005
Native American	61	328	2,105	3,074	3,452	9,020	15,709
Others	1,804	30,980	99,658	147,692	270,194	550,328	598,490
Hispanic	5,248	66,368	167,090	276,799	458,204	973,709	966,219
Color	8,251	116,606	372,202	753,852	789,596	2,040,507	2,502,202
White (%)	94.09	78.88	73.56	65.55	57.67	66.55	74.69
Black (%)	2.79	10.32	11.64	22.76	16.49	17.03	13.17
Asian (%)	0.59	2.09	5.55	3.16	6.29	4.53	4.50
Native American (%)	0.08	0.09	0.19	0.17	0.25	0.19	0.20
Others (%)	2.44	8.61	9.07	8.35	19.31	11.71	7.44
Hispanic (%)	7.11	18.45	15.20	15.65	32.75	20.71	12.01
Color (%)	11.18	32.42	33.86	42.61	56.44	43.41	31.10
Female (%)	49.78	50.04	51.05	51.73	50.75	51.12	50.87
Old (65+ years) (%)	9.59	9.56	10.11	12.86	9.74	10.99	13.07

(continued)

Table B.14 (continued)

	0–10	11–20	21–30	31–40	41–50	0–50	Outside
Kid (<5 years) (%)	7.06	7.99	7.05	6.45	7.75	7.10	6.35
Native-born (%)	96.74	89.36	86.80	87.02	75.06	83.74	87.91
Renter housing units (%)	15.50	17.87	18.43	22.83	30.04	23.43	30.15
College degree or higher (%)	21.83	22.93	34.85	30.84	24.95	29.32	30.82
Unemployment (%)	8.57	8.40	7.86	9.81	9.78	9.21	8.23
Poverty (%)	5.75	8.51	7.22	10.60	12.77	10.22	14.08
Mean household income ($)	79,156	76,132	89,143	81,429	71,974	80,096	72,559
Index	14	14	14	14	14	14	14

B.15 LaSalle County Station, Illinois

See Fig. B.15.
See Table B.15.

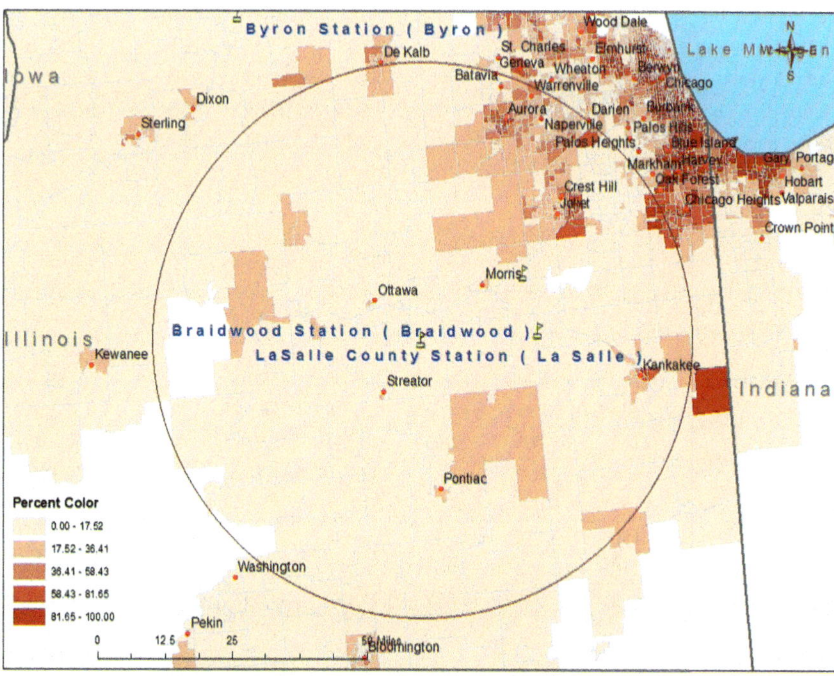

Fig. B.15 Spatial distribution of percent color among the populations surrounding LaSalle County Station, Illinois, in 2010

Table B.15 Demographic composition of population, as sorted by distance from LaSalle County Station, Illinois, in 1990, 2000, and 2010 *Source* (Kyne 2014)

	0–10	11–20	21–30	31–40	41–50	0–50	Outside
Year 1990							
Tract	3	25	36	93	78	235	2,888
Tract area (sq. mile)	188	1,133	1,545	1,931	978	5,775	52,139
Total population	10,831	94,231	110,499	244,400	223,305	683,266	10,747,317
White	10,725	91,717	106,268	207,604	191,201	607,515	8,350,398
Black	23	1,048	2,243	25,300	14,318	42,932	1,647,925
Asian	41	208	499	1,737	3,508	5,993	278,935
Native American	39	263	213	509	419	1,443	22,625
Others	3	996	1,276	9,272	13,844	25,391	447,414
Hispanic	95	2,607	2,535	16,504	26,543	48,284	830,404
Color	190	4,144	5,476	43,685	44,387	97,882	2,761,975
White (%)	99.02	97.33	96.17	84.94	85.62	88.91	77.70
Black (%)	0.21	1.11	2.03	10.35	6.41	6.28	15.33
Asian (%)	0.38	0.22	0.45	0.71	1.57	0.88	2.60
Native American (%)	0.36	0.28	0.19	0.21	0.19	0.21	0.21
Others (%)	0.03	1.06	1.15	3.79	6.20	3.72	4.16
Hispanic (%)	0.88	2.77	2.29	6.75	11.89	7.07	7.73
Color (%)	1.75	4.40	4.96	17.87	19.88	14.33	25.70
Female (%)	50.08	50.89	50.44	50.26	50.43	50.43	51.54
Old (65+ years) (%)	14.63	16.04	13.80	12.27	8.73	11.92	12.59
Kid (<5 years) (%)	8.20	8.41	8.68	9.09	11.34	9.65	8.84
Native-born (%)	99.34	98.37	98.23	95.85	92.61	95.58	91.42
Renter housing units (%)	21.94	26.92	24.16	26.20	27.94	26.45	33.77
Education (%)	8.43	10.72	11.30	14.14	24.83	16.46	21.33
Unemployment (%)	9.15	6.92	5.78	5.61	4.55	5.49	6.71
Poverty (%)	9.59	10.36	8.02	8.04	6.66	7.93	12.16
Mean household income ($)	52,577	55,557	58,868	64,361	75,609	65,522	67,207
Year 2000							
Total population	11,165	99,235	125,264	318,005	333,166	886,835	11,532,458
White	10,930	94,847	118,096	261,283	265,495	750,651	8,372,913
Black	11	1,763	2,616	30,533	22,255	57,178	1,807,441
Asian	57	372	525	4,327	12,869	18,150	409,101
Native American	17	131	257	917	782	2,104	28,303
Others	150	2,125	3,775	20,992	31,724	58,766	914,686
Hispanic	110	4,512	5,403	35,981	57,740	103,746	1,425,395
Color	312	7,297	9,867	74,183	97,297	188,956	3,806,929

(continued)

Table B.15 (continued)

	0–10	11–20	21–30	31–40	41–50	0–50	Outside
White (%)	97.90	95.58	94.28	82.16	79.69	84.64	72.60
Black (%)	0.10	1.78	2.09	9.60	6.68	6.45	15.67
Asian (%)	0.51	0.37	0.42	1.36	3.86	2.05	3.55
Native American (%)	0.15	0.13	0.21	0.29	0.23	0.24	0.25
Others (%)	1.34	2.14	3.01	6.60	9.52	6.63	7.93
Hispanic (%)	0.99	4.55	4.31	11.31	17.33	11.70	12.36
Color (%)	2.79	7.35	7.88	23.33	29.20	21.31	33.01
Female (%)	50.70	50.20	50.29	49.90	50.17	50.10	51.17
Old (65+ years) (%)	14.46	15.33	12.88	10.83	6.96	10.22	12.21
Kid (<5 years) (%)	7.59	7.55	7.95	10.00	11.25	9.87	8.39
Native-born (%)	98.82	97.76	97.46	93.08	86.44	91.80	87.37
Renter housing units (%)	18.67	25.08	20.14	21.21	21.00	21.42	31.43
College degree or higher (%)	10.20	13.30	15.28	20.05	36.60	24.42	26.18
Unemployment (%)	4.59	5.74	4.35	4.90	4.03	4.58	6.16
Poverty (%)	7.13	8.01	6.68	6.86	5.28	6.36	11.01
Mean household income ($)	61,628	63,822	69,120	76,227	97,065	80,932	76,366
Year 2010							
Total population	12,195	107,361	146,245	446,375	425,689	1,137,865	11,607,494
White	11,724	101,516	134,898	350,749	297,318	896,205	8,240,475
Black	0	1,681	3,965	43,871	30,892	80,409	1,779,896
Asian	61	553	1,142	12,663	31,579	45,998	528,829
Native American	2	96	235	476	1,098	1,907	22,822
Others	408	3,515	6,005	38,616	64,802	113,346	1,035,472
Hispanic	568	6,639	10,533	84,372	100,559	202,671	1,737,257
Color	905	10,609	17,192	148,063	169,611	346,380	4,196,329
White (%)	96.14	94.56	92.24	78.58	69.84	78.76	70.99
Black (%)	0.00	1.57	2.71	9.83	7.26	7.07	15.33
Asian (%)	0.50	0.52	0.78	2.84	7.42	4.04	4.56
Native American (%)	0.02	0.09	0.16	0.11	0.26	0.17	0.20
Others (%)	3.35	3.27	4.11	8.65	15.22	9.96	8.92
Hispanic (%)	4.66	6.18	7.20	18.90	23.62	17.81	14.97
Color (%)	7.42	9.88	11.76	33.17	39.84	30.44	36.15
Female (%)	52.04	50.15	50.02	49.94	50.43	50.18	51.04
Old (65+ years) (%)	16.06	14.84	12.25	9.74	7.61	9.82	12.55
Kid (<5 years) (%)	6.10	6.19	6.40	8.16	8.21	7.74	6.52

(continued)

Table B.15 (continued)

	0–10	11–20	21–30	31–40	41–50	0–50	Outside
Native-born (%)	98.82	96.70	96.44	89.10	82.12	88.25	86.19
Renter housing units (%)	18.09	23.37	16.60	18.12	20.98	19.50	28.62
College degree or higher (%)	15.79	16.19	19.23	24.75	40.28	28.65	30.43
Unemployment (%)	9.64	9.62	8.12	8.14	7.37	8.00	8.65
Poverty (%)	7.94	9.59	7.89	8.38	7.87	8.23	13.07
Mean household income ($)	62,441	64,620	72,170	76,444	94,070	80,807	74,629
Index	15	15	15	15	15	15	15

B.16 Quad Cities Nuclear Power Station, Illinois

See Fig. B.16.
See Table B.16.

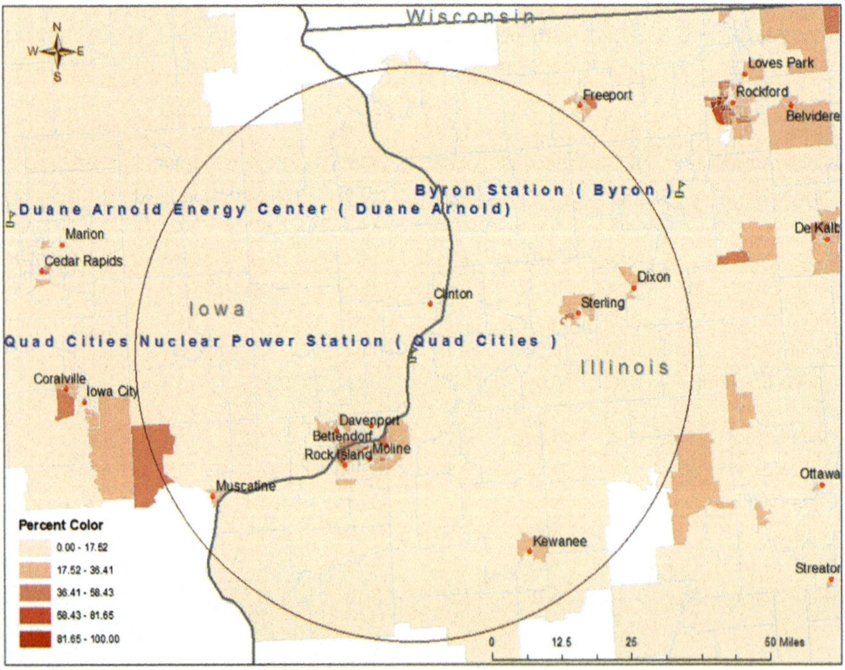

Fig. B.16 Spatial distribution of percent color among the populations surrounding Quad Cities Nuclear Power Station, Illinois, in 2010

Table B.16 Demographic composition of population, as sorted by distance from Quad Cities Nuclear Power Station, Illinois, in 1990, 2000, and 2010 *Source* (Kyne 2014)

	0–10	11–20	21–30	31–40	41–50	0–50	Outside
Year 1990							
Tract	7	63	51	22	11	154	3,794
Tract area (sq. mile)	272	909	1,628	2,047	968	5,824	108,363
Total population	33,742	226,327	169,415	72,924	37,153	539,561	13,667,777
White	33,015	209,709	156,924	70,767	36,474	506,889	11,136,128
Black	438	10,299	8,520	485	406	20,148	1,717,967
Asian	68	1,568	985	233	82	2,936	306,319
Native American	131	723	526	126	57	1,563	30,316
Others	88	4,030	2,460	1,318	129	8,025	477,042
Hispanic	322	8,889	5,783	2,342	298	17,634	891,696
Color	967	21,410	15,555	3,191	910	42,033	2,927,212
White (%)	97.85	92.66	92.63	97.04	98.17	93.94	81.48
Black (%)	1.30	4.55	5.03	0.67	1.09	3.73	12.57
Asian (%)	0.20	0.69	0.58	0.32	0.22	0.54	2.24
Native American (%)	0.39	0.32	0.31	0.17	0.15	0.29	0.22
Others (%)	0.26	1.78	1.45	1.81	0.35	1.49	3.49
Hispanic (%)	0.95	3.93	3.41	3.21	0.80	3.27	6.52
Color (%)	2.87	9.46	9.18	4.38	2.45	7.79	21.42
Female (%)	51.39	51.71	51.57	51.37	52.25	51.64	51.49
Old (65+ years) (%)	14.19	12.95	14.99	16.09	19.39	14.54	13.04
Kid (<5 years) (%)	8.30	8.98	8.53	8.19	7.60	8.59	8.80
Native-born (%)	99.26	97.77	98.16	98.80	99.05	98.21	92.79
Renter housing units (%)	23.73	31.72	27.53	24.88	22.96	28.35	32.41
Education (%)	14.43	19.45	12.39	10.91	11.04	15.14	20.43
Unemployment (%)	6.74	5.82	6.35	5.15	6.23	5.98	6.23
Poverty (%)	9.75	11.78	13.14	10.25	11.54	11.85	11.82
Mean household income ($)	54,959	56,525	51,892	52,878	47,269	53,843	64,541
Year 2000							
Total population	32,764	234,762	169,147	74,209	36,753	547,635	14,797,982
White	31,489	209,459	151,031	70,901	35,260	498,140	11,375,161
Black	518	11,027	9,844	426	372	22,187	1,902,190
Asian	128	3,049	1,040	318	89	4,624	458,605
Native American	127	967	620	158	31	1,903	37,767
Others	500	10,261	6,613	2,408	999	20,781	1,024,259
Hispanic	641	14,572	8,323	3,413	805	27,754	1,582,888
Color	1,666	32,245	21,690	4,776	1,809	62,186	4,146,997

(continued)

Table B.16 (continued)

	0–10	11–20	21–30	31–40	41–50	0–50	Outside
White (%)	96.11	89.22	89.29	95.54	95.94	90.96	76.87
Black (%)	1.58	4.70	5.82	0.57	1.01	4.05	12.85
Asian (%)	0.39	1.30	0.61	0.43	0.24	0.84	3.10
Native American (%)	0.39	0.41	0.37	0.21	0.08	0.35	0.26
Others (%)	1.53	4.37	3.91	3.24	2.72	3.79	6.92
Hispanic (%)	1.96	6.21	4.92	4.60	2.19	5.07	10.70
Color (%)	5.08	13.74	12.82	6.44	4.92	11.36	28.02
Female (%)	50.63	51.44	50.94	50.54	51.92	51.14	51.07
Old (65+ years) (%)	14.70	13.27	15.27	16.16	19.54	14.79	12.53
Kid (<5 years) (%)	7.30	8.19	7.48	7.26	7.14	7.72	8.38
Native-born (%)	98.97	96.08	97.43	98.06	98.56	97.10	89.16
Renter housing units (%)	21.03	28.94	24.52	21.59	19.99	25.45	29.93
College degree or higher (%)	15.78	22.23	14.76	13.29	14.54	17.75	25.41
Unemployment (%)	5.03	5.07	6.04	4.35	5.18	5.27	5.69
Poverty (%)	9.14	9.69	10.96	8.13	9.21	9.80	10.41
Mean household income ($)	61,306	63,441	59,077	62,467	54,935	61,258	74,051
Year 2010							
Total population	32,223	237,903	165,804	71,555	36,903	544,388	15,217,238
White	30,401	205,786	145,270	67,243	35,820	484,520	11,425,570
Black	760	14,681	10,773	830	355	27,399	1,915,588
Asian	209	3,790	2,016	271	147	6,433	619,718
Native American	95	514	379	236	108	1,332	33,222
Others	758	13,132	7,366	2,975	473	24,704	1,223,140
Hispanic	791	17,435	11,421	4,714	1,316	35,677	2,041,183
Color	2,259	40,532	26,967	6,918	2,078	78,754	4,783,553
White (%)	94.35	86.50	87.62	93.97	97.07	89.00	75.08
Black (%)	2.36	6.17	6.50	1.16	0.96	5.03	12.59
Asian (%)	0.65	1.59	1.22	0.38	0.40	1.18	4.07
Native American (%)	0.29	0.22	0.23	0.33	0.29	0.24	0.22
Others (%)	2.35	5.52	4.44	4.16	1.28	4.54	8.04
Hispanic (%)	2.45	7.33	6.89	6.59	3.57	6.55	13.41
Color (%)	7.01	17.04	16.26	9.67	5.63	14.47	31.44
Female (%)	50.30	50.85	51.29	50.48	50.36	50.87	50.90
Old (65+ years) (%)	16.25	14.07	16.30	17.65	20.15	15.76	12.68

(continued)

Table B.16 (continued)

	0–10	11–20	21–30	31–40	41–50	0–50	Outside
Kid (<5 years) (%)	5.06	6.93	5.87	5.84	5.57	6.26	6.63
Native-born (%)	98.56	95.59	96.30	98.06	98.25	96.49	87.89
Renter housing units (%)	19.74	27.48	23.01	19.68	21.90	24.22	27.34
College degree or higher (%)	18.22	27.15	18.02	18.18	18.35	21.99	29.44
Unemployment (%)	6.68	6.05	6.66	6.06	6.83	6.33	7.99
Poverty (%)	10.01	11.82	12.54	9.91	13.11	11.76	12.46
Mean household income ($)	64,349	64,003	57,579	61,077	52,949	60,911	72,883
Index	16	16	16	16	16	16	16

B.17 Duane Arnold Energy Center, Iowa

See Fig. B.17.
See Table B.17.

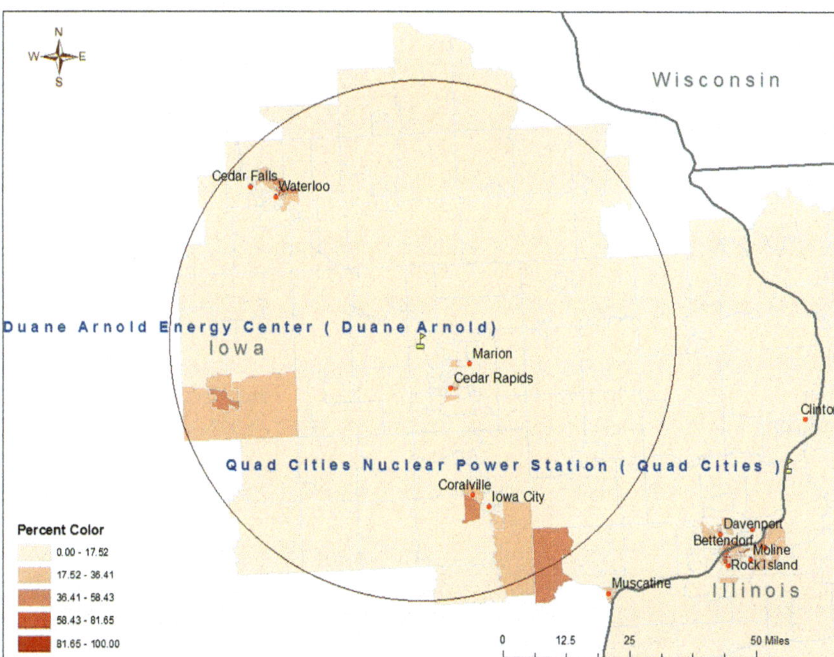

Fig. B.17 Spatial distribution of percent color among the populations surrounding Duane Arnold Energy Center, Iowa, in 2010

Table B.17 Demographic composition of population, as sorted by distance from Duane Arnold Energy Center, Iowa, in 1990, 2000, and 2010 *Source* (Kyne 2014)

	0–10	11–20	21–30	31–40	41–50	0–50	Outside
Year 1990							
Tract	22	27	16	50	14	129	696
Tract area (sq. mile)	291	973	1,505	1,955	1,071	5,795	50,478
Total population	77,265	102,461	63,200	176,101	43,929	462,956	2,313,799
White	75,417	98,804	62,432	164,558	40,575	441,786	2,243,318
Black	824	2,413	408	6,638	2,978	13,261	33,997
Asian	734	762	160	4,094	165	5,915	18,412
Native American	86	230	91	214	111	732	7,079
Others	202	253	112	596	98	1,261	11,001
Hispanic	608	749	253	1,645	368	3,623	27,019
Color	2,266	4,164	887	12,611	3,604	23,532	85,856
White (%)	97.61	96.43	98.78	93.45	92.36	95.43	96.95
Black (%)	1.07	2.36	0.65	3.77	6.78	2.86	1.47
Asian (%)	0.95	0.74	0.25	2.32	0.38	1.28	0.80
Native American (%)	0.11	0.22	0.14	0.12	0.25	0.16	0.31
Others (%)	0.26	0.25	0.18	0.34	0.22	0.27	0.48
Hispanic (%)	0.79	0.73	0.40	0.93	0.84	0.78	1.17
Color (%)	2.93	4.06	1.40	7.16	8.20	5.08	3.71
Female (%)	51.08	51.68	50.13	51.19	52.49	51.26	51.64
Old (65+ years) (%)	10.29	14.19	13.74	12.00	16.72	12.88	15.83
Kid (<5 years) (%)	8.79	8.56	8.74	8.29	8.86	8.55	8.41
Native-born (%)	98.32	98.51	99.31	96.73	99.37	97.99	98.53
Renter housing units (%)	27.88	28.05	21.76	38.08	26.53	30.86	27.31
Education (%)	24.31	17.43	15.93	27.42	10.30	21.33	15.98
Unemployment (%)	4.77	6.26	4.56	4.34	5.73	4.99	4.44
Poverty (%)	6.15	11.26	8.76	16.43	16.17	12.45	11.29
Mean household income ($)	64,629	59,144	56,997	54,324	43,093	56,429	51,501
Year 2000							
Total population	90,277	114,979	72,354	186,824	46,027	510,461	2,415,863
White	85,844	107,818	70,570	168,664	41,008	473,904	2,275,833
Black	1,314	3,197	579	8,230	3,678	16,998	42,760
Asian	1,439	1,308	413	4,945	222	8,327	27,651
Native American	154	268	148	281	99	950	8,313
Others	1,520	2,392	649	4,701	1,020	10,282	61,306
Hispanic	1,414	1,461	708	4,204	1,076	8,863	72,638
Color	5,260	7,901	2,253	20,135	5,477	41,026	172,272

(continued)

Table B.17 (continued)

	0–10	11–20	21–30	31–40	41–50	0–50	Outside
White (%)	95.09	93.77	97.53	90.28	89.10	92.84	94.20
Black (%)	1.46	2.78	0.80	4.41	7.99	3.33	1.77
Asian (%)	1.59	1.14	0.57	2.65	0.48	1.63	1.14
Native American (%)	0.17	0.23	0.20	0.15	0.22	0.19	0.34
Others (%)	1.68	2.08	0.90	2.52	2.22	2.01	2.54
Hispanic (%)	1.57	1.27	0.98	2.25	2.34	1.74	3.01
Color (%)	5.83	6.87	3.11	10.78	11.90	8.04	7.13
Female (%)	50.58	50.85	49.45	50.91	51.65	50.70	51.04
Old (65+ years) (%)	11.06	13.60	12.66	12.01	15.76	12.63	15.39
Kid (<5 years) (%)	8.73	8.04	7.80	7.14	8.49	7.84	7.73
Native-born (%)	97.19	97.91	98.57	94.72	98.18	96.73	96.92
Renter housing units (%)	24.75	25.96	19.09	36.47	25.05	28.60	25.20
College degree or higher (%)	30.93	22.61	22.16	31.56	13.63	26.36	20.15
Unemployment (%)	2.81	4.06	3.83	4.03	4.88	3.85	4.25
Poverty (%)	4.50	8.06	5.36	13.97	12.80	9.61	9.03
Mean household income ($)	74,393	66,328	71,347	59,890	52,523	64,832	60,361
Year 2010							
Total population	95,814	126,017	80,322	194,292	47,275	543,720	2,472,547
White	87,913	115,401	76,896	169,723	41,260	491,193	2,282,217
Black	2,763	5,088	1,110	10,902	3,415	23,278	59,404
Asian	2,385	1,349	894	6,412	522	11,562	39,762
Native American	208	306	147	571	82	1,314	8,511
Others	2,545	3,873	1,275	6,684	1,996	16,373	82,653
Hispanic	2,409	2,752	1,355	7,108	2,208	15,832	121,100
Color	9,183	12,036	4,358	28,005	6,892	60,474	259,124
White (%)	91.75	91.58	95.73	87.35	87.28	90.34	92.30
Black (%)	2.88	4.04	1.38	5.61	7.22	4.28	2.40
Asian (%)	2.49	1.07	1.11	3.30	1.10	2.13	1.61
Native American (%)	0.22	0.24	0.18	0.29	0.17	0.24	0.34
Others (%)	2.66	3.07	1.59	3.44	4.22	3.01	3.34
Hispanic (%)	2.51	2.18	1.69	3.66	4.67	2.91	4.90
Color (%)	9.58	9.55	5.43	14.41	14.58	11.12	10.48
Female (%)	50.94	50.39	50.06	50.62	52.25	50.68	50.62
Old (65+ years) (%)	12.32	13.38	13.03	11.81	15.80	12.79	15.25
Kid (<5 years) (%)	6.50	6.89	6.79	6.14	6.87	6.54	6.57

(continued)

Table B.17 (continued)

	0–10	11–20	21–30	31–40	41–50	0–50	Outside
Native-born (%)	96.80	97.63	97.99	93.62	96.52	96.01	95.84
Renter housing units (%)	24.20	25.15	17.72	33.91	24.68	27.01	24.03
College degree or higher (%)	32.74	25.13	29.23	33.28	17.39	29.22	23.51
Unemployment (%)	4.98	5.08	3.50	5.23	5.86	4.95	5.35
Poverty (%)	7.50	10.89	6.60	17.73	12.70	12.22	11.46
Mean household income ($)	73,105	63,563	72,294	59,068	53,298	64,054	61,340
Index	17	17	17	17	17	17	17

B.18 Wolf Creek Generating Station, Kansas

See Fig. B.18.
See Table B.18.

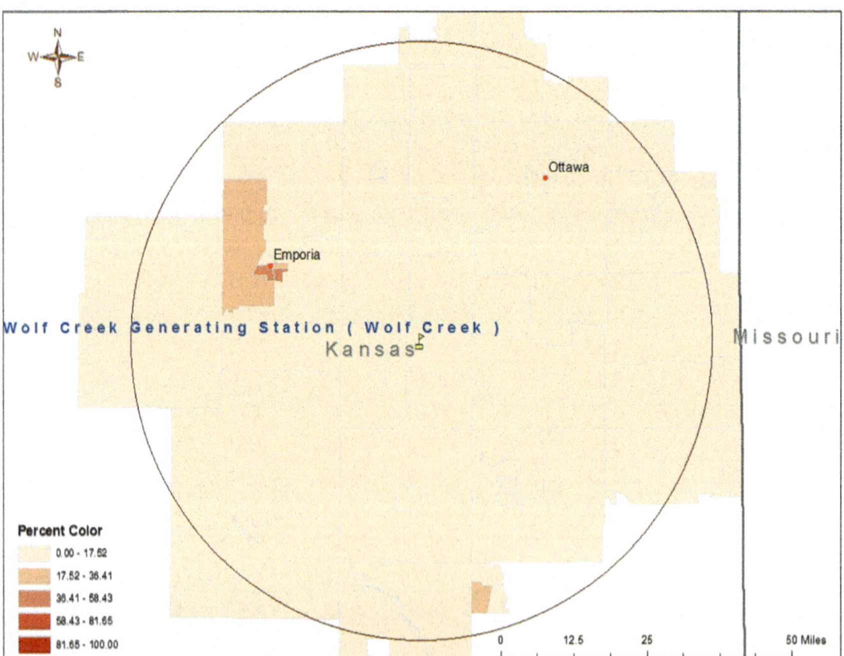

Fig. B.18 Spatial distribution of percent color among the populations surrounding Wolf Creek Generating Station, Kansas, in 2010

Table B.18 Demographic composition of population, as sorted by distance from Wolf Creek Generating Station, Kansas, in 1990, 2000, and 2010 *Source* (Kyne 2014)

	0–10	11–20	21–30	31–40	41–50	0–50	Outside
Year 1990							
Tract	3	1	14	11	10	39	731
Tract area (sq. mile)	654	452	1,774	1,996	1,096	5,973	76,306
Total population	8,404	3,028	51,599	39,484	31,274	133,789	2,343,798
White	8,330	2,999	48,322	38,212	30,382	128,245	2,105,651
Black	11	7	894	369	236	1,517	140,440
Asian	2	0	728	215	51	996	30,120
Native American	59	21	342	323	300	1,045	22,208
Others	2	0	1,319	361	303	1,985	45,371
Hispanic	38	9	1,976	827	563	3,413	86,881
Color	110	34	3,904	1,670	1,142	6,860	276,841
White (%)	99.12	99.04	93.65	96.78	97.15	95.86	89.84
Black (%)	0.13	0.23	1.73	0.93	0.75	1.13	5.99
Asian (%)	0.02	0.00	1.41	0.54	0.16	0.74	1.29
Native American (%)	0.70	0.69	0.66	0.82	0.96	0.78	0.95
Others (%)	0.02	0.00	2.56	0.91	0.97	1.48	1.94
Hispanic (%)	0.45	0.30	3.83	2.09	1.80	2.55	3.71
Color (%)	1.31	1.12	7.57	4.23	3.65	5.13	11.81
Female (%)	50.51	50.59	51.28	51.56	51.67	51.39	50.94
Old (65+ years) (%)	19.61	16.88	15.79	17.35	17.87	17.00	13.64
Kid (<5 years) (%)	8.67	8.72	8.99	9.20	7.79	8.75	9.18
Native-born (%)	99.80	99.60	97.70	98.82	99.41	98.61	97.40
Renter housing units (%)	20.26	11.69	29.62	23.62	19.84	24.57	29.28
Education (%)	13.46	6.07	14.00	16.39	12.25	14.07	21.50
Unemployment (%)	5.29	4.64	5.55	4.63	4.78	5.06	4.68
Poverty (%)	8.94	16.69	15.22	10.29	11.45	12.49	11.42
Mean household income ($)	47,226	46,237	42,044	50,760	48,175	46,456	56,668
Year 2000							
Total population	8,865	3,072	53,774	42,012	33,274	140,997	2,547,421
White	8,521	3,002	47,328	40,169	31,879	130,899	2,181,220
Black	22	5	975	361	237	1,600	148,984
Asian	70	0	444	153	120	787	45,193
Native American	80	27	377	301	185	970	23,753
Others	172	35	4,656	1,024	847	6,734	148,278
Hispanic	122	32	5,967	1,240	686	8,047	178,252
Color	398	90	8,620	2,568	1,783	13,459	440,471

(continued)

Table B.18 (continued)

	0–10	11–20	21–30	31–40	41–50	0–50	Outside
White (%)	96.12	97.72	88.01	95.61	95.81	92.84	85.62
Black (%)	0.25	0.16	1.81	0.86	0.71	1.13	5.85
Asian (%)	0.79	0.00	0.83	0.36	0.36	0.56	1.77
Native American (%)	0.90	0.88	0.70	0.72	0.56	0.69	0.93
Others (%)	1.94	1.14	8.66	2.44	2.55	4.78	5.82
Hispanic (%)	1.38	1.04	11.10	2.95	2.06	5.71	7.00
Color (%)	4.49	2.93	16.03	6.11	5.36	9.55	17.29
Female (%)	50.98	49.67	50.51	50.61	50.97	50.66	50.61
Old (65+ years) (%)	16.29	16.93	14.14	15.91	16.08	15.32	13.11
Kid (<5 years) (%)	7.22	8.01	8.33	7.62	7.03	7.73	8.44
Native-born (%)	99.21	99.77	94.20	98.23	99.28	97.04	94.87
Renter housing units (%)	19.56	12.41	29.61	23.73	18.70	24.23	28.43
College degree or higher (%)	20.11	7.76	16.20	20.51	16.91	17.77	26.25
Unemployment (%)	4.44	5.59	5.79	3.02	4.40	4.54	4.23
Poverty (%)	6.61	14.90	14.79	8.33	9.42	11.06	9.83
Mean household income ($)	57,364	43,779	48,027	55,755	59,166	53,502	65,521
Year 2010							
Total population	8,587	2,978	50,347	43,830	32,722	138,464	2,670,865
White	8,291	2,924	45,173	41,125	30,724	128,237	2,263,549
Black	60	0	919	375	410	1,764	160,004
Asian	11	3	883	161	135	1,193	66,551
Native American	20	4	140	135	218	517	23,905
Others	205	47	3,232	2,034	1,235	6,753	156,856
Hispanic	146	0	6,116	2,551	1,116	9,929	266,179
Color	410	54	9,338	4,151	2,887	16,840	571,537
White (%)	96.55	98.19	89.72	93.83	93.89	92.61	84.75
Black (%)	0.70	0.00	1.83	0.86	1.25	1.27	5.99
Asian (%)	0.13	0.10	1.75	0.37	0.41	0.86	2.49
Native American (%)	0.23	0.13	0.28	0.31	0.67	0.37	0.90
Others (%)	2.39	1.58	6.42	4.64	3.77	4.88	5.87
Hispanic (%)	1.70	0.00	12.15	5.82	3.41	7.17	9.97
Color (%)	4.77	1.81	18.55	9.47	8.82	12.16	21.40
Female (%)	50.22	50.20	50.30	50.78	51.64	50.76	50.47
Old (65+ years) (%)	16.75	15.51	14.35	15.35	16.72	15.40	12.97
Kid (<5 years) (%)	5.69	5.71	6.79	6.76	6.36	6.59	7.12

(continued)

Table B.18 (continued)

	0–10	11–20	21–30	31–40	41–50	0–50	Outside
Native-born (%)	99.10	99.66	93.10	98.00	98.68	96.48	93.55
Renter housing units (%)	20.07	18.17	28.18	24.73	18.05	24.02	27.72
College degree or higher (%)	19.98	11.92	17.69	22.44	21.77	20.23	29.83
Unemployment (%)	4.06	6.07	6.77	5.91	7.54	6.50	5.98
Poverty (%)	11.47	9.04	19.85	9.94	12.78	14.25	12.34
Mean household income ($)	58,379	58,687	46,664	55,506	59,414	53,425	65,608
Index	18	18	18	18	18	18	18

B.19 River Bend Station, Louisiana

See Fig. B.19.
See Table B.19.

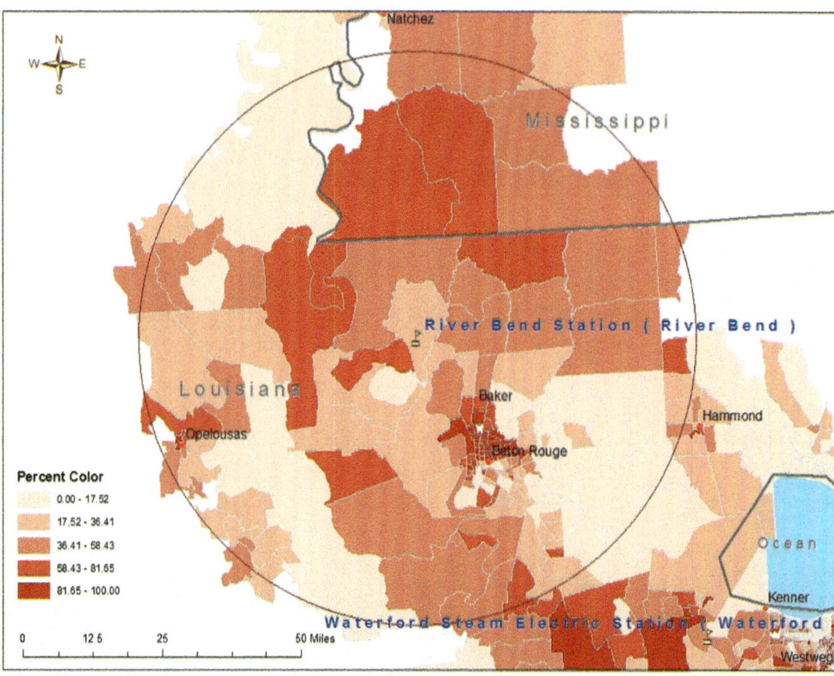

Fig. B.19 Spatial distribution of percent color among the populations surrounding River Bend Station, Louisiana, in 2010

Table B.19 Demographic composition of population, as sorted by distance from River Bend Station, Louisiana, in 1990, 2000, and 2010 *Source* (Kyne 2014)

	0–10	11–20	21–30	31–40	41–50	0–50	Outside
Year 1990							
Tract	5	20	79	38	19	161	1,651
Tract area (sq. mile)	239	810	1,545	2,308	1,966	6,869	93,941
Total population	16,273	88,612	319,426	156,491	75,780	656,582	6,136,605
White	8,767	45,501	202,277	118,508	57,002	432,055	4,040,847
Black	7,455	42,749	111,543	36,150	17,981	215,878	1,998,642
Asian	13	151	4,004	1,219	313	5,700	46,681
Native American	32	172	544	313	91	1,152	27,777
Others	5	41	1,058	298	385	1,787	22,669
Hispanic	106	378	4,552	2,522	952	8,510	96,837
Color	7,594	43,373	120,438	40,155	19,451	231,011	2,162,469
White (%)	53.87	51.35	63.33	75.73	75.22	65.80	65.85
Black (%)	45.81	48.24	34.92	23.10	23.73	32.88	32.57
Asian (%)	0.08	0.17	1.25	0.78	0.41	0.87	0.76
Native American (%)	0.20	0.19	0.17	0.20	0.12	0.18	0.45
Others (%)	0.03	0.05	0.33	0.19	0.51	0.27	0.37
Hispanic (%)	0.65	0.43	1.43	1.61	1.26	1.30	1.58
Color (%)	46.67	48.95	37.70	25.66	25.67	35.18	35.24
Female (%)	51.88	50.64	51.42	50.80	51.04	51.13	52.12
Old (65+ years) (%)	12.70	8.83	10.48	7.56	9.43	9.50	11.84
Kid (<5 years) (%)	9.78	9.52	9.00	10.40	10.30	9.57	9.45
Native-born (%)	99.25	99.47	97.70	98.35	99.16	98.30	98.43
Renter housing units (%)	21.57	22.69	35.18	22.75	16.08	28.33	28.27
Education (%)	14.43	13.21	23.71	20.13	6.78	19.33	15.21
Unemployment (%)	10.11	10.78	8.67	7.16	9.50	8.69	9.24
Poverty (%)	31.64	21.85	22.06	18.63	22.07	21.44	24.49
Mean household income ($)	43,827	49,517	53,801	57,401	44,976	52,902	46,170
Year 2000							
Total population	18,311	92,984	340,810	195,870	89,839	737,814	6,575,820
White	10,720	43,256	194,414	145,696	69,082	463,168	4,138,149
Black	7,377	48,128	133,868	44,762	18,486	252,621	2,225,382
Asian	39	447	6,596	2,276	599	9,957	65,300
Native American	32	218	917	684	247	2,098	35,571
Others	139	939	5,012	2,457	1,398	9,945	111,443
Hispanic	184	706	5,708	3,566	1,966	12,130	133,514
Color	7,672	50,016	149,958	52,517	21,789	281,952	2,508,726
White (%)	58.54	46.52	57.04	74.38	76.90	62.78	62.93

(continued)

Table B.19 (continued)

	0–10	11–20	21–30	31–40	41–50	0–50	Outside
Black (%)	40.29	51.76	39.28	22.85	20.58	34.24	33.84
Asian (%)	0.21	0.48	1.94	1.16	0.67	1.35	0.99
Native American (%)	0.17	0.23	0.27	0.35	0.27	0.28	0.54
Others (%)	0.76	1.01	1.47	1.25	1.56	1.35	1.69
Hispanic (%)	1.00	0.76	1.67	1.82	2.19	1.64	2.03
Color (%)	41.90	53.79	44.00	26.81	24.25	38.21	38.15
Female (%)	51.65	50.13	51.33	50.77	50.56	50.94	51.77
Old (65+ years) (%)	12.76	10.40	10.81	8.28	8.93	9.91	12.00
Kid (<5 years) (%)	8.40	8.25	8.10	9.31	9.30	8.60	8.56
Native-born (%)	98.84	99.25	96.61	97.59	98.29	97.47	97.92
Renter housing units (%)	20.20	21.13	35.73	20.05	13.80	26.86	27.31
College degree or higher (%)	17.59	14.70	26.43	23.46	9.62	21.90	17.59
Unemployment (%)	6.48	7.97	6.65	4.98	6.28	6.31	7.47
Poverty (%)	20.68	19.06	19.97	13.56	16.26	17.69	19.98
Mean household income ($)	59,551	54,976	59,157	66,307	55,222	60,088	54,041
Year 2010							
Total population	18,786	96,927	357,465	232,127	109,517	814,822	6,557,109
White	11,393	40,274	188,276	170,165	81,788	491,896	4,089,088
Black	7,162	54,179	153,312	53,841	24,051	292,545	2,201,410
Asian	0	634	8,892	3,462	688	13,676	80,277
Native American	102	146	666	599	429	1,942	39,020
Others	129	1,694	6,319	4,060	2,561	14,763	147,314
Hispanic	185	1,113	10,586	8,029	4,462	24,375	218,213
Color	7,570	57,006	176,777	67,723	30,399	339,475	2,598,980
White (%)	60.65	41.55	52.67	73.31	74.68	60.37	62.36
Black (%)	38.12	55.90	42.89	23.19	21.96	35.90	33.57
Asian (%)	0.00	0.65	2.49	1.49	0.63	1.68	1.22
Native American (%)	0.54	0.15	0.19	0.26	0.39	0.24	0.60
Others (%)	0.69	1.75	1.77	1.75	2.34	1.81	2.25
Hispanic (%)	0.98	1.15	2.96	3.46	4.07	2.99	3.33
Color (%)	40.30	58.81	49.45	29.17	27.76	41.66	39.64
Female (%)	52.88	51.40	51.13	50.97	50.15	51.02	51.29
Old (65+ years) (%)	15.09	11.85	10.81	9.84	10.42	10.70	12.49
Kid (<5 years) (%)	6.58	6.35	6.50	7.53	7.84	6.96	6.97
Native-born (%)	98.62	98.59	95.66	96.54	97.75	96.61	97.05

(continued)

Table B.19 (continued)

	0–10	11–20	21–30	31–40	41–50	0–50	Outside
Renter housing units (%)	19.16	23.15	34.97	21.79	16.54	27.05	26.36
College degree or higher (%)	20.53	17.62	27.62	27.44	16.55	24.70	19.82
Unemployment (%)	6.83	9.89	7.52	5.23	6.92	7.04	8.65
Poverty (%)	19.02	17.10	19.60	13.40	14.88	16.86	19.68
Mean household income ($)	58,682	55,685	60,344	74,007	65,867	64,356	56,114
Index	19	19	19	19	19	19	19

B.20 Waterford Steam Electric Station, Louisiana

See Fig. B.20.
See Table B.20.

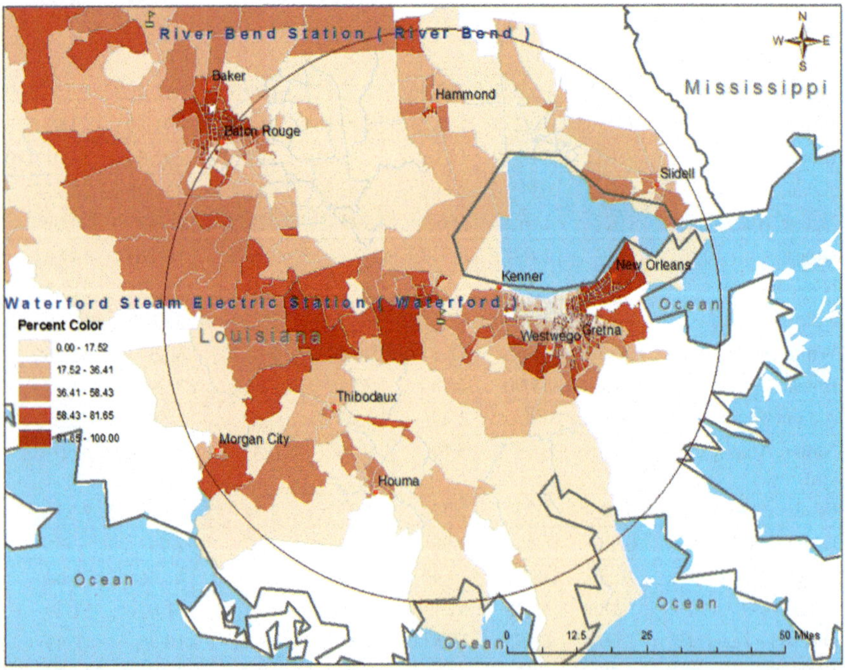

Fig. B.20 Spatial distribution of percent color among the populations surrounding Waterford Steam Electric Station, Louisiana, in 2010

Table B.20 Demographic composition of population, as sorted by distance from Waterford Steam Electric Station, Louisiana, in 1990, 2000, and 2010 *Source* (Kyne 2014)

	0–10	11–20	21–30	31–40	41–50	0–50	Outside
Year 1990							
Tract	18	100	242	80	47	487	661
Tract area (sq. mile)	285	1,096	1,177	1,766	1,944	6,269	46,109
Total population	61,978	333,717	763,682	298,467	197,977	1,655,821	2,564,155
White	40,357	274,614	386,278	226,292	170,125	1,097,666	1,742,354
Black	20,849	48,108	357,404	62,343	24,157	512,861	785,800
Asian	296	5,475	9,836	6,254	1,524	23,385	16,289
Native American	136	749	2,853	2,733	1,459	7,930	12,144
Others	349	4,766	7,301	860	729	14,005	7,539
Hispanic	1,396	19,558	26,034	6,628	3,396	57,012	33,593
Color	22,566	73,483	393,883	77,755	30,459	598,146	845,481
White (%)	65.12	82.29	50.58	75.82	85.93	66.29	67.95
Black (%)	33.64	14.42	46.80	20.89	12.20	30.97	30.65
Asian (%)	0.48	1.64	1.29	2.10	0.77	1.41	0.64
Native American (%)	0.22	0.22	0.37	0.92	0.74	0.48	0.47
Others (%)	0.56	1.43	0.96	0.29	0.37	0.85	0.29
Hispanic (%)	2.25	5.86	3.41	2.22	1.72	3.44	1.31
Color (%)	36.41	22.02	51.58	26.05	15.39	36.12	32.97
Female (%)	51.74	52.18	52.85	51.61	50.68	52.19	51.72
Old (65+ years) (%)	7.31	11.17	11.61	8.56	7.57	10.33	11.56
Kid (<5 years) (%)	11.75	8.46	9.54	10.40	10.08	9.62	9.64
Native-born (%)	98.36	94.57	96.42	97.67	98.23	96.56	98.81
Renter housing units (%)	19.18	33.05	40.95	23.78	24.57	33.90	27.11
Education (%)	13.01	20.94	18.23	14.47	20.85	18.29	14.70
Unemployment (%)	8.76	6.24	11.10	8.58	6.25	8.89	10.18
Poverty (%)	17.11	12.82	26.88	20.48	15.63	21.13	25.18
Mean household income ($)	55,028	58,954	48,406	51,604	57,855	52,518	45,751
Year 2000							
Total population	67,869	332,732	773,325	361,234	243,092	1,778,252	2,690,724
White	38,815	256,157	345,306	272,187	198,830	1,111,295	1,744,669
Black	27,092	55,593	392,160	71,744	33,852	580,441	864,125
Asian	487	7,807	13,364	7,014	3,393	32,065	24,806
Native American	190	929	3,404	3,386	2,407	10,316	15,517
Others	1,298	12,241	19,082	6,921	4,629	44,171	41,571
Hispanic	1,917	23,706	27,164	8,274	4,915	65,976	41,878
Color	30,064	91,533	442,258	94,042	47,430	705,327	969,301

(continued)

Table B.20 (continued)

	0–10	11–20	21–30	31–40	41–50	0–50	Outside
White (%)	57.19	76.99	44.65	75.35	81.79	62.49	64.84
Black (%)	39.92	16.71	50.71	19.86	13.93	32.64	32.11
Asian (%)	0.72	2.35	1.73	1.94	1.40	1.80	0.92
Native American (%)	0.28	0.28	0.44	0.94	0.99	0.58	0.58
Others (%)	1.91	3.68	2.47	1.92	1.90	2.48	1.54
Hispanic (%)	2.82	7.12	3.51	2.29	2.02	3.71	1.56
Color (%)	44.30	27.51	57.19	26.03	19.51	39.66	36.02
Female (%)	51.71	52.04	52.54	51.57	50.73	51.97	51.47
Old (65+ years) (%)	8.97	13.20	11.16	9.36	8.75	10.77	12.14
Kid (<5 years) (%)	9.42	7.38	8.44	8.92	8.84	8.43	8.57
Native-born (%)	97.50	92.87	96.05	97.39	97.33	95.96	98.37
Renter housing units (%)	17.56	33.31	39.92	21.67	21.71	31.96	26.66
College degree or higher (%)	14.85	24.28	20.70	19.75	24.23	21.49	16.86
Unemployment (%)	6.65	5.10	8.25	6.09	4.67	6.60	7.83
Poverty (%)	15.44	11.96	23.37	15.13	11.22	17.56	21.03
Mean household income ($)	61,806	66,376	54,632	64,561	68,526	61,006	52,411
Year 2010							
Total population	73,071	323,442	591,275	387,069	281,307	1,656,164	2,773,776
White	38,350	237,178	287,520	291,984	219,223	1,074,255	1,744,819
Black	32,527	63,335	263,676	73,780	48,454	481,772	923,913
Asian	574	8,478	15,003	8,347	5,405	37,807	30,819
Native American	181	1,573	3,806	3,460	2,002	11,022	16,330
Others	1,439	12,878	21,270	9,498	6,223	51,308	57,895
Hispanic	3,236	34,819	34,793	16,044	11,403	100,295	72,459
Color	37,128	110,324	321,320	105,225	70,306	644,303	1,074,365
White (%)	52.48	73.33	48.63	75.43	77.93	64.86	62.90
Black (%)	44.51	19.58	44.59	19.06	17.22	29.09	33.31
Asian (%)	0.79	2.62	2.54	2.16	1.92	2.28	1.11
Native American (%)	0.25	0.49	0.64	0.89	0.71	0.67	0.59
Others (%)	1.97	3.98	3.60	2.45	2.21	3.10	2.09
Hispanic (%)	4.43	10.77	5.88	4.14	4.05	6.06	2.61
Color (%)	50.81	34.11	54.34	27.19	24.99	38.90	38.73
Female (%)	51.35	51.57	51.12	50.70	51.19	51.13	51.11
Old (65+ years) (%)	10.40	14.22	10.90	10.76	10.64	11.45	12.53
Kid (<5 years) (%)	7.37	6.07	6.53	7.51	6.90	6.77	6.95
Native-born (%)	97.42	90.11	94.18	96.47	95.77	94.33	97.70

(continued)

Table B.20 (continued)

	0–10	11–20	21–30	31–40	41–50	0–50	Outside
Renter housing units (%)	18.38	30.61	32.12	21.35	22.60	27.40	27.21
College degree or higher (%)	18.22	26.23	23.21	24.42	26.94	24.53	18.72
Unemployment (%)	7.79	6.65	9.02	6.04	6.08	7.30	8.04
Poverty (%)	14.27	12.27	20.20	13.86	11.87	15.47	19.76
Mean household income ($)	64,391	69,301	60,215	71,477	72,135	66,827	55,925
Index	20	20	20	20	20	20	20

B.21 Calvert Cliffs Nuclear Power Plant, Maryland

See Fig. B.21.
See Table B.21.

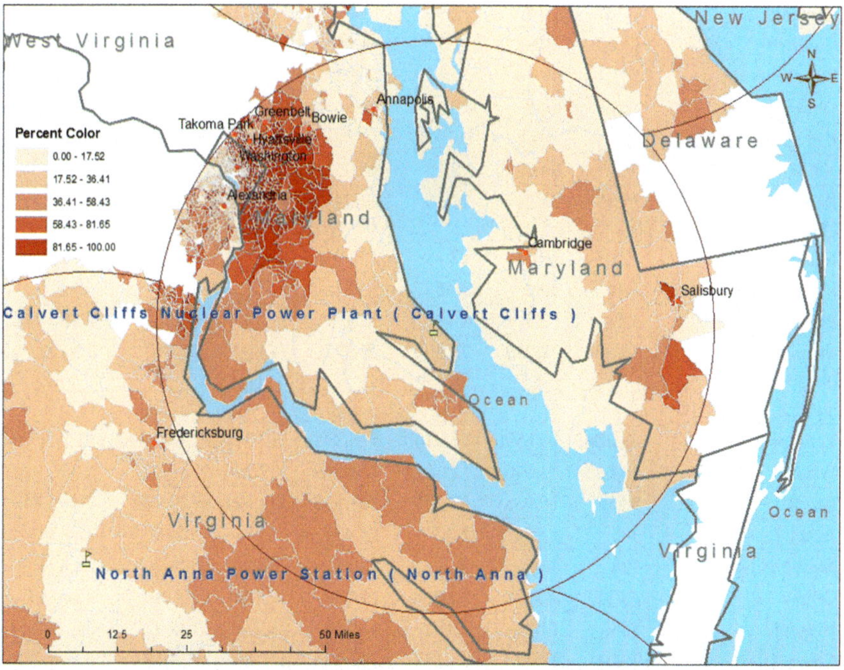

Fig. B.21 Spatial distribution of percent color among the populations surrounding Calvert Cliffs Nuclear Power Plant, Maryland, in 2010

Table B.21 Demographic composition of population, as sorted by distance from Calvert Cliffs Nuclear Power Plant, Maryland, in 1990, 2000, and 2010 *Source* (Kyne 2014)

	0–10	11–20	21–30	31–40	41–50	0–50	Outside
Year 1990							
Tract	10	26	45	152	205	438	3,054
Tract area (sq. mile)	258	907	1,720	2,244	953	6,082	49,167
Total population	33,438	83,023	146,292	507,912	689,904	1,460,569	10,115,140
White	28,854	67,156	116,513	263,709	316,745	792,977	7,576,252
Black	4,261	14,418	27,106	229,112	351,427	626,324	2,125,403
Asian	123	781	1,406	10,206	13,885	26,401	281,292
Native American	134	329	766	2,079	2,023	5,331	26,866
Others	73	312	510	2,824	5,831	9,550	105,271
Hispanic	369	965	2,030	8,148	14,457	25,969	280,745
Color	4,892	16,452	30,948	248,204	379,863	680,359	2,696,288
White (%)	86.29	80.89	79.64	51.92	45.91	54.29	74.90
Black (%)	12.74	17.37	18.53	45.11	50.94	42.88	21.01
Asian (%)	0.37	0.94	0.96	2.01	2.01	1.81	2.78
Native American (%)	0.40	0.40	0.52	0.41	0.29	0.36	0.27
Others (%)	0.22	0.38	0.35	0.56	0.85	0.65	1.04
Hispanic (%)	1.10	1.16	1.39	1.60	2.10	1.78	2.78
Color (%)	14.63	19.82	21.15	48.87	55.06	46.58	26.66
Female (%)	50.31	49.89	50.62	51.92	51.85	51.61	51.30
Old (65+ years) (%)	10.21	8.65	10.18	8.96	9.04	9.13	11.11
Kid (<5 years) (%)	10.24	10.46	9.19	8.78	9.12	9.11	8.56
Native-born (%)	98.31	98.02	98.15	96.05	95.12	95.99	93.82
Renter housing units (%)	15.45	25.21	22.50	29.94	40.72	33.78	32.80
Education (%)	19.08	15.68	17.18	23.79	25.58	23.43	26.12
Unemployment (%)	3.46	3.80	3.45	3.92	5.85	4.75	4.52
Poverty (%)	5.13	7.55	6.40	5.59	11.14	8.38	9.96
Mean household income ($)	79,006	71,759	77,800	82,773	74,909	77,867	72,122
Year 2000							
Total population	48,206	100,590	168,160	576,295	712,341	1,605,592	11,341,468
White	40,685	79,046	121,814	238,569	307,224	787,338	7,895,918
Black	5,858	16,935	39,155	308,555	359,421	729,924	2,465,540
Asian	335	1,854	2,379	11,437	16,471	32,476	454,384
Native American	123	404	933	2,017	2,090	5,567	34,484
Others	1,215	2,317	3,891	15,744	27,149	50,316	491,113
Hispanic	858	1,622	3,182	13,528	28,757	47,947	551,446
Color	7,931	22,428	48,137	343,484	417,201	839,181	3,698,281
White (%)	84.40	78.58	72.44	41.40	43.13	49.04	69.62

(continued)

Table B.21 (continued)

	0–10	11–20	21–30	31–40	41–50	0–50	Outside
Black (%)	12.15	16.84	23.28	53.54	50.46	45.46	21.74
Asian (%)	0.69	1.84	1.41	1.98	2.31	2.02	4.01
Native American (%)	0.26	0.40	0.55	0.35	0.29	0.35	0.30
Others (%)	2.52	2.30	2.31	2.73	3.81	3.13	4.33
Hispanic (%)	1.78	1.61	1.89	2.35	4.04	2.99	4.86
Color (%)	16.45	22.30	28.63	59.60	58.57	52.27	32.61
Female (%)	50.02	50.05	51.33	52.49	52.20	52.01	51.31
Old (65+ years) (%)	9.88	9.08	10.88	9.82	10.18	10.05	11.44
Kid (<5 years) (%)	8.69	8.43	8.19	8.18	8.56	8.38	7.82
Native-born (%)	97.64	97.37	97.27	95.05	92.12	94.20	90.57
Renter housing units (%)	14.78	23.59	20.31	26.40	36.83	29.98	31.08
College degree or higher (%)	23.27	20.55	21.49	27.35	30.72	27.68	31.13
Unemployment (%)	3.66	3.96	3.73	4.53	6.92	5.41	4.63
Poverty (%)	4.56	7.26	6.63	6.38	13.15	9.40	9.63
Mean household income ($)	85,546	79,782	83,435	87,376	80,277	83,268	79,389
Year 2010							
Total population	48,285	129,775	197,255	603,123	741,007	1,719,445	12,403,132
White	40,337	100,056	122,696	222,718	322,224	808,031	8,290,535
Black	6,042	21,710	63,121	340,523	360,158	791,554	2,714,879
Asian	520	3,147	3,469	14,220	21,726	43,082	707,414
Native American	23	486	725	1,640	1,374	4,248	38,531
Others	1,363	4,376	7,244	24,022	35,525	72,530	651,773
Hispanic	1,656	4,126	7,535	32,128	60,124	105,569	950,644
Color	9,084	32,728	79,157	397,726	453,351	972,046	4,624,980
White (%)	83.54	77.10	62.20	36.93	43.48	46.99	66.84
Black (%)	12.51	16.73	32.00	56.46	48.60	46.04	21.89
Asian (%)	1.08	2.42	1.76	2.36	2.93	2.51	5.70
Native American (%)	0.05	0.37	0.37	0.27	0.19	0.25	0.31
Others (%)	2.82	3.37	3.67	3.98	4.79	4.22	5.25
Hispanic (%)	3.43	3.18	3.82	5.33	8.11	6.14	7.66
Color (%)	18.81	25.22	40.13	65.94	61.18	56.53	37.29
Female (%)	51.42	50.04	51.53	52.66	52.23	52.11	51.17
Old (65+ years) (%)	10.91	9.76	11.81	11.67	10.92	11.20	11.94
Kid (<5 years) (%)	6.43	6.46	6.30	6.31	7.12	6.67	6.36
Native-born (%)	97.18	95.85	95.67	92.11	88.36	91.33	87.70

(continued)

Table B.21 (continued)

	0–10	11–20	21–30	31–40	41–50	0–50	Outside
Renter housing units (%)	14.68	21.89	17.88	24.03	33.89	27.27	29.22
College degree or higher (%)	28.94	26.46	27.30	31.36	35.86	32.39	35.62
Unemployment (%)	5.21	4.63	5.80	7.38	8.59	7.45	6.19
Poverty (%)	5.45	6.30	5.84	6.02	12.20	8.66	10.10
Mean household income ($)	97,216	96,429	94,526	94,529	89,320	92,445	85,728
Index	21	21	21	21	21	21	21

B.22 Pilgrim Nuclear Power Station, Massachusetts

See Fig. B.22.
See Table B.22.

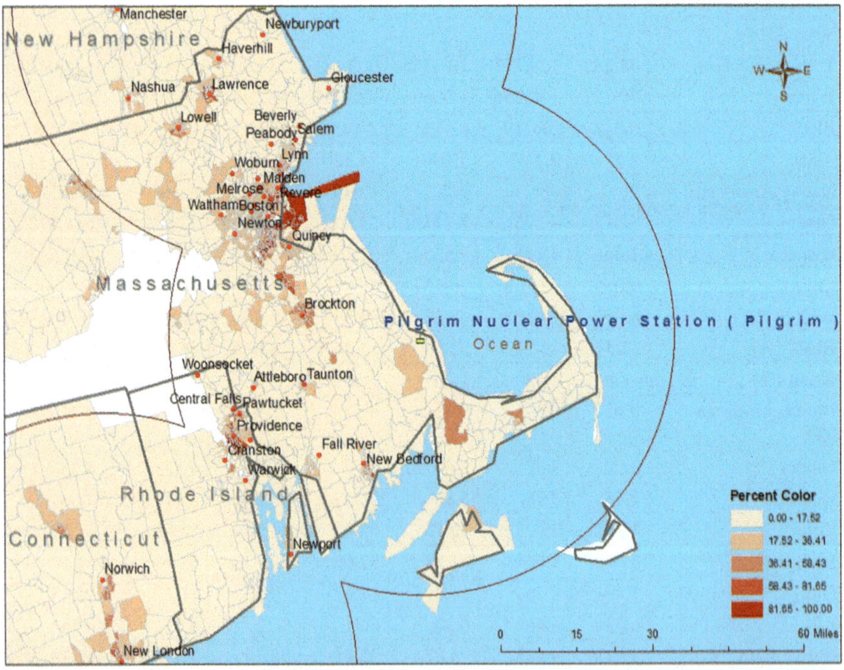

Fig. B.22 Spatial distribution of percent color among the populations surrounding Pilgrim Nuclear Power Station, Massachusetts, in 2010

Table B.22 Demographic composition of population, as sorted by distance from Pilgrim Nuclear Power Station, Massachusetts, in 1990, 2000, and 2010 *Source* (Kyne 2014)

	0–10	11–20	21–30	31–40	41–50	0–50	Outside
Year 1990							
Tract	15	50	187	332	163	747	975
Tract area (sq. mile)	453	539	1,710	1,222	470	4,394	7,705
Total population	59,169	208,525	750,365	1,188,509	701,079	2,907,647	4,112,244
White	57,619	201,256	698,882	943,095	632,029	2,532,881	3,797,948
Black	811	2,846	25,199	156,125	31,051	216,032	118,952
Asian	274	1,191	7,875	44,492	24,456	78,288	80,075
Native American	104	511	2,124	3,011	1,387	7,137	9,710
Others	359	2,710	16,291	41,801	12,138	73,299	105,539
Hispanic	626	2,506	17,890	75,151	30,322	126,495	193,291
Color	1,996	8,929	62,260	275,155	84,735	433,075	396,824
White (%)	97.38	96.51	93.14	79.35	90.15	87.11	92.36
Black (%)	1.37	1.36	3.36	13.14	4.43	7.43	2.89
Asian (%)	0.46	0.57	1.05	3.74	3.49	2.69	1.95
Native American (%)	0.18	0.25	0.28	0.25	0.20	0.25	0.24
Others (%)	0.61	1.30	2.17	3.52	1.73	2.52	2.57
Hispanic (%)	1.06	1.20	2.38	6.32	4.33	4.35	4.70
Color (%)	3.37	4.28	8.30	23.15	12.09	14.89	9.65
Female (%)	51.08	50.53	52.38	52.32	52.75	52.29	51.84
Old (65+ years) (%)	12.51	11.12	15.39	13.82	14.45	14.16	13.60
Kid (<5 years) (%)	8.57	9.23	8.27	7.60	7.06	7.78	8.38
Native-born (%)	97.01	97.06	91.29	84.78	86.08	87.90	92.29
Renter housing units (%)	21.03	17.10	28.72	47.90	49.74	40.59	34.25
Education (%)	26.33	22.91	20.77	28.34	29.29	26.20	26.51
Unemployment (%)	6.43	6.59	7.67	7.41	6.30	7.12	6.41
Poverty (%)	5.06	4.65	8.32	12.37	9.78	9.99	8.35
Mean household income ($)	81,690	78,555	68,836	69,897	70,109	70,453	75,141
Year 2000							
Total population	66,086	240,422	797,241	1,245,095	729,608	3,078,452	4,318,964
White	62,913	228,761	695,065	900,117	597,650	2,484,506	3,771,399
Black	1,216	3,360	36,136	164,871	37,894	243,477	138,916
Asian	511	1,103	14,892	73,288	42,926	132,720	131,627
Native American	170	575	2,548	3,986	1,991	9,270	11,159
Others	1,270	6,622	48,596	102,868	49,122	208,478	265,864
Hispanic	1,052	2,640	26,337	114,589	51,723	196,341	321,451
Color	3,714	12,982	114,602	388,475	153,948	673,721	667,906
White (%)	95.20	95.15	87.18	72.29	81.91	80.71	87.32

(continued)

Table B.22 (continued)

	0–10	11–20	21–30	31–40	41–50	0–50	Outside
Black (%)	1.84	1.40	4.53	13.24	5.19	7.91	3.22
Asian (%)	0.77	0.46	1.87	5.89	5.88	4.31	3.05
Native American (%)	0.26	0.24	0.32	0.32	0.27	0.30	0.26
Others (%)	1.92	2.75	6.10	8.26	6.73	6.77	6.16
Hispanic (%)	1.59	1.10	3.30	9.20	7.09	6.38	7.44
Color (%)	5.62	5.40	14.37	31.20	21.10	21.89	15.46
Female (%)	50.80	50.94	52.18	52.04	52.11	51.98	51.76
Old (65+ years) (%)	12.04	11.66	15.82	13.28	13.88	13.93	13.51
Kid (<5 years) (%)	8.25	8.39	7.35	6.94	6.41	7.06	7.79
Native-born (%)	96.34	97.02	89.91	80.50	81.17	84.73	90.23
Renter housing units (%)	19.20	15.72	27.35	47.42	50.00	39.75	32.92
College degree or higher (%)	32.42	28.93	25.66	34.84	35.21	32.01	32.21
Unemployment (%)	4.16	3.80	4.89	5.81	4.55	5.08	4.44
Poverty (%)	5.12	4.87	8.61	12.91	10.64	10.45	9.18
Mean household income ($)	88,648	87,939	74,599	78,444	77,034	77,983	82,556
Year 2010							
Total population	69,440	253,462	807,574	1,269,310	730,272	3,130,058	4,403,427
White	66,147	240,502	675,606	888,055	574,404	2,444,714	3,711,042
Black	1,030	4,137	69,729	176,485	52,918	304,299	180,086
Asian	636	2,597	21,965	94,481	55,279	174,958	194,577
Native American	88	424	2,102	3,353	1,325	7,292	9,813
Others	1,539	5,802	38,172	106,936	46,346	198,795	307,909
Hispanic	940	4,268	38,503	145,939	77,598	267,248	442,159
Color	3,847	15,811	153,254	439,092	199,623	811,627	889,914
White (%)	95.26	94.89	83.66	69.96	78.66	78.10	84.28
Black (%)	1.48	1.63	8.63	13.90	7.25	9.72	4.09
Asian (%)	0.92	1.02	2.72	7.44	7.57	5.59	4.42
Native American (%)	0.13	0.17	0.26	0.26	0.18	0.23	0.22
Others (%)	2.22	2.29	4.73	8.42	6.35	6.35	6.99
Hispanic (%)	1.35	1.68	4.77	11.50	10.63	8.54	10.04
Color (%)	5.54	6.24	18.98	34.59	27.34	25.93	20.21
Female (%)	51.51	50.62	52.20	51.93	51.81	51.86	51.51
Old (65+ years) (%)	14.78	13.20	15.96	12.97	13.20	13.85	13.46
Kid (<5 years) (%)	5.57	5.62	5.65	5.52	5.31	5.52	5.78
Native-born (%)	95.38	96.43	87.33	78.68	77.04	82.34	88.12

(continued)

Table B.22 (continued)

	0–10	11–20	21–30	31–40	41–50	0–50	Outside
Renter housing units (%)	17.06	12.97	25.54	43.61	46.54	36.53	29.55
College degree or higher (%)	38.47	34.91	30.19	40.59	40.83	37.41	37.02
Unemployment (%)	7.94	6.86	7.84	8.55	7.01	7.85	7.13
Poverty (%)	6.16	5.17	9.70	14.39	12.51	11.80	10.01
Mean household income ($)	99,048	91,855	77,390	83,902	78,785	81,903	85,439
Index	22	22	22	22	22	22	22

B.23 Donald C. Cook Nuclear Power Plant, Michigan

See Fig. B.23.
See Table B.23.

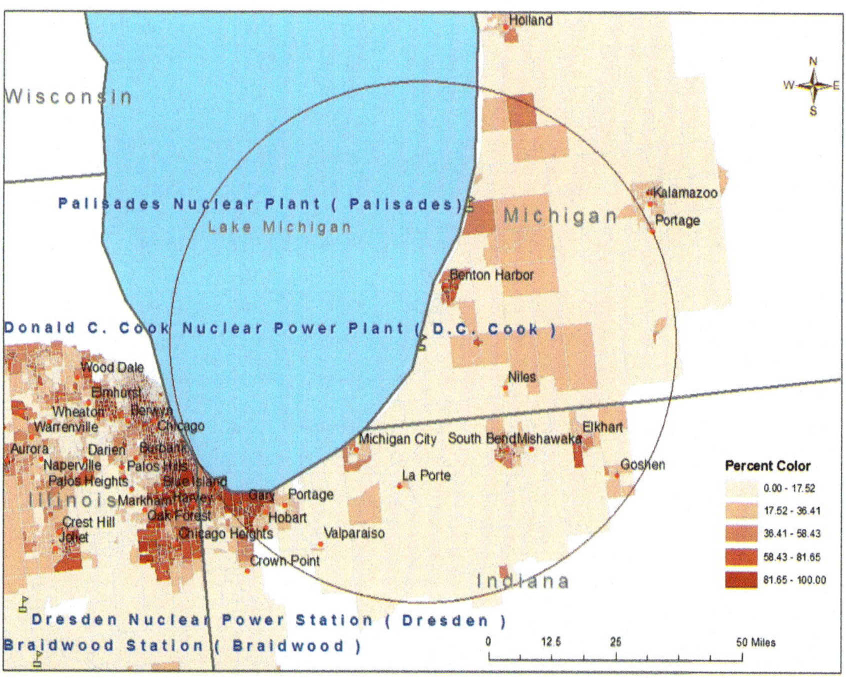

Fig. B.23 Spatial distribution of percent color among the populations surrounding Donald C. Cook Nuclear Power Plant, Michigan, in 2010

Table B.23 Demographic composition of population, as sorted by distance from Donald C. Cook Nuclear Power Plant, Michigan, in 1990, 2000, and 2010 *Source* (Kyne 2014)

	0–10	11–20	21–30	31–40	41–50	0–50	Outside
Year 1990							
Tract	16	29	99	58	25	227	4,097
Tract area (sq. mile)	183	1,396	1,350	1,322	1,602	5,853	127,280
Total population	51,729	92,894	336,028	234,769	94,097	809,517	14,029,827
White	47,625	69,660	291,514	218,719	92,300	719,818	12,061,910
Black	3,217	21,282	36,324	12,332	349	73,504	1,645,230
Asian	415	911	2,654	1,351	405	5,736	133,729
Native American	228	484	1,675	1,055	284	3,726	69,686
Others	242	558	3,864	1,305	767	6,736	119,247
Hispanic	715	1,481	7,428	3,360	1,945	14,929	270,349
Color	4,567	24,109	48,082	18,102	2,988	97,848	2,111,729
White (%)	92.07	74.99	86.75	93.16	98.09	88.92	85.97
Black (%)	6.22	22.91	10.81	5.25	0.37	9.08	11.73
Asian (%)	0.80	0.98	0.79	0.58	0.43	0.71	0.95
Native American (%)	0.44	0.52	0.50	0.45	0.30	0.46	0.50
Others (%)	0.47	0.60	1.15	0.56	0.82	0.83	0.85
Hispanic (%)	1.38	1.59	2.21	1.43	2.07	1.84	1.93
Color (%)	8.83	25.95	14.31	7.71	3.18	12.09	15.05
Female (%)	51.53	52.32	51.44	50.68	51.16	51.29	51.49
Old (65+ years) (%)	14.59	13.27	13.95	12.38	11.66	13.19	12.09
Kid (<5 years) (%)	8.00	9.47	8.64	8.92	8.33	8.74	8.92
Native-born (%)	95.96	96.72	97.26	98.04	98.13	97.44	96.94
Renter housing units (%)	20.07	32.12	25.73	22.87	24.93	25.18	26.43
Education (%)	23.07	14.44	16.07	14.79	15.23	15.89	16.73
Unemployment (%)	4.51	8.99	6.69	5.50	4.18	6.14	7.36
Poverty (%)	6.53	20.40	11.42	8.89	7.10	10.90	12.29
Mean household income ($)	65,530	48,129	55,111	59,657	61,153	56,980	60,707
Year 2000							
Total population	53,837	91,502	352,593	260,465	103,421	861,818	15,157,111
White	47,338	65,745	288,402	230,900	98,274	730,659	12,547,017
Black	4,383	21,303	41,933	14,403	888	82,910	1,823,262
Asian	726	825	3,897	1,717	751	7,916	228,532
Native American	232	468	1,478	1,060	178	3,416	74,594
Others	1,154	3,164	16,885	12,377	3,339	36,919	483,704
Hispanic	968	3,191	18,086	13,997	5,227	41,469	491,229
Color	6,979	27,345	71,825	34,868	7,995	149,012	2,843,870

(continued)

Table B.23 (continued)

	0–10	11–20	21–30	31–40	41–50	0–50	Outside
White (%)	87.93	71.85	81.79	88.65	95.02	84.78	82.78
Black (%)	8.14	23.28	11.89	5.53	0.86	9.62	12.03
Asian (%)	1.35	0.90	1.11	0.66	0.73	0.92	1.51
Native American (%)	0.43	0.51	0.42	0.41	0.17	0.40	0.49
Others (%)	2.14	3.46	4.79	4.75	3.23	4.28	3.19
Hispanic (%)	1.80	3.49	5.13	5.37	5.05	4.81	3.24
Color (%)	12.96	29.88	20.37	13.39	7.73	17.29	18.76
Female (%)	51.26	51.95	51.21	50.26	50.87	50.96	50.99
Old (65+ years) (%)	15.62	13.65	13.48	12.73	12.01	13.23	12.26
Kid (<5 years) (%)	6.79	8.62	8.43	8.35	7.81	8.25	8.25
Native-born (%)	94.90	94.87	95.65	95.85	96.27	95.66	95.56
Renter housing units (%)	17.59	29.20	25.12	22.22	25.08	24.19	24.54
College degree or higher (%)	27.78	16.57	19.55	17.23	19.98	19.13	20.97
Unemployment (%)	3.78	6.66	5.78	4.46	3.92	5.11	5.46
Poverty (%)	6.41	17.38	11.27	8.00	7.13	10.12	10.14
Mean household income ($)	77,581	54,391	61,725	66,732	69,938	64,439	69,314
Year 2010							
Total population	53,507	87,062	350,164	272,079	110,916	873,728	15,496,357
White	46,784	61,852	276,966	238,535	101,458	725,595	12,631,003
Black	4,612	18,965	44,796	15,928	2,018	86,319	1,887,649
Asian	835	1,680	5,462	2,382	1,313	11,672	331,635
Native American	114	540	1,504	1,140	269	3,567	65,422
Others	1,162	4,025	21,436	14,094	5,858	46,575	580,648
Hispanic	920	4,051	27,500	23,200	9,943	65,614	719,270
Color	7,226	27,040	87,212	46,689	14,616	182,783	3,260,870
White (%)	87.44	71.04	79.10	87.67	91.47	83.05	81.51
Black (%)	8.62	21.78	12.79	5.85	1.82	9.88	12.18
Asian (%)	1.56	1.93	1.56	0.88	1.18	1.34	2.14
Native American (%)	0.21	0.62	0.43	0.42	0.24	0.41	0.42
Others (%)	2.17	4.62	6.12	5.18	5.28	5.33	3.75
Hispanic (%)	1.72	4.65	7.85	8.53	8.96	7.51	4.64
Color (%)	13.50	31.06	24.91	17.16	13.18	20.92	21.04
Female (%)	50.95	51.34	50.92	50.30	50.76	50.75	50.90
Old (65+ years) (%)	17.24	14.80	13.27	13.37	12.70	13.62	12.99
Kid (<5 years) (%)	4.93	7.16	6.59	6.94	6.83	6.69	6.39
Native-born (%)	94.45	94.38	94.47	94.70	95.73	94.69	94.64

(continued)

Table B.23 (continued)

	0–10	11–20	21–30	31–40	41–50	0–50	Outside
Renter housing units (%)	16.06	26.68	23.67	20.87	23.10	22.54	23.20
College degree or higher (%)	32.49	19.90	21.45	19.87	22.27	21.63	24.14
Unemployment (%)	7.35	11.87	10.39	9.21	6.76	9.49	10.31
Poverty (%)	7.33	22.06	16.58	12.51	11.80	14.66	14.32
Mean household income ($)	72,172	52,262	56,423	61,769	64,423	59,720	62,918
Index	23	23	23	23	23	23	23

B.24 Fermi, Michigan

See Fig. B.24.
See Table B.24.

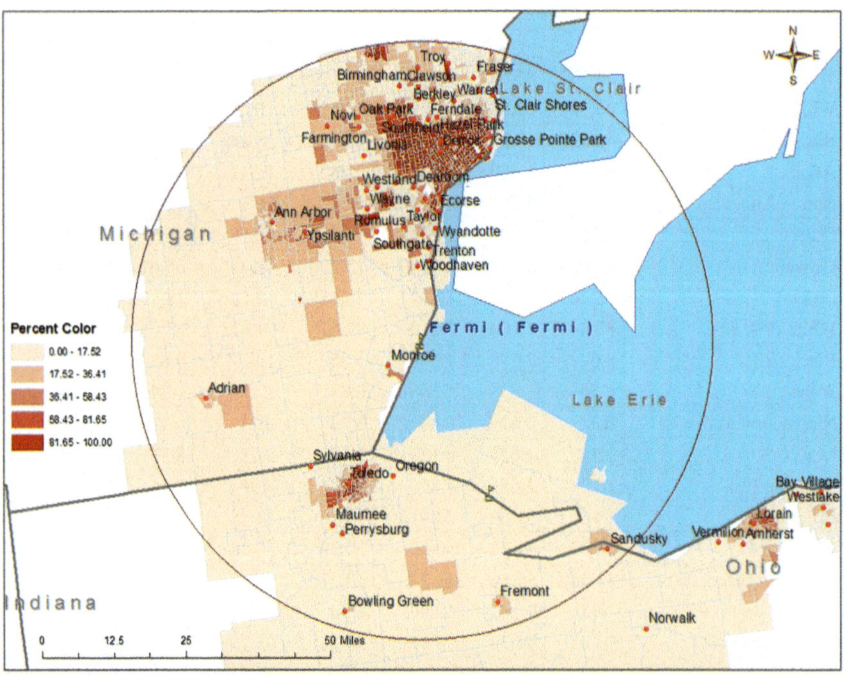

Fig. B.24 Spatial distribution of percent color among the populations surrounding Fermi, Michigan, in 2010

Table B.24 Demographic composition of population, as sorted by distance from Fermi, Michigan, in 1990, 2000, and 2010 *Source* (Kyne 2014)

	0–10	11–20	21–30	31–40	41–50	0–50	Outside
Year 1990							
Tract	24	73	401	510	304	1,312	4,453
Tract area (sq. mile)	268	597	1,148	1,353	1,186	4,552	136,987
Total population	78,888	248,261	1,327,704	1,809,704	1,025,409	4,489,966	15,652,356
White	76,095	235,400	1,034,834	1,036,411	963,855	3,346,595	13,937,582
Black	1,403	7,292	248,652	732,429	32,397	1,022,173	1,419,063
Asian	457	2,963	14,281	28,509	21,279	67,489	124,623
Native American	447	1,389	5,683	5,418	3,260	16,197	65,054
Others	486	1,217	24,239	6,972	4,623	37,537	105,995
Hispanic	1,150	5,517	49,793	21,155	13,959	91,574	230,325
Color	3,418	17,086	317,166	785,641	70,749	1,194,060	1,835,817
White (%)	96.46	94.82	77.94	57.27	94.00	74.53	89.04
Black (%)	1.78	2.94	18.73	40.47	3.16	22.77	9.07
Asian (%)	0.58	1.19	1.08	1.58	2.08	1.50	0.80
Native American (%)	0.57	0.56	0.43	0.30	0.32	0.36	0.42
Others (%)	0.62	0.49	1.83	0.39	0.45	0.84	0.68
Hispanic (%)	1.46	2.22	3.75	1.17	1.36	2.04	1.47
Color (%)	4.33	6.88	23.89	43.41	6.90	26.59	11.73
Female (%)	51.45	51.29	51.73	52.82	51.57	52.11	51.53
Old (65+ years) (%)	10.51	10.80	12.42	12.49	11.94	12.21	12.55
Kid (<5 years) (%)	9.64	8.67	9.24	9.35	7.99	8.98	8.84
Native-born (%)	97.96	96.17	94.89	94.91	92.87	94.56	97.63
Renter housing units (%)	26.56	26.21	32.87	34.49	22.80	30.81	27.52
Education (%)	9.95	11.47	14.30	20.78	24.52	19.06	16.60
Unemployment (%)	7.53	6.77	9.32	11.08	5.27	8.82	6.94
Poverty (%)	10.79	7.94	15.87	18.61	5.32	14.03	12.45
Mean household income ($)	62,352	67,987	57,573	62,446	86,966	66,870	58,074
Year 2000							
Total population	82,873	253,915	1,335,661	1,795,730	1,074,800	4,542,979	16,748,605
White	78,028	230,699	950,111	928,797	958,790	3,146,425	14,454,440
Black	1,809	12,187	274,917	764,888	45,291	1,099,092	1,590,990
Asian	904	3,206	25,274	48,427	39,750	117,561	194,704
Native American	281	1,099	5,941	5,218	3,104	15,643	72,198
Others	1,857	6,721	79,402	48,435	27,849	164,264	436,267
Hispanic	1,761	6,768	82,990	28,604	21,716	141,839	394,210
Color	6,018	27,347	423,634	880,992	128,772	1,466,763	2,482,414

(continued)

Table B.24 (continued)

	0–10	11–20	21–30	31–40	41–50	0–50	Outside
White (%)	94.15	90.86	71.13	51.72	89.21	69.26	86.30
Black (%)	2.18	4.80	20.58	42.59	4.21	24.19	9.50
Asian (%)	1.09	1.26	1.89	2.70	3.70	2.59	1.16
Native American (%)	0.34	0.43	0.44	0.29	0.29	0.34	0.43
Others (%)	2.24	2.65	5.94	2.70	2.59	3.62	2.60
Hispanic (%)	2.12	2.67	6.21	1.59	2.02	3.12	2.35
Color (%)	7.26	10.77	31.72	49.06	11.98	32.29	14.82
Female (%)	50.56	51.18	51.13	52.39	51.23	51.64	51.12
Old (65+ years) (%)	11.00	12.17	11.99	12.08	13.69	12.42	12.91
Kid (<5 years) (%)	8.62	7.84	8.90	8.33	7.50	8.28	8.03
Native-born (%)	97.56	96.22	92.13	93.59	89.79	92.48	96.89
Renter housing units (%)	22.85	23.05	31.25	32.06	22.06	28.78	25.57
College degree or higher (%)	12.33	14.83	18.55	26.17	30.79	24.22	20.64
Unemployment (%)	4.04	4.58	6.63	7.42	4.00	6.10	5.15
Poverty (%)	8.68	6.88	14.10	14.94	5.52	11.89	10.20
Mean household income ($)	71,669	75,830	64,286	72,876	93,245	75,436	66,546
Year 2010							
Total population	85,991	260,959	1,282,033	1,651,761	1,087,491	4,368,235	17,096,884
White	79,329	227,803	883,946	846,961	924,248	2,962,287	14,531,779
Black	2,739	19,590	286,121	688,136	79,305	1,075,891	1,716,965
Asian	813	4,921	34,409	67,618	54,853	162,614	271,620
Native American	311	707	5,284	4,136	2,674	13,112	64,175
Others	2,799	7,938	72,273	44,910	26,411	154,331	512,344
Hispanic	3,820	10,636	99,397	35,478	28,598	177,929	578,502
Color	9,645	39,995	448,258	825,355	181,050	1,504,303	2,907,823
White (%)	92.25	87.29	68.95	51.28	84.99	67.81	85.00
Black (%)	3.19	7.51	22.32	41.66	7.29	24.63	10.04
Asian (%)	0.95	1.89	2.68	4.09	5.04	3.72	1.59
Native American (%)	0.36	0.27	0.41	0.25	0.25	0.30	0.38
Others (%)	3.25	3.04	5.64	2.72	2.43	3.53	3.00
Hispanic (%)	4.44	4.08	7.75	2.15	2.63	4.07	3.38
Color (%)	11.22	15.33	34.96	49.97	16.65	34.44	17.01
Female (%)	50.95	51.06	51.33	52.19	51.57	51.69	50.92
Old (65+ years) (%)	11.80	13.21	11.59	12.65	14.61	12.84	13.69
Kid (<5 years) (%)	6.60	5.86	7.13	5.94	5.72	6.24	6.24
Native-born (%)	97.38	95.77	91.08	93.08	88.39	91.57	96.12

(continued)

Table B.24 (continued)

	0–10	11–20	21–30	31–40	41–50	0–50	Outside
Renter housing units (%)	21.89	21.54	29.16	28.53	21.56	26.50	24.44
College degree or higher (%)	15.75	17.73	21.56	29.98	34.33	27.67	23.75
Unemployment (%)	12.01	11.06	13.54	13.95	9.65	12.48	9.31
Poverty (%)	10.97	10.67	18.74	19.35	8.67	15.82	14.13
Mean household income ($)	62,923	64,973	56,406	65,310	80,715	66,641	61,932
Index	24	24	24	24	24	24	24

B.25 Palisades Nuclear Plant, Michigan

See Fig. B.25.
See Table B.25.

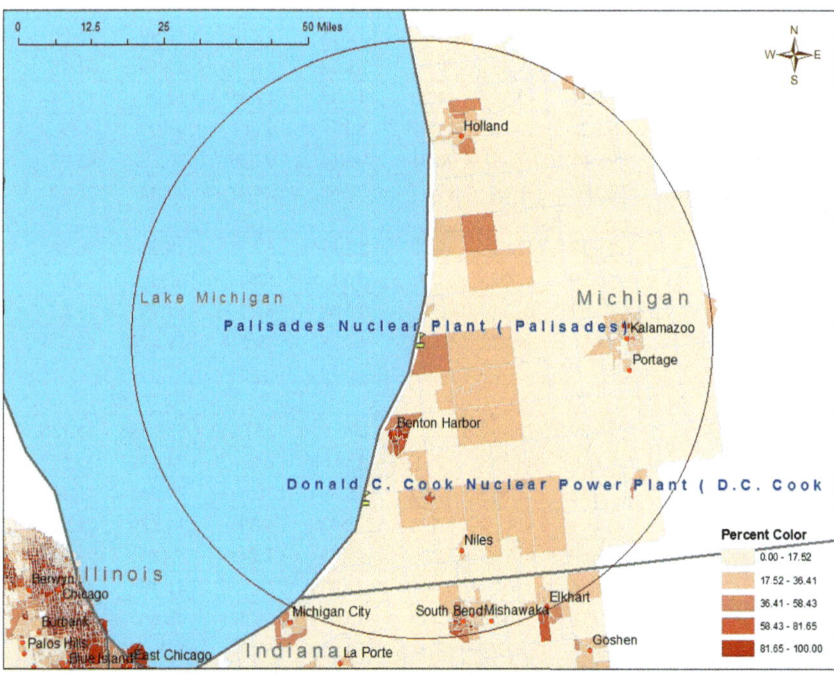

Fig. B.25 Spatial distribution of percent color among the populations surrounding Palisades Nuclear Plant, Michigan, in 2010

Table B.25 Demographic composition of population, as sorted by distance from Palisades Nuclear Plant, Michigan, in 1990, 2000, and 2010 *Source* (Kyne 2014)

	0–10	11–20	21–30	31–40	41–50	0–50	Outside
Year 1990							
Tract	8	23	31	70	94	226	4,098
Tract area (sq. mile)	183	904	1,852	2,037	1,810	6,785	126,348
Total population	34,133	73,590	124,582	263,660	298,779	794,744	14,044,600
White	29,981	50,126	117,323	236,091	268,462	701,983	12,079,745
Black	3,282	21,648	3,627	17,125	24,118	69,800	1,648,934
Asian	120	238	1,343	4,475	2,713	8,889	130,576
Native American	355	511	922	1,100	817	3,705	69,707
Others	396	1,061	1,368	4,881	2,673	10,379	115,604
Hispanic	866	1,755	3,427	9,283	5,736	21,067	264,211
Color	4,603	24,090	9,266	31,926	33,340	103,225	2,106,352
White (%)	87.84	68.12	94.17	89.54	89.85	88.33	86.01
Black (%)	9.62	29.42	2.91	6.50	8.07	8.78	11.74
Asian (%)	0.35	0.32	1.08	1.70	0.91	1.12	0.93
Native American (%)	1.04	0.69	0.74	0.42	0.27	0.47	0.50
Others (%)	1.16	1.44	1.10	1.85	0.89	1.31	0.82
Hispanic (%)	2.54	2.38	2.75	3.52	1.92	2.65	1.88
Color (%)	13.49	32.74	7.44	12.11	11.16	12.99	15.00
Female (%)	51.97	52.34	50.92	51.62	51.37	51.50	51.48
Old (65+ years) (%)	14.01	13.72	11.70	11.45	12.65	12.26	12.14
Kid (<5 years) (%)	9.32	10.22	9.04	8.87	9.26	9.19	8.89
Native-born (%)	98.11	97.49	96.31	96.16	97.20	96.78	96.98
Renter housing units (%)	22.57	29.10	19.47	28.68	24.21	25.34	26.42
Education (%)	11.48	12.24	19.02	22.37	17.35	18.51	16.58
Unemployment (%)	10.29	11.92	5.70	5.71	5.56	6.33	7.35
Poverty (%)	17.69	24.71	8.72	12.08	8.87	11.79	12.23
Mean household income ($)	48,562	46,222	60,703	61,482	59,214	58,476	60,617
Year 2000							
Total population	34,594	75,167	138,356	297,216	337,253	882,586	15,136,343
White	29,507	48,833	125,892	254,089	286,676	744,997	12,532,679
Black	2,782	21,865	4,489	18,094	30,033	77,263	1,828,909
Asian	234	414	1,593	7,642	4,356	14,239	222,209
Native American	266	416	782	1,473	1,249	4,186	73,824
Others	1,806	3,632	5,598	15,931	14,939	41,906	478,717
Hispanic	2,331	5,015	6,449	18,411	16,461	48,667	484,031
Color	6,075	28,575	15,446	50,823	57,374	158,293	2,834,589
White (%)	85.30	64.97	90.99	85.49	85.00	84.41	82.80

(continued)

Table B.25 (continued)

	0–10	11–20	21–30	31–40	41–50	0–50	Outside
Black (%)	8.04	29.09	3.24	6.09	8.91	8.75	12.08
Asian (%)	0.68	0.55	1.15	2.57	1.29	1.61	1.47
Native American (%)	0.77	0.55	0.57	0.50	0.37	0.47	0.49
Others (%)	5.22	4.83	4.05	5.36	4.43	4.75	3.16
Hispanic (%)	6.74	6.67	4.66	6.19	4.88	5.51	3.20
Color (%)	17.56	38.02	11.16	17.10	17.01	17.94	18.73
Female (%)	51.73	51.05	50.50	51.35	51.24	51.17	50.98
Old (65+ years) (%)	14.38	12.76	12.15	11.85	11.89	12.09	12.32
Kid (<5 years) (%)	7.86	9.12	7.87	8.36	8.77	8.48	8.24
Native-born (%)	96.31	95.88	94.85	94.64	95.49	95.17	95.59
Renter housing units (%)	20.17	26.86	16.83	27.74	23.50	24.03	24.55
College degree or higher (%)	12.03	14.23	23.19	26.41	21.76	22.47	20.78
Unemployment (%)	6.73	7.58	4.35	5.96	5.15	5.55	5.44
Poverty (%)	12.35	19.52	7.88	10.56	9.21	10.46	10.12
Mean household income ($)	54,881	54,445	72,629	67,498	66,009	66,082	69,223
Year 2010							
Total population	33,432	70,910	144,272	307,925	343,061	899,600	15,470,485
White	27,653	46,185	130,813	257,360	284,755	746,766	12,609,832
Black	2,575	19,388	4,432	19,533	33,182	79,110	1,894,858
Asian	178	473	2,457	9,599	5,582	18,289	325,018
Native American	141	424	621	1,540	1,136	3,862	65,127
Others	2,885	4,440	5,949	19,893	18,406	51,573	575,650
Hispanic	4,704	5,157	6,855	25,218	23,368	65,302	719,582
Color	8,086	26,811	16,906	62,170	70,304	184,277	3,259,376
White (%)	82.71	65.13	90.67	83.58	83.00	83.01	81.51
Black (%)	7.70	27.34	3.07	6.34	9.67	8.79	12.25
Asian (%)	0.53	0.67	1.70	3.12	1.63	2.03	2.10
Native American (%)	0.42	0.60	0.43	0.50	0.33	0.43	0.42
Others (%)	8.63	6.26	4.12	6.46	5.37	5.73	3.72
Hispanic (%)	14.07	7.27	4.75	8.19	6.81	7.26	4.65
Color (%)	24.19	37.81	11.72	20.19	20.49	20.48	21.07
Female (%)	51.47	51.13	50.42	51.04	50.98	50.94	50.89
Old (65+ years) (%)	15.42	13.63	13.61	12.55	12.13	12.75	13.04
Kid (<5 years) (%)	6.50	6.94	5.81	6.59	6.98	6.64	6.39
Native-born (%)	92.75	95.91	94.86	93.89	94.61	94.44	94.66
Renter housing units (%)	18.33	24.20	16.34	26.27	21.41	22.34	23.22

(continued)

Table B.25 (continued)

	0–10	11–20	21–30	31–40	41–50	0–50	Outside
College degree or higher (%)	14.82	17.06	27.43	28.89	23.98	25.28	23.94
Unemployment (%)	10.22	13.00	8.50	10.17	9.06	9.69	10.30
Poverty (%)	18.61	25.65	11.36	15.27	12.91	14.70	14.32
Mean household income ($)	53,186	50,337	66,579	60,247	62,056	60,861	62,857
Index	25	25	25	25	25	25	25

B.26 Monticello Nuclear Generating Plant, Minnesota

See Fig. B.26.
See Table B.26.

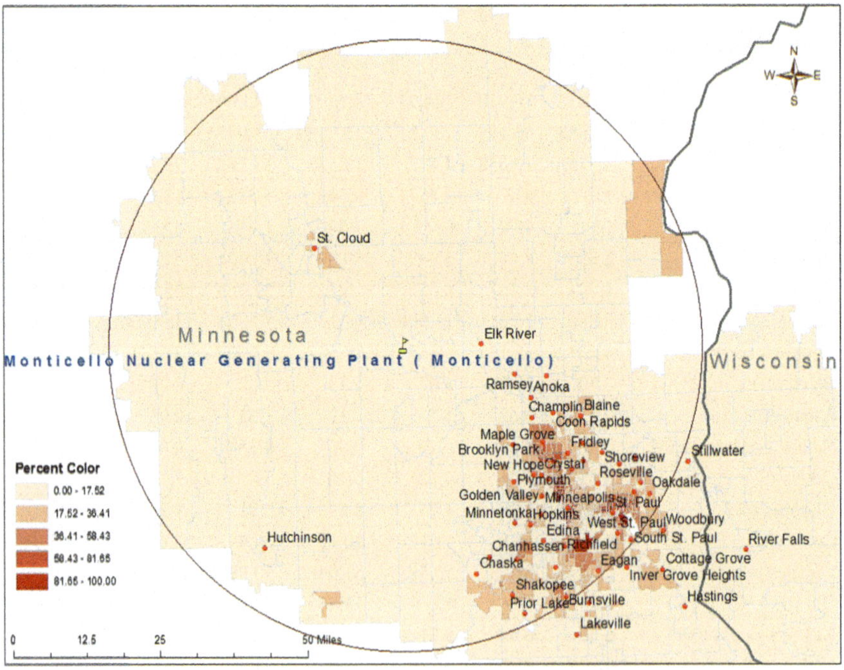

Fig. B.26 Spatial distribution of percent color among the populations surrounding Monticello Nuclear Generating Plant, Minnesota, in 2010

Table B.26 Demographic composition of population, as sorted by distance from Monticello Nuclear Generating Plant, Minnesota, in 1990, 2000, and 2010 *Source* (Kyne 2014)

	0–10	11–20	21–30	31–40	41–50	0–50	Outside
Year 1990							
Tract	8	24	122	245	51	450	888
Tract area (sq. mile)	308	760	1,613	1,377	1,241	5,299	81,637
Total population	28,856	82,718	439,709	807,901	185,450	1,544,634	2,830,480
White	28,631	81,457	426,650	719,344	175,183	1,431,265	2,701,930
Black	42	261	4,947	48,309	5,365	58,924	35,869
Asian	94	356	4,736	23,171	3,307	31,664	45,085
Native American	79	520	2,368	13,003	1,007	16,977	32,523
Others	13	125	1,006	4,034	591	5,769	15,059
Hispanic	188	401	2,974	9,976	1,643	15,182	34,474
Color	386	1,545	15,178	93,784	11,259	122,152	146,780
White (%)	99.22	98.48	97.03	89.04	94.46	92.66	95.46
Black (%)	0.15	0.32	1.13	5.98	2.89	3.81	1.27
Asian (%)	0.33	0.43	1.08	2.87	1.78	2.05	1.59
Native American (%)	0.27	0.63	0.54	1.61	0.54	1.10	1.15
Others (%)	0.05	0.15	0.23	0.50	0.32	0.37	0.53
Hispanic (%)	0.65	0.48	0.68	1.23	0.89	0.98	1.22
Color (%)	1.34	1.87	3.45	11.61	6.07	7.91	5.19
Female (%)	49.66	49.04	50.39	51.30	51.98	50.97	50.98
Old (65+ years) (%)	7.45	7.73	6.65	11.19	12.97	9.86	13.93
Kid (<5 years) (%)	11.94	10.47	10.59	8.88	8.73	9.49	9.19
Native-born (%)	99.30	99.18	98.50	95.58	97.29	96.88	97.71
Renter housing units (%)	15.57	14.87	23.41	36.35	25.43	30.32	22.45
Education (%)	13.52	14.36	20.86	29.45	30.59	26.32	19.38
Unemployment (%)	5.42	5.56	4.51	5.17	3.68	4.82	5.34
Poverty (%)	7.14	6.15	6.94	10.74	5.43	8.70	11.05
Mean household income ($)	62,847	67,078	73,253	69,922	77,801	71,518	56,617
Year 2000							
Total population	42,437	118,777	540,296	870,778	209,265	1,781,553	3,137,926
White	41,520	115,648	497,519	705,497	187,732	1,547,916	2,854,208
Black	136	684	14,845	75,739	6,908	98,312	69,545
Asian	226	542	12,965	38,960	6,324	59,017	81,952
Native American	142	453	2,903	8,721	1,284	13,503	41,065
Others	423	1,456	12,058	41,839	7,014	62,790	91,171
Hispanic	471	1,274	8,322	35,882	7,105	53,054	88,732
Color	1,204	3,867	46,861	179,550	24,834	256,316	322,491

(continued)

Table B.26 (continued)

	0–10	11–20	21–30	31–40	41–50	0–50	Outside
White (%)	97.84	97.37	92.08	81.02	89.71	86.89	90.96
Black (%)	0.32	0.58	2.75	8.70	3.30	5.52	2.22
Asian (%)	0.53	0.46	2.40	4.47	3.02	3.31	2.61
Native American (%)	0.33	0.38	0.54	1.00	0.61	0.76	1.31
Others (%)	1.00	1.23	2.23	4.80	3.35	3.52	2.91
Hispanic (%)	1.11	1.07	1.54	4.12	3.40	2.98	2.83
Color (%)	2.84	3.26	8.67	20.62	11.87	14.39	10.28
Female (%)	49.82	49.18	50.15	50.49	51.67	50.42	50.56
Old (65+ years) (%)	6.90	7.27	7.58	10.81	12.85	9.74	13.38
Kid (<5 years) (%)	10.34	9.98	8.97	7.85	8.08	8.42	7.85
Native-born (%)	98.88	98.66	95.66	90.44	93.44	93.12	95.60
Renter housing units (%)	12.93	12.94	20.98	35.10	22.67	27.79	21.03
College degree or higher (%)	17.31	19.96	28.03	35.98	36.59	32.31	24.69
Unemployment (%)	3.02	2.83	3.16	4.16	3.04	3.61	4.32
Poverty (%)	3.65	4.35	4.74	9.65	4.69	7.08	8.43
Mean household income ($)	76,311	82,793	86,842	81,397	89,604	83,920	68,262
Year 2010							
Total population	56,001	163,032	591,309	898,525	241,302	1,950,169	3,291,745
White	53,442	154,601	511,554	704,342	204,643	1,628,582	2,911,244
Black	322	2,718	32,957	96,531	10,693	143,221	113,944
Asian	248	2,726	26,447	47,382	13,747	90,550	115,787
Native American	310	550	2,322	6,891	1,748	11,821	43,549
Others	1,679	2,437	18,029	43,379	10,471	75,995	107,221
Hispanic	1,504	2,787	18,029	53,937	13,668	89,925	143,791
Color	3,430	10,491	90,337	226,762	44,376	375,396	469,677
White (%)	95.43	94.83	86.51	78.39	84.81	83.51	88.44
Black (%)	0.57	1.67	5.57	10.74	4.43	7.34	3.46
Asian (%)	0.44	1.67	4.47	5.27	5.70	4.64	3.52
Native American (%)	0.55	0.34	0.39	0.77	0.72	0.61	1.32
Others (%)	3.00	1.49	3.05	4.83	4.34	3.90	3.26
Hispanic (%)	2.69	1.71	3.05	6.00	5.66	4.61	4.37
Color (%)	6.12	6.43	15.28	25.24	18.39	19.25	14.27
Female (%)	49.17	49.35	50.35	50.44	51.81	50.46	50.32
Old (65+ years) (%)	7.32	8.45	9.48	10.91	12.93	10.42	13.82
Kid (<5 years) (%)	8.07	8.59	6.99	6.70	7.22	7.05	6.54
Native-born (%)	98.28	97.05	92.38	88.58	89.81	90.87	94.26

(continued)

Table B.26 (continued)

	0–10	11–20	21–30	31–40	41–50	0–50	Outside
Renter housing units (%)	13.58	12.56	20.95	34.26	21.69	26.72	21.13
College degree or higher (%)	22.50	25.62	31.34	40.69	41.11	36.27	28.58
Unemployment (%)	5.83	6.43	6.56	7.14	5.84	6.71	6.18
Poverty (%)	5.41	6.30	8.05	13.12	7.15	10.05	10.91
Mean household income ($)	78,611	81,582	84,060	80,013	88,509	82,333	68,448
Index	26	26	26	26	26	26	26

B.27 Prairie Island Nuclear Generating Plant, Minnesota

See Fig. B.27.
See Table B.27.

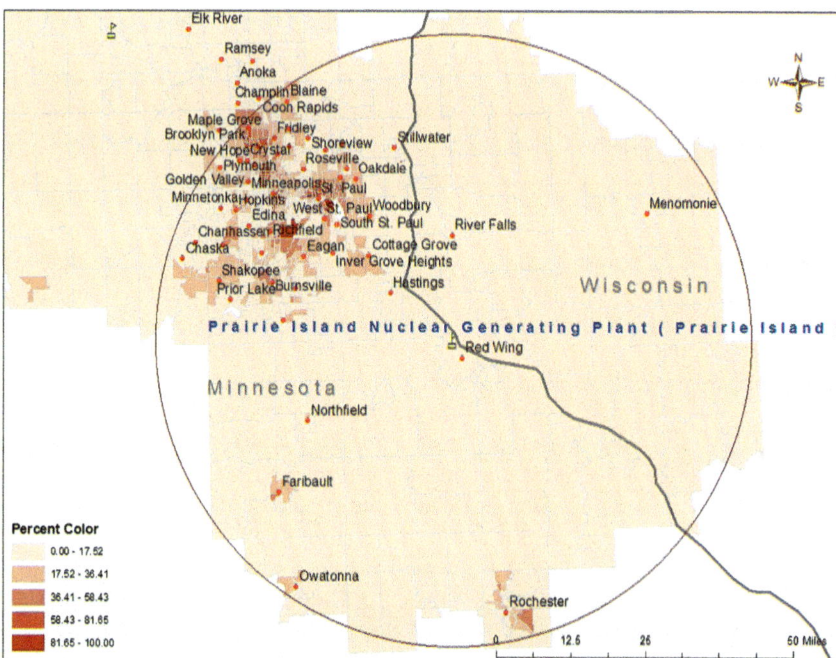

Fig. B.27 Spatial distribution of percent color among the populations surrounding Prairie Island Nuclear Generating Plant, Minnesota, in 2010

Table B.27 Demographic composition of population, as sorted by distance from Prairie Island Nuclear Generating Plant, Minnesota, in 1990, 2000, and 2010 *Source* (Kyne 2014)

	0–10	11–20	21–30	31–40	41–50	0–50	Outside
Year 1990							
Tract	5	22	107	260	56	450	2,297
Tract area (sq. mile)	226	1,126	1,319	2,217	756	5,643	146,789
Total population	22,549	81,536	306,309	877,923	177,005	1,465,322	7,801,560
White	22,086	80,351	295,746	772,878	170,339	1,341,400	7,306,105
Black	58	274	3,477	47,572	1,163	52,544	286,553
Asian	83	534	4,878	36,398	4,569	46,462	83,347
Native American	261	162	877	13,085	533	14,918	74,299
Others	61	221	1,325	7,978	399	9,984	51,206
Hispanic	165	544	4,003	18,221	1,505	24,438	112,824
Color	557	1,526	13,167	114,195	7,742	137,187	554,420
White (%)	97.95	98.55	96.55	88.03	96.23	91.54	93.65
Black (%)	0.26	0.34	1.14	5.42	0.66	3.59	3.67
Asian (%)	0.37	0.65	1.59	4.15	2.58	3.17	1.07
Native American (%)	1.16	0.20	0.29	1.49	0.30	1.02	0.95
Others (%)	0.27	0.27	0.43	0.91	0.23	0.68	0.66
Hispanic (%)	0.73	0.67	1.31	2.08	0.85	1.67	1.45
Color (%)	2.47	1.87	4.30	13.01	4.37	9.36	7.11
Female (%)	50.66	50.01	50.79	51.50	51.46	51.25	50.99
Old (65+ years) (%)	13.81	9.28	7.62	11.91	10.24	10.69	13.34
Kid (<5 years) (%)	8.85	9.84	10.75	9.37	9.93	9.74	8.99
Native-born (%)	99.28	98.84	97.90	94.95	96.42	96.02	97.74
Renter housing units (%)	24.76	21.14	22.60	35.53	23.94	30.80	26.84
Education (%)	13.88	18.20	25.73	27.04	30.90	26.58	18.38
Unemployment (%)	4.20	4.53	3.90	5.00	3.72	4.57	5.29
Poverty (%)	8.19	6.51	5.10	11.59	6.29	9.25	10.70
Mean household income ($)	57,921	67,105	74,654	62,450	75,699	66,560	58,426
Year 2000							
Total population	24,295	96,933	412,332	946,237	207,032	1,686,829	8,596,325
White	23,302	93,933	378,717	744,661	189,405	1,430,018	7,745,659
Black	189	645	9,394	73,590	4,404	88,222	379,990
Asian	87	717	10,698	60,980	7,255	79,737	145,886
Native American	563	360	1,681	10,828	689	14,121	90,108
Others	153	1,284	11,850	56,168	5,274	74,729	234,684
Hispanic	243	1,036	11,383	51,412	4,356	68,430	264,405
Color	1,141	3,574	39,255	220,614	19,805	284,389	970,444

(continued)

Table B.27 (continued)

	0–10	11–20	21–30	31–40	41–50	0–50	Outside
White (%)	95.91	96.91	91.85	78.70	91.49	84.78	90.10
Black (%)	0.78	0.67	2.28	7.78	2.13	5.23	4.42
Asian (%)	0.36	0.74	2.59	6.44	3.50	4.73	1.70
Native American (%)	2.32	0.37	0.41	1.14	0.33	0.84	1.05
Others (%)	0.63	1.32	2.87	5.94	2.55	4.43	2.73
Hispanic (%)	1.00	1.07	2.76	5.43	2.10	4.06	3.08
Color (%)	4.70	3.69	9.52	23.31	9.57	16.86	11.29
Female (%)	49.94	50.20	50.73	50.87	50.90	50.79	50.53
Old (65+ years) (%)	13.66	9.54	7.86	10.95	10.98	10.16	13.08
Kid (<5 years) (%)	7.17	7.99	9.41	8.13	8.25	8.44	7.77
Native-born (%)	98.80	98.41	95.55	89.64	93.35	92.17	96.25
Renter housing units (%)	21.93	18.39	18.94	34.58	21.17	28.27	25.59
College degree or higher (%)	19.01	23.68	34.26	32.57	36.37	32.75	23.28
Unemployment (%)	3.20	3.68	2.93	4.38	3.36	3.84	4.49
Poverty (%)	5.92	4.92	3.75	10.37	5.46	7.76	8.43
Mean household income ($)	70,523	79,048	90,123	71,653	85,950	78,059	68,873
Year 2010							
Total population	25,260	106,574	472,356	962,120	233,785	1,800,095	9,079,766
White	23,992	101,144	409,258	729,796	208,712	1,472,902	7,981,475
Black	433	1,329	19,120	96,949	8,253	126,084	479,743
Asian	94	1,590	22,719	74,248	10,663	109,314	221,018
Native American	428	325	1,723	9,440	659	12,575	91,984
Others	313	2,186	19,536	51,687	5,498	79,220	305,546
Hispanic	431	2,946	22,211	73,753	8,799	108,140	436,125
Color	1,518	7,437	75,956	275,734	31,748	392,393	1,355,202
White (%)	94.98	94.90	86.64	75.85	89.28	81.82	87.90
Black (%)	1.71	1.25	4.05	10.08	3.53	7.00	5.28
Asian (%)	0.37	1.49	4.81	7.72	4.56	6.07	2.43
Native American (%)	1.69	0.30	0.36	0.98	0.28	0.70	1.01
Others (%)	1.24	2.05	4.14	5.37	2.35	4.40	3.37
Hispanic (%)	1.71	2.76	4.70	7.67	3.76	6.01	4.80
Color (%)	6.01	6.98	16.08	28.66	13.58	21.80	14.93
Female (%)	51.14	50.26	50.74	50.57	50.99	50.66	50.34
Old (65+ years) (%)	15.68	10.71	9.34	10.85	12.35	10.71	13.44
Kid (<5 years) (%)	5.76	6.17	7.13	6.81	6.95	6.86	6.43
Native-born (%)	98.77	97.31	92.71	87.49	92.28	90.22	95.06

(continued)

Table B.27 (continued)

	0–10	11–20	21–30	31–40	41–50	0–50	Outside
Renter housing units (%)	17.84	17.08	19.44	33.74	19.84	27.23	24.63
College degree or higher (%)	21.96	27.51	39.29	36.87	40.08	37.16	26.81
Unemployment (%)	4.56	6.12	5.59	7.58	4.90	6.58	6.51
Poverty (%)	7.82	6.84	5.61	15.15	7.58	11.04	11.14
Mean household income ($)	67,433	77,993	89,852	70,134	85,429	77,474	67,642
Index	27	27	27	27	27	27	27

B.28 Grand Gulf Nuclear Station, Mississippi

See Fig. B.28.
See Table B.28.

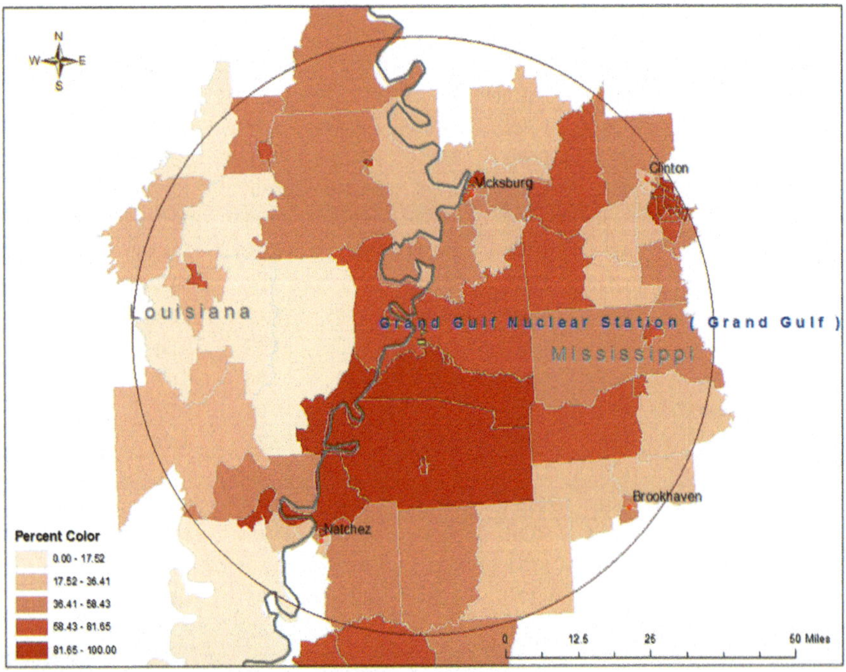

Fig. B.28 Spatial distribution of percent color among the populations surrounding Grand Gulf Nuclear Station, Mississippi, in 2010

Table B.28 Demographic composition of population, as sorted by distance from Grand Gulf Nuclear Station, Mississippi, in 1990, 2000, and 2010 *Source* (Kyne 2014)

	0–10	11–20	21–30	31–40	41–50	0–50	Outside
Year 1990							
Tract	4	4	14	25	9	56	1,756
Tract area (sq. mile)	662	1,150	763	2,442	1,059	6,076	94,733
Total population	13,756	15,718	48,711	96,720	39,711	214,616	6,578,571
White	3,256	6,666	25,633	50,723	24,195	110,473	4,362,429
Black	10,469	8,994	22,648	45,635	15,440	103,186	2,111,334
Asian	8	45	227	158	15	453	51,928
Native American	24	13	105	104	37	283	28,646
Others	0	0	95	103	26	224	24,232
Hispanic	72	30	224	529	204	1,059	104,288
Color	10,518	9,075	23,128	46,308	15,685	104,714	2,288,766
White (%)	23.67	42.41	52.62	52.44	60.93	51.47	66.31
Black (%)	76.10	57.22	46.49	47.18	38.88	48.08	32.09
Asian (%)	0.06	0.29	0.47	0.16	0.04	0.21	0.79
Native American (%)	0.17	0.08	0.22	0.11	0.09	0.13	0.44
Others (%)	0.00	0.00	0.20	0.11	0.07	0.10	0.37
Hispanic (%)	0.52	0.19	0.46	0.55	0.51	0.49	1.59
Color (%)	76.46	57.74	47.48	47.88	39.50	48.79	34.79
Female (%)	53.61	52.33	53.40	52.87	52.58	52.94	51.99
Old (65+ years) (%)	12.63	11.79	15.02	13.15	14.85	13.75	11.54
Kid (<5 years) (%)	9.02	10.13	8.99	8.86	9.51	9.11	9.47
Native-born (%)	99.32	99.15	99.31	99.66	99.79	99.55	98.38
Renter housing units (%)	21.23	19.14	27.89	22.99	21.65	23.49	28.43
Education (%)	16.33	9.97	17.11	13.12	9.99	13.42	15.68
Unemployment (%)	18.18	16.81	8.62	11.04	11.56	11.36	9.12
Poverty (%)	43.18	37.24	26.62	30.76	35.34	31.84	23.95
Mean household income ($)	34,522	35,052	44,723	40,505	34,894	39,738	47,036
Year 2000							
Total population	14,443	16,522	49,879	99,218	38,825	218,887	7,094,747
White	3,120	6,426	23,774	50,055	22,303	105,678	4,495,639
Black	11,183	9,801	25,198	47,955	15,904	110,041	2,367,962
Asian	12	106	223	347	246	934	74,323
Native American	7	75	60	187	164	493	37,176
Others	121	114	623	676	207	1,741	119,647
Hispanic	154	142	655	711	334	1,996	143,648
Color	11,338	10,138	26,371	49,479	16,768	114,094	2,676,584

(continued)

Table B.28 (continued)

	0–10	11–20	21–30	31–40	41–50	0–50	Outside
White (%)	21.60	38.89	47.66	50.45	57.44	48.28	63.37
Black (%)	77.43	59.32	50.52	48.33	40.96	50.27	33.38
Asian (%)	0.08	0.64	0.45	0.35	0.63	0.43	1.05
Native American (%)	0.05	0.45	0.12	0.19	0.42	0.23	0.52
Others (%)	0.84	0.69	1.25	0.68	0.53	0.80	1.69
Hispanic (%)	1.07	0.86	1.31	0.72	0.86	0.91	2.02
Color (%)	78.50	61.36	52.87	49.87	43.19	52.12	37.73
Female (%)	52.45	52.54	52.89	52.02	51.79	52.24	51.67
Old (65+ years) (%)	11.43	10.96	13.42	13.71	15.01	13.52	11.74
Kid (<5 years) (%)	7.75	9.73	8.61	8.14	8.50	8.41	8.57
Native-born (%)	99.10	99.04	99.04	99.22	99.04	99.12	97.83
Renter housing units (%)	17.67	19.85	27.73	23.14	21.98	23.37	27.38
College degree or higher (%)	18.53	12.78	18.40	14.98	10.50	14.99	18.11
Unemployment (%)	16.44	11.41	7.63	8.83	9.09	9.23	7.30
Poverty (%)	32.23	30.18	20.87	24.85	29.21	25.56	19.57
Mean household income ($)	43,808	41,487	53,859	48,053	40,263	47,253	54,864
Year 2010							
Total population	12,168	14,097	49,182	96,251	38,838	210,536	7,161,395
White	2,337	5,095	21,415	49,010	19,965	97,822	4,483,162
Black	9,732	8,840	26,662	46,050	18,424	109,708	2,384,247
Asian	13	47	417	308	50	835	93,118
Native American	7	0	41	189	115	352	40,610
Others	79	115	647	694	284	1,819	160,258
Hispanic	70	213	855	1,075	402	2,615	239,973
Color	9,855	9,116	28,272	47,840	19,273	114,356	2,824,099
White (%)	19.21	36.14	43.54	50.92	51.41	46.46	62.60
Black (%)	79.98	62.71	54.21	47.84	47.44	52.11	33.29
Asian (%)	0.11	0.33	0.85	0.32	0.13	0.40	1.30
Native American (%)	0.06	0.00	0.08	0.20	0.30	0.17	0.57
Others (%)	0.65	0.82	1.32	0.72	0.73	0.86	2.24
Hispanic (%)	0.58	1.51	1.74	1.12	1.04	1.24	3.35
Color (%)	80.99	64.67	57.48	49.70	49.62	54.32	39.44
Female (%)	51.97	51.78	51.38	51.57	48.87	51.07	51.26
Old (65+ years) (%)	12.91	12.72	13.18	14.68	13.85	13.95	12.25
Kid (<5 years) (%)	5.81	6.36	6.84	6.50	7.21	6.66	6.97

(continued)

Table B.28 (continued)

	0–10	11–20	21–30	31–40	41–50	0–50	Outside
Native-born (%)	99.81	99.16	98.42	98.83	99.72	98.98	96.94
Renter housing units (%)	18.07	23.22	29.74	23.24	25.42	24.82	26.48
College degree or higher (%)	16.38	17.28	21.43	16.74	11.63	16.94	20.45
Unemployment (%)	10.22	10.85	9.73	10.95	14.90	11.28	8.39
Poverty (%)	35.13	30.44	23.21	24.33	30.24	26.14	19.18
Mean household income ($)	34,624	41,988	49,130	49,040	47,656	47,561	57,294
Index	28	28	28	28	28	28	28

B.29 Callaway Plant, Missouri

See Fig. B.29.
See Table B.29.

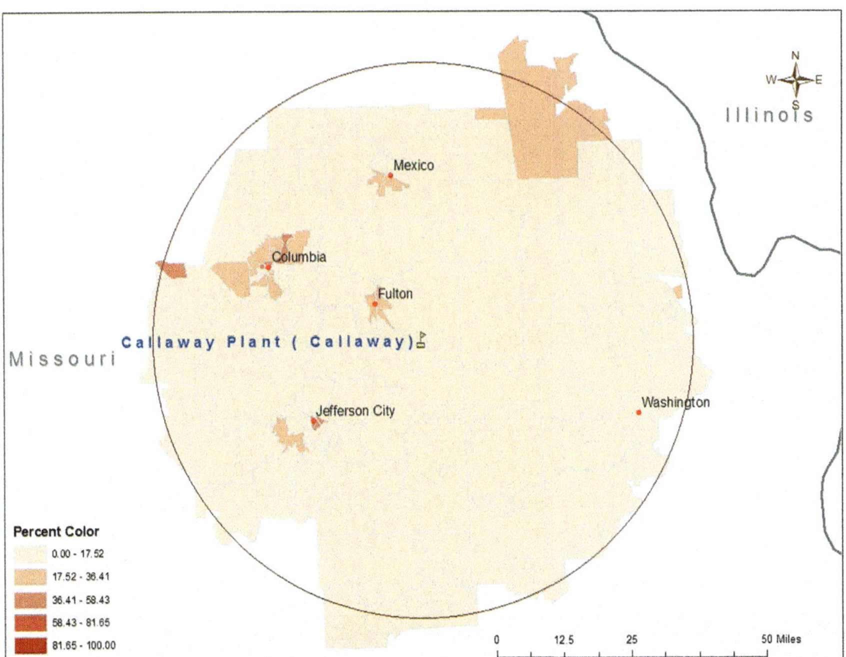

Fig. B.29 Spatial distribution of percent color among the populations surrounding Callaway Plant, Missouri, in 2010

Table B.29 Demographic composition of population, as sorted by distance from Callaway Plant, Missouri, in 1990, 2000, and 2010 *Source* (Kyne 2014)

	0–10	11–20	21–30	31–40	41–50	0–50	Outside
Year 1990							
Tract	2	9	26	39	5	81	1,312
Tract area (sq. mile)	237	1,093	1,412	2,286	724	5,752	63,955
Total population	8,124	32,061	101,712	148,640	16,808	307,345	4,809,724
White	7,439	30,918	94,401	136,270	16,279	285,307	4,201,530
Black	612	910	6,188	8,558	410	16,678	530,170
Asian	33	68	552	2,966	50	3,669	35,912
Native American	31	113	360	446	22	972	21,349
Others	6	59	209	401	45	720	20,729
Hispanic	67	138	803	1,259	77	2,344	58,089
Color	746	1,231	7,858	13,200	574	23,609	643,604
White (%)	91.57	96.43	92.81	91.68	96.85	92.83	87.35
Black (%)	7.53	2.84	6.08	5.76	2.44	5.43	11.02
Asian (%)	0.41	0.21	0.54	2.00	0.30	1.19	0.75
Native American (%)	0.38	0.35	0.35	0.30	0.13	0.32	0.44
Others (%)	0.07	0.18	0.21	0.27	0.27	0.23	0.43
Hispanic (%)	0.82	0.43	0.79	0.85	0.46	0.76	1.21
Color (%)	9.18	3.84	7.73	8.88	3.42	7.68	13.38
Female (%)	49.80	51.20	50.43	50.88	50.40	50.71	51.91
Old (65+ years) (%)	17.68	14.44	13.14	11.06	15.97	12.55	14.12
Kid (<5 years) (%)	7.40	8.97	8.23	8.44	8.64	8.41	8.67
Native-born (%)	98.89	99.45	98.72	97.05	99.20	98.02	98.39
Renter housing units (%)	25.95	16.96	29.42	32.79	12.97	28.63	27.80
Education (%)	11.90	12.72	22.83	24.24	7.51	21.20	17.59
Unemployment (%)	5.00	3.76	4.82	4.50	6.66	4.65	6.26
Poverty (%)	11.68	9.69	10.61	14.59	14.00	12.63	13.38
Mean household income ($)	44,564	50,121	56,184	51,029	43,086	52,107	55,064
Year 2000							
Total population	10,726	37,518	118,776	171,258	19,448	357,726	5,237,485
White	9,210	35,677	106,388	152,475	18,223	321,973	4,424,979
Black	1,219	969	8,427	11,200	737	22,552	599,535
Asian	125	92	1,256	3,584	10	5,067	58,433
Native American	25	200	400	700	91	1,416	24,784
Others	144	585	2,306	3,299	384	6,718	129,754
Hispanic	73	284	1,467	2,364	127	4,315	112,058
Color	1,550	1,994	13,199	20,214	1,344	38,301	869,073

(continued)

Table B.29 (continued)

	0–10	11–20	21–30	31–40	41–50	0–50	Outside
White (%)	85.87	95.09	89.57	89.03	93.70	90.01	84.49
Black (%)	11.36	2.58	7.09	6.54	3.79	6.30	11.45
Asian (%)	1.17	0.25	1.06	2.09	0.05	1.42	1.12
Native American (%)	0.23	0.53	0.34	0.41	0.47	0.40	0.47
Others (%)	1.34	1.56	1.94	1.93	1.97	1.88	2.48
Hispanic (%)	0.68	0.76	1.24	1.38	0.65	1.21	2.14
Color (%)	14.45	5.31	11.11	11.80	6.91	10.71	16.59
Female (%)	40.42	51.01	50.11	50.89	53.41	50.47	51.48
Old (65+ years) (%)	13.48	12.97	12.16	10.60	14.57	11.67	13.64
Kid (<5 years) (%)	4.61	7.85	7.48	7.84	6.55	7.56	7.97
Native-born (%)	98.83	99.15	97.90	96.56	99.20	97.49	97.28
Renter housing units (%)	26.33	17.13	30.02	31.51	13.21	28.32	26.61
College degree or higher (%)	13.57	15.21	27.73	29.39	9.14	25.59	21.32
Unemployment (%)	3.91	4.08	4.24	4.73	3.87	4.44	5.36
Poverty (%)	8.77	8.00	11.03	12.64	10.50	11.40	11.76
Mean household income ($)	50,488	58,609	62,757	59,772	51,768	60,034	62,384
Year 2010							
Total population	11,642	39,493	129,790	191,791	20,201	392,917	5,529,397
White	10,081	37,582	112,854	166,198	19,103	345,818	4,592,404
Black	899	945	10,775	13,007	582	26,208	652,854
Asian	120	254	1,772	5,220	6	7,372	89,976
Native American	38	45	381	754	25	1,243	21,586
Others	504	667	4,008	6,612	485	12,276	172,577
Hispanic	333	541	3,417	4,809	308	9,408	189,262
Color	1,728	2,394	18,442	27,658	1,333	51,555	1,051,943
White (%)	86.59	95.16	86.95	86.66	94.56	88.01	83.05
Black (%)	7.72	2.39	8.30	6.78	2.88	6.67	11.81
Asian (%)	1.03	0.64	1.37	2.72	0.03	1.88	1.63
Native American (%)	0.33	0.11	0.29	0.39	0.12	0.32	0.39
Others (%)	4.33	1.69	3.09	3.45	2.40	3.12	3.12
Hispanic (%)	2.86	1.37	2.63	2.51	1.52	2.39	3.42
Color (%)	14.84	6.06	14.21	14.42	6.60	13.12	19.02
Female (%)	43.81	50.72	50.70	50.97	53.54	50.78	51.13
Old (65+ years) (%)	11.40	13.54	12.62	11.04	15.68	12.06	13.84
Kid (<5 years) (%)	4.22	6.42	6.59	6.16	6.72	6.30	6.55
Native-born (%)	99.30	98.89	96.92	95.34	99.70	96.56	96.33

(continued)

Table B.29 (continued)

	0–10	11–20	21–30	31–40	41–50	0–50	Outside
Renter housing units (%)	29.76	17.43	30.41	31.18	12.55	28.49	26.07
College degree or higher (%)	15.86	19.76	30.90	33.30	10.69	29.22	24.77
Unemployment (%)	6.11	6.54	5.05	6.10	7.20	5.85	7.56
Poverty (%)	16.14	9.70	14.30	15.34	12.41	14.29	13.95
Mean household income ($)	57,079	58,081	61,274	60,457	50,873	59,960	61,327
Index	29	29	29	29	29	29	29

B.30 Cooper Nuclear Station, Nebraska

See Fig. B.30.
See Table B.30.

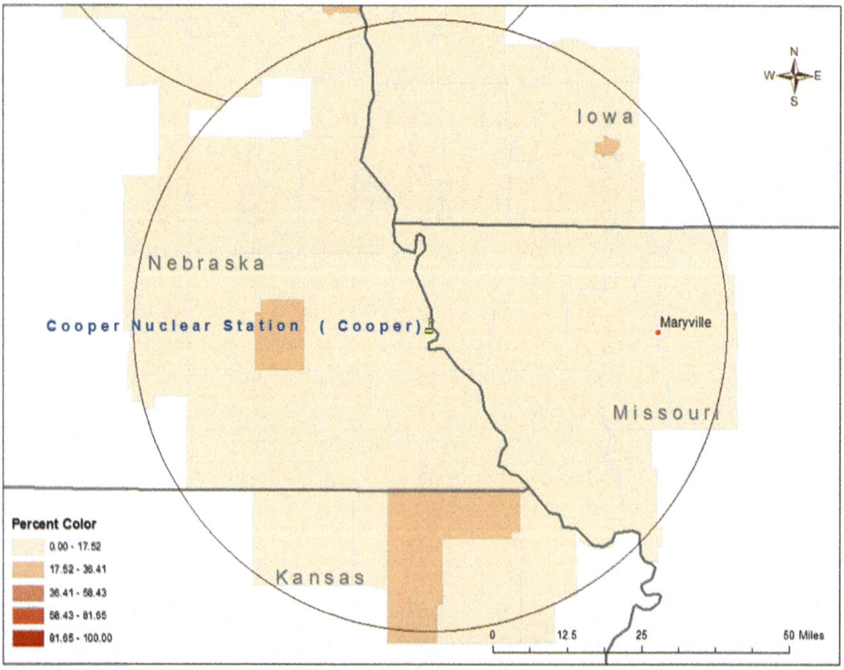

Fig. B.30 Spatial distribution of percent color among the populations surrounding Cooper Nuclear Station, Nebraska, in 2010

Table B.30 Demographic composition of population, as sorted by distance from Cooper Nuclear Station, Nebraska, in 1990, 2000, and 2010 *Source* (Kyne 2014)

	0–10	11–20	21–30	31–40	41–50	0–50	Outside
Year 1990							
Tract	3	2	10	14	9	38	3,482
Tract area (sq. mile)	670	554	1,395	2,259	1,290	6,168	279,438
Total population	11,078	6,798	28,334	40,875	30,314	117,399	11,832,396
White	10,993	6,521	28,042	39,923	29,763	115,242	10,771,594
Black	36	121	8	191	107	463	792,724
Asian	4	23	94	87	67	275	107,311
Native American	45	91	100	607	285	1,128	64,822
Others	0	42	90	66	91	289	95,907
Hispanic	26	47	281	341	187	882	215,579
Color	111	291	483	1,198	637	2,720	1,174,239
White (%)	99.23	95.93	98.97	97.67	98.18	98.16	91.03
Black (%)	0.32	1.78	0.03	0.47	0.35	0.39	6.70
Asian (%)	0.04	0.34	0.33	0.21	0.22	0.23	0.91
Native American (%)	0.41	1.34	0.35	1.49	0.94	0.96	0.55
Others (%)	0.00	0.62	0.32	0.16	0.30	0.25	0.81
Hispanic (%)	0.23	0.69	0.99	0.83	0.62	0.75	1.82
Color (%)	1.00	4.28	1.70	2.93	2.10	2.32	9.92
Female (%)	50.41	51.21	52.66	51.81	50.12	51.41	51.52
Old (65+ years) (%)	20.05	18.05	22.90	22.09	16.21	20.34	14.24
Kid (<5 years) (%)	7.10	8.68	7.85	7.71	8.95	8.06	8.78
Native-born (%)	99.58	99.35	99.34	99.31	99.15	99.30	98.16
Renter housing units (%)	28.30	28.07	25.15	23.16	23.77	24.59	28.50
Education (%)	16.66	14.17	11.89	12.85	12.15	12.87	18.46
Unemployment (%)	2.82	3.46	5.06	3.72	3.87	3.98	5.15
Poverty (%)	13.18	18.28	12.73	14.12	12.91	13.62	12.22
Mean household income ($)	45,263	39,940	42,655	43,994	45,163	43,834	54,371
Year 2000							
Total population	10,205	6,117	27,836	40,863	30,847	115,868	12,805,348
White	10,040	5,854	26,739	39,308	29,413	111,354	11,231,241
Black	24	138	46	236	319	763	899,101
Asian	58	4	248	103	194	607	166,650
Native American	27	50	220	730	546	1,573	74,034
Others	56	71	583	486	373	1,569	434,324
Hispanic	50	25	578	408	566	1,627	476,418
Color	201	273	1,399	1,792	1,733	5,398	1,784,914

(continued)

Table B.30 (continued)

	0–10	11–20	21–30	31–40	41–50	0–50	Outside
White (%)	98.38	95.70	96.06	96.19	95.35	96.10	87.71
Black (%)	0.24	2.26	0.17	0.58	1.03	0.66	7.02
Asian (%)	0.57	0.07	0.89	0.25	0.63	0.52	1.30
Native American (%)	0.26	0.82	0.79	1.79	1.77	1.36	0.58
Others (%)	0.55	1.16	2.09	1.19	1.21	1.35	3.39
Hispanic (%)	0.49	0.41	2.08	1.00	1.83	1.40	3.72
Color (%)	1.97	4.46	5.03	4.39	5.62	4.66	13.94
Female (%)	51.54	49.84	51.53	51.59	49.04	50.80	51.06
Old (65+ years) (%)	19.63	18.54	20.37	20.71	16.17	19.21	13.73
Kid (<5 years) (%)	5.47	5.72	7.01	7.04	7.26	6.88	8.04
Native-born (%)	99.33	99.79	97.93	99.19	98.84	98.84	96.48
Renter housing units (%)	25.06	24.13	24.08	21.84	23.24	23.17	27.28
College degree or higher (%)	20.44	18.28	14.41	17.15	15.44	16.39	22.70
Unemployment (%)	5.61	3.81	4.27	3.47	5.63	4.44	4.58
Poverty (%)	11.85	12.37	10.58	9.70	9.73	10.25	10.50
Mean household income ($)	51,308	46,785	50,783	52,927	54,282	52,257	62,562
Year 2010							
Total population	9,770	5,155	27,379	38,925	29,884	111,113	13,435,922
White	9,501	4,976	26,004	36,883	28,213	105,577	11,585,698
Black	63	1	352	169	648	1,233	1,000,265
Asian	17	5	86	231	86	425	246,824
Native American	56	53	113	790	313	1,325	71,402
Others	133	120	824	852	624	2,553	531,733
Hispanic	114	37	1,430	738	598	2,917	760,764
Color	342	195	2,460	2,343	2,146	7,486	2,308,154
White (%)	97.25	96.53	94.98	94.75	94.41	95.02	86.23
Black (%)	0.64	0.02	1.29	0.43	2.17	1.11	7.44
Asian (%)	0.17	0.10	0.31	0.59	0.29	0.38	1.84
Native American (%)	0.57	1.03	0.41	2.03	1.05	1.19	0.53
Others (%)	1.36	2.33	3.01	2.19	2.09	2.30	3.96
Hispanic (%)	1.17	0.72	5.22	1.90	2.00	2.63	5.66
Color (%)	3.50	3.78	8.98	6.02	7.18	6.74	17.18
Female (%)	51.28	50.79	49.32	52.35	47.44	50.12	50.80
Old (65+ years) (%)	18.54	19.96	20.09	20.74	15.93	19.06	13.75
Kid (<5 years) (%)	5.94	5.24	5.95	5.72	6.36	5.95	6.75

(continued)

Table B.30 (continued)

	0–10	11–20	21–30	31–40	41–50	0–50	Outside
Native-born (%)	98.88	99.17	97.13	98.63	98.68	98.32	95.36
Renter housing units (%)	24.19	20.13	22.92	21.22	20.22	21.63	26.45
College degree or higher (%)	22.21	19.79	16.22	18.04	17.92	17.98	26.22
Unemployment (%)	5.28	6.84	5.08	4.60	4.63	4.90	6.33
Poverty (%)	13.21	12.57	13.80	11.43	9.83	11.80	12.85
Mean household income ($)	54,322	52,507	49,662	52,796	58,011	53,419	62,412
Index	30	30	30	30	30	30	30

Fort Calhoun Station, Nebraska

See Fig. B.31.
See Table B.31.

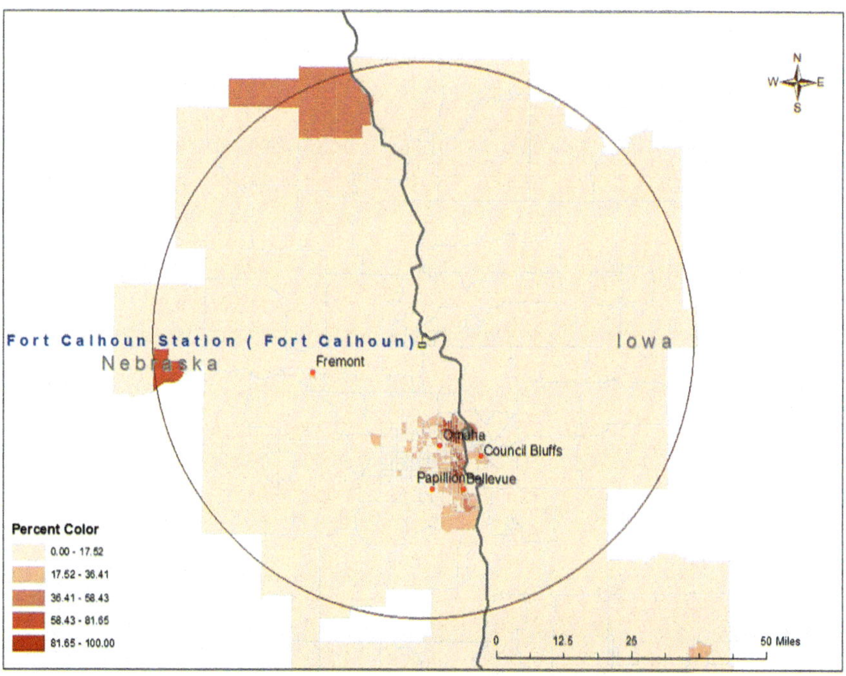

Fig. B.31 Spatial distribution of percent color among the populations surrounding Fort Calhoun Station, Nebraska, in 2010

Table B.31 Demographic composition of population, as sorted by distance from Fort Calhoun Station, Nebraska, in 1990, 2000, and 2010 *Source* (Kyne 2014)

	0–10	11–20	21–30	31–40	41–50	0–50	Outside
Year 1990							
Tract	5	91	142	24	7	269	1,088
Tract area (sq. mile)	173	1,121	1,468	1,809	1,015	5,587	128,033
Total population	15,687	243,551	385,024	67,593	19,265	731,120	3,624,019
White	15,520	197,208	365,490	65,201	19,059	662,478	3,503,625
Black	57	40,882	8,879	1,314	14	51,146	53,236
Asian	14	2,585	4,064	551	77	7,291	29,598
Native American	50	1,411	1,855	281	66	3,663	16,713
Others	46	1,469	4,735	239	49	6,538	20,853
Hispanic	101	3,404	11,444	772	77	15,798	49,936
Color	203	48,090	25,988	2,847	233	77,361	148,684
White (%)	98.94	80.97	94.93	96.46	98.93	90.61	96.68
Black (%)	0.36	16.79	2.31	1.94	0.07	7.00	1.47
Asian (%)	0.09	1.06	1.06	0.82	0.40	1.00	0.82
Native American (%)	0.32	0.58	0.48	0.42	0.34	0.50	0.46
Others (%)	0.29	0.60	1.23	0.35	0.25	0.89	0.58
Hispanic (%)	0.64	1.40	2.97	1.14	0.40	2.16	1.38
Color (%)	1.29	19.75	6.75	4.21	1.21	10.58	4.10
Female (%)	52.09	52.19	51.25	50.24	49.94	51.45	51.46
Old (65+ years) (%)	14.94	11.82	10.76	14.22	14.82	11.63	15.56
Kid (<5 years) (%)	7.93	9.77	9.53	9.20	9.15	9.53	8.52
Native-born (%)	99.18	98.00	97.56	98.55	99.16	97.87	98.45
Renter housing units (%)	23.19	33.51	33.46	25.47	18.76	32.13	28.23
Education (%)	15.67	24.03	20.98	16.33	11.87	21.20	16.89
Unemployment (%)	3.45	4.78	3.80	3.61	4.57	4.13	4.23
Poverty (%)	7.06	12.19	8.33	7.98	11.99	9.65	11.71
Mean household income ($)	57,733	60,822	58,746	51,719	47,843	58,519	51,287
Year 2000							
Total population	17,820	273,372	428,422	73,370	20,776	813,760	3,823,827
White	17,428	212,241	385,082	69,991	20,304	705,046	3,578,478
Black	66	45,594	11,847	1,090	15	58,612	68,581
Asian	77	4,722	5,996	477	76	11,348	46,429
Native American	38	1,945	2,326	310	107	4,726	19,958
Others	211	8,868	23,181	1,488	274	34,022	110,387
Hispanic	215	8,080	31,824	1,619	429	42,167	133,206
Color	451	64,820	56,609	4,233	724	126,837	302,171

(continued)

Table B.31 (continued)

	0–10	11–20	21–30	31–40	41–50	0–50	Outside
White (%)	97.80	77.64	89.88	95.39	97.73	86.64	93.58
Black (%)	0.37	16.68	2.77	1.49	0.07	7.20	1.79
Asian (%)	0.43	1.73	1.40	0.65	0.37	1.39	1.21
Native American (%)	0.21	0.71	0.54	0.42	0.52	0.58	0.52
Others (%)	1.18	3.24	5.41	2.03	1.32	4.18	2.89
Hispanic (%)	1.21	2.96	7.43	2.21	2.06	5.18	3.48
Color (%)	2.53	23.71	13.21	5.77	3.48	15.59	7.90
Female (%)	51.27	51.74	50.59	50.30	49.71	50.94	50.88
Old (65+ years) (%)	14.14	11.27	11.04	14.03	14.54	11.54	15.03
Kid (<5 years) (%)	8.26	8.82	8.82	7.81	7.57	8.68	7.76
Native-born (%)	98.23	96.35	94.38	97.90	98.89	95.56	96.61
Renter housing units (%)	23.07	33.28	32.30	22.68	17.59	31.18	26.57
College degree or higher (%)	21.39	29.81	25.81	20.84	14.79	26.29	21.26
Unemployment (%)	3.44	4.42	3.38	2.92	3.62	3.69	4.00
Poverty (%)	6.08	10.53	7.60	5.89	7.76	8.40	9.55
Mean household income ($)	66,484	70,483	69,512	64,129	59,642	69,052	59,757
Year 2010							
Total population	18,933	298,263	474,429	78,563	20,775	890,963	3,924,429
White	18,330	224,840	418,617	73,280	20,297	755,364	3,605,903
Black	100	48,836	13,921	1,680	74	64,611	96,057
Asian	79	7,326	8,489	771	67	16,732	65,425
Native American	8	1,343	2,108	515	10	3,984	21,492
Others	416	15,918	31,294	2,317	327	50,272	135,552
Hispanic	464	14,601	55,721	3,011	461	74,258	214,645
Color	824	81,377	87,845	7,015	855	177,916	445,849
White (%)	96.82	75.38	88.24	93.28	97.70	84.78	91.88
Black (%)	0.53	16.37	2.93	2.14	0.36	7.25	2.45
Asian (%)	0.42	2.46	1.79	0.98	0.32	1.88	1.67
Native American (%)	0.04	0.45	0.44	0.66	0.05	0.45	0.55
Others (%)	2.20	5.34	6.60	2.95	1.57	5.64	3.45
Hispanic (%)	2.45	4.90	11.74	3.83	2.22	8.33	5.47
Color (%)	4.35	27.28	18.52	8.93	4.12	19.97	11.36
Female (%)	50.77	51.68	50.42	49.39	50.43	50.76	50.55
Old (65+ years) (%)	13.95	11.02	11.06	14.57	15.10	11.51	14.90
Kid (<5 years) (%)	6.68	7.48	7.81	6.84	6.44	7.56	6.63

(continued)

Table B.31 (continued)

	0–10	11–20	21–30	31–40	41–50	0–50	Outside
Native-born (%)	98.66	94.66	92.13	97.17	98.48	93.71	95.54
Renter housing units (%)	19.96	31.75	30.01	20.20	16.82	29.19	25.29
College degree or higher (%)	26.38	34.75	30.89	25.69	20.46	31.34	24.42
Unemployment (%)	3.89	6.53	5.72	4.34	4.77	5.82	5.07
Poverty (%)	5.66	13.83	10.39	7.25	7.61	11.09	11.82
Mean household income ($)	69,650	69,389	68,860	69,785	65,129	69,048	60,642
Index	31	31	31	31	31	31	31

B.32 Seabrook Station, New Hampshire

See Fig. B.32.
See Table B.32.

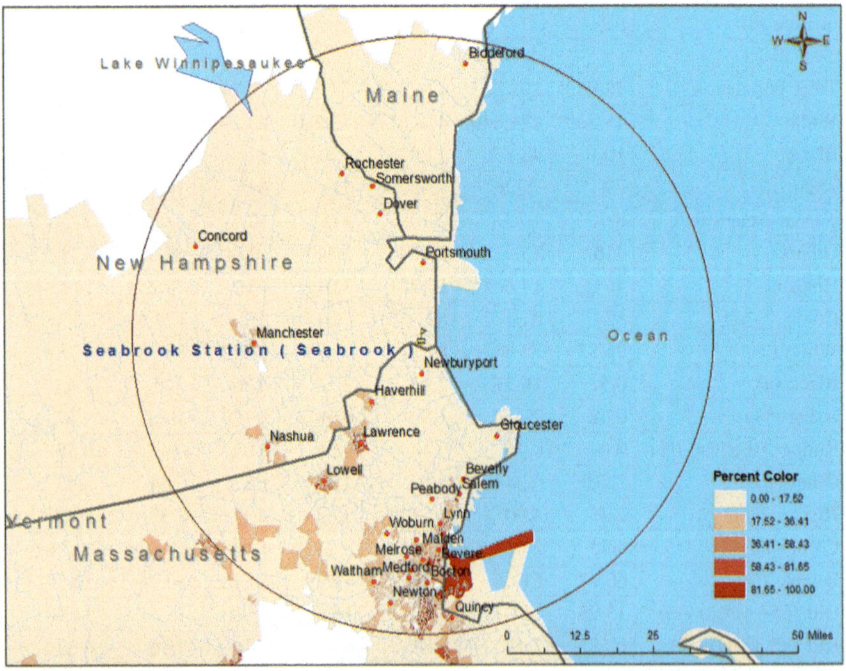

Fig. B.32 Spatial distribution of percent color among the populations surrounding Seabrook Station, New Hampshire, in 2010

Table B.32 Demographic composition of population, as sorted by distance from Seabrook Station, New Hampshire, in 1990, 2000, and 2010 *Source* (Kyne 2014)

	0–10	11–20	21–30	31–40	41–50	0–50	Outside
Year 1990							
Tract	30	78	180	271	309	868	1,263
Tract area (sq. mile)	270	650	1,185	1,256	835	4,195	51,088
Total population	101,342	310,511	755,938	1,146,601	1,151,223	3,465,615	4,887,988
White	100,218	296,305	703,646	1,067,227	891,135	3,058,531	4,648,230
Black	336	4,414	9,885	28,791	164,244	207,670	101,835
Asian	410	2,935	17,657	30,075	56,099	107,176	49,462
Native American	180	635	1,688	2,037	2,630	7,170	14,233
Others	189	6,229	23,044	18,449	37,142	85,053	74,211
Hispanic	522	12,218	42,624	41,738	69,970	167,072	127,409
Color	1,564	18,932	67,906	100,792	288,926	478,120	301,834
White (%)	98.89	95.42	93.08	93.08	77.41	88.25	95.09
Black (%)	0.33	1.42	1.31	2.51	14.27	5.99	2.08
Asian (%)	0.40	0.95	2.34	2.62	4.87	3.09	1.01
Native American (%)	0.18	0.20	0.22	0.18	0.23	0.21	0.29
Others (%)	0.19	2.01	3.05	1.61	3.23	2.45	1.52
Hispanic (%)	0.52	3.93	5.64	3.64	6.08	4.82	2.61
Color (%)	1.54	6.10	8.98	8.79	25.10	13.80	6.18
Female (%)	51.54	51.11	51.61	51.97	52.50	51.98	51.65
Old (65+ years) (%)	12.60	11.43	11.89	13.44	12.86	12.71	13.66
Kid (<5 years) (%)	8.52	9.20	9.10	8.13	6.95	8.06	8.53
Native-born (%)	97.14	94.89	92.15	90.00	83.87	89.08	94.42
Renter housing units (%)	25.23	29.89	34.54	39.42	51.12	41.14	27.93
Education (%)	27.42	26.24	23.87	27.40	36.97	29.73	22.67
Unemployment (%)	6.53	6.33	7.05	6.42	6.67	6.63	6.64
Poverty (%)	5.22	6.51	8.73	7.54	12.17	9.15	8.68
Mean household income ($)	76,264	77,196	74,512	76,767	78,690	76,972	66,827
Year 2000							
Total population	117,140	343,887	816,708	1,222,632	1,183,397	3,683,764	5,176,042
White	114,111	318,611	726,005	1,046,964	837,270	3,042,961	4,745,048
Black	547	3,407	12,745	42,309	164,634	223,642	128,546
Asian	714	5,064	25,765	59,376	91,949	182,868	81,452
Native American	114	802	1,683	2,728	3,734	9,061	16,425
Others	1,641	16,016	50,488	71,230	85,848	225,223	204,580
Hispanic	1,173	19,973	64,582	84,956	90,951	261,635	194,841
Color	3,676	31,844	113,423	213,537	380,831	743,311	513,643
White (%)	97.41	92.65	88.89	85.63	70.75	82.60	91.67
Black (%)	0.47	0.99	1.56	3.46	13.91	6.07	2.48
Asian (%)	0.61	1.47	3.15	4.86	7.77	4.96	1.57

(continued)

Table B.32 (continued)

	0–10	11–20	21–30	31–40	41–50	0–50	Outside
Native American (%)	0.10	0.23	0.21	0.22	0.32	0.25	0.32
Others (%)	1.40	4.66	6.18	5.83	7.25	6.11	3.95
Hispanic (%)	1.00	5.81	7.91	6.95	7.69	7.10	3.76
Color (%)	3.14	9.26	13.89	17.47	32.18	20.18	9.92
Female (%)	51.63	51.48	51.42	51.50	52.23	51.72	51.54
Old (65+ years) (%)	12.95	11.78	12.06	13.43	12.32	12.60	14.05
Kid (<5 years) (%)	7.55	7.86	8.19	7.50	6.38	7.33	7.42
Native-born (%)	96.51	93.65	90.45	85.08	79.70	85.70	93.48
Renter housing units (%)	24.22	27.98	33.65	39.02	51.35	40.52	26.60
College degree or higher (%)	34.62	32.39	28.85	33.10	44.35	35.75	27.72
Unemployment (%)	2.72	3.33	4.20	3.84	5.31	4.32	4.59
Poverty (%)	5.03	6.79	8.20	8.18	12.80	9.40	9.02
Mean household income ($)	88,305	86,147	81,831	84,032	89,198	85,616	72,897
Year 2010							
Total population	120,876	362,044	847,854	1,247,097	1,189,344	3,767,215	5,351,485
White	117,238	322,258	722,244	1,017,788	809,101	2,988,629	4,810,793
Black	648	4,687	18,987	60,905	173,556	258,783	191,609
Asian	1,300	8,066	37,582	85,282	120,580	252,810	127,288
Native American	168	601	1,059	1,834	3,280	6,942	15,479
Others	1,522	26,432	67,982	81,288	82,827	260,051	206,316
Hispanic	1,724	31,377	87,065	128,598	109,397	358,161	279,195
Color	4,811	47,970	154,265	297,323	424,698	929,067	699,721
White (%)	96.99	89.01	85.18	81.61	68.03	79.33	89.90
Black (%)	0.54	1.29	2.24	4.88	14.59	6.87	3.58
Asian (%)	1.08	2.23	4.43	6.84	10.14	6.71	2.38
Native American (%)	0.14	0.17	0.12	0.15	0.28	0.18	0.29
Others (%)	1.26	7.30	8.02	6.52	6.96	6.90	3.86
Hispanic (%)	1.43	8.67	10.27	10.31	9.20	9.51	5.22
Color (%)	3.98	13.25	18.19	23.84	35.71	24.66	13.08
Female (%)	51.61	51.32	51.29	51.44	52.13	51.62	51.27
Old (65+ years) (%)	14.87	12.10	12.67	13.31	12.02	12.69	14.42
Kid (<5 years) (%)	5.10	5.68	6.24	6.01	5.03	5.69	5.53
Native-born (%)	95.50	92.00	88.50	81.16	77.79	83.25	92.07

(continued)

Table B.32 (continued)

	0–10	11–20	21–30	31–40	41–50	0–50	Outside
Renter housing units (%)	21.22	24.51	29.55	36.23	47.37	36.85	24.00
College degree or higher (%)	41.35	37.48	34.01	37.87	50.28	40.93	32.13
Unemployment (%)	5.19	6.10	6.70	6.94	7.32	6.87	7.12
Poverty (%)	5.02	7.26	8.89	9.48	14.91	10.65	10.28
Mean household income ($)	93,745	89,716	83,699	86,038	93,567	88,585	75,537
Index	32	32	32	32	32	32	32

B.33 Hope Creek Generating Station, New Jersey

See Fig. B.33.
See Table B.33.

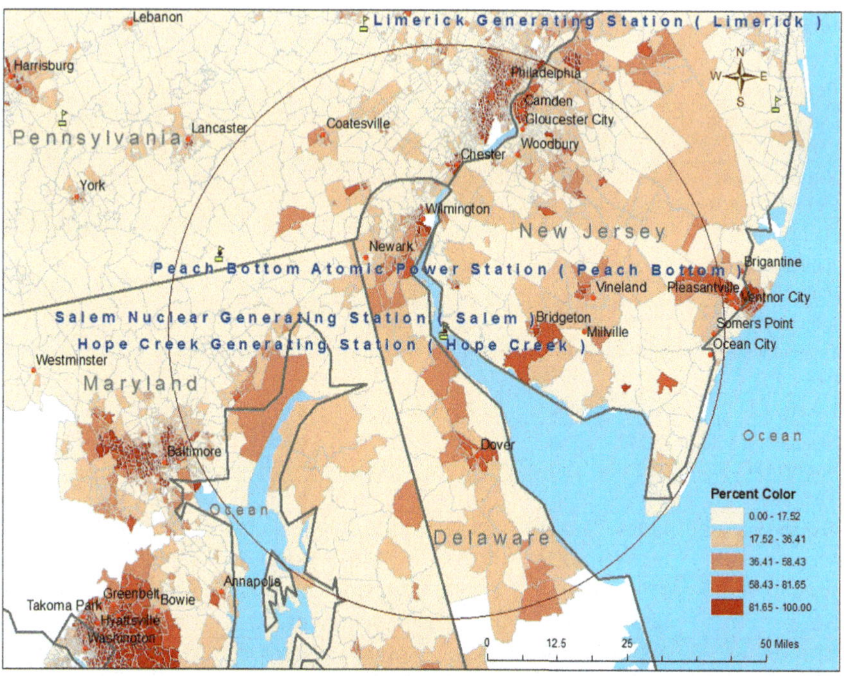

Fig. B.33 Spatial distribution of percent color among the populations surrounding Hope Creek Generating Station, New Jersey, in 2010

Table B.33 Demographic composition of population, as sorted by distance from Hope Creek Generating Station, New Jersey, in 1990, 2000, and 2010 *Source* (Kyne 2014)

	0–10	11–20	21–30	31–40	41–50	0–50	Outside
Year 1990							
Tract	15	99	188	398	397	1,097	5,755
Tract area (sq. mile)	332	848	1,870	1,653	982	5,684	63,987
Total population	34,738	325,339	650,001	1,509,343	1,533,210	4,052,631	21,006,748
White	26,049	267,177	518,830	1,150,923	916,694	2,879,673	17,710,736
Black	7,987	48,281	106,188	313,228	501,262	976,946	2,447,042
Asian	84	4,030	7,329	32,446	36,439	80,328	470,992
Native American	195	1,590	1,675	2,893	3,206	9,559	37,510
Others	436	4,250	15,972	9,857	75,608	106,123	340,396
Hispanic	745	8,137	29,376	26,399	104,396	169,053	906,894
Color	8,919	61,688	143,035	372,107	639,329	1,225,078	3,809,682
White (%)	74.99	82.12	79.82	76.25	59.79	71.06	84.31
Black (%)	22.99	14.84	16.34	20.75	32.69	24.11	11.65
Asian (%)	0.24	1.24	1.13	2.15	2.38	1.98	2.24
Native American (%)	0.56	0.49	0.26	0.19	0.21	0.24	0.18
Others (%)	1.26	1.31	2.46	0.65	4.93	2.62	1.62
Hispanic (%)	2.14	2.50	4.52	1.75	6.81	4.17	4.32
Color (%)	25.68	18.96	22.01	24.65	41.70	30.23	18.14
Female (%)	49.36	51.39	51.77	51.98	53.18	52.33	51.74
Old (65+ years) (%)	11.86	10.72	12.32	13.60	13.69	13.18	13.91
Kid (<5 years) (%)	8.96	8.88	8.91	8.54	9.02	8.81	8.14
Native-born (%)	98.26	96.87	96.30	95.35	94.63	95.38	92.93
Renter housing units (%)	30.02	29.00	27.68	29.50	31.40	29.92	29.17
Education (%)	10.36	15.11	22.28	20.25	23.75	21.41	21.86
Unemployment (%)	5.87	4.49	4.87	6.05	7.66	6.31	5.36
Poverty (%)	12.17	9.15	9.07	11.24	16.74	12.82	8.76
Mean household income ($)	58,742	63,687	73,575	65,371	69,621	68,084	71,689
Year 2000							
Total population	40,224	383,484	709,421	1,525,610	1,531,056	4,189,795	22,585,696
White	29,785	282,207	543,558	1,073,697	840,955	2,770,202	17,791,120
Black	8,705	72,859	117,520	349,576	513,013	1,061,673	2,894,328
Asian	485	7,311	13,372	54,540	54,175	129,883	803,103
Native American	196	1,703	2,536	3,194	4,365	11,994	44,266
Others	1,067	19,398	32,414	44,612	118,557	216,048	1,052,874
Hispanic	1,085	24,730	45,000	43,697	151,190	265,702	1,506,994

(continued)

Table B.33 (continued)

	0–10	11–20	21–30	31–40	41–50	0–50	Outside
Color	10,870	111,838	185,101	471,678	735,636	1,515,123	5,522,464
White (%)	74.05	73.59	76.62	70.38	54.93	66.12	78.77
Black (%)	21.64	19.00	16.57	22.91	33.51	25.34	12.81
Asian (%)	1.21	1.91	1.88	3.57	3.54	3.10	3.56
Native American (%)	0.49	0.44	0.36	0.21	0.29	0.29	0.20
Others (%)	2.65	5.06	4.57	2.92	7.74	5.16	4.66
Hispanic (%)	2.70	6.45	6.34	2.86	9.87	6.34	6.67
Color (%)	27.02	29.16	26.09	30.92	48.05	36.16	24.45
Female (%)	49.04	50.85	51.57	52.16	53.08	52.25	51.57
Old (65+ years) (%)	10.23	10.57	12.63	13.54	13.15	12.94	14.13
Kid (<5 years) (%)	8.76	8.16	7.87	7.66	8.11	7.92	7.60
Native-born (%)	96.97	94.25	94.79	93.50	92.82	93.57	89.91
Renter housing units (%)	24.46	28.56	25.96	29.79	32.45	29.99	28.20
College degree or higher (%)	16.76	19.11	27.10	24.28	28.36	25.70	26.70
Unemployment (%)	5.18	5.61	6.25	6.50	8.81	7.16	5.20
Poverty (%)	10.07	9.70	9.23	12.21	17.88	13.52	8.93
Mean household income ($)	69,448	68,663	81,529	70,201	73,843	73,276	78,009
Year 2010							
Total population	47,275	427,702	768,301	1,578,254	1,551,169	4,372,701	23,539,282
White	34,531	289,805	580,602	1,052,228	830,412	2,787,578	17,769,170
Black	10,020	100,007	130,603	385,252	507,839	1,133,721	3,240,060
Asian	914	14,427	20,250	75,950	72,566	184,107	1,187,353
Native American	167	1,863	2,476	2,891	3,727	11,124	46,053
Others	1,643	21,600	34,370	61,933	136,625	256,171	1,296,646
Hispanic	2,217	44,395	67,069	79,274	196,806	389,761	2,231,784
Color	13,744	165,952	229,283	565,115	790,951	1,765,045	6,974,797
White (%)	73.04	67.76	75.57	66.67	53.53	63.75	75.49
Black (%)	21.20	23.38	17.00	24.41	32.74	25.93	13.76
Asian (%)	1.93	3.37	2.64	4.81	4.68	4.21	5.04
Native American (%)	0.35	0.44	0.32	0.18	0.24	0.25	0.20
Others (%)	3.48	5.05	4.47	3.92	8.81	5.86	5.51
Hispanic (%)	4.69	10.38	8.73	5.02	12.69	8.91	9.48
Color (%)	29.07	38.80	29.84	35.81	50.99	40.37	29.63
Female (%)	47.94	51.08	51.51	51.85	52.80	52.01	51.29
Old (65+ years) (%)	10.70	10.83	13.13	12.82	12.69	12.61	14.15

(continued)

Table B.33 (continued)

	0–10	11–20	21–30	31–40	41–50	0–50	Outside
Kid (<5 years) (%)	6.41	6.90	6.35	6.24	6.54	6.44	6.01
Native-born (%)	95.99	90.38	92.84	91.43	90.66	91.35	87.58
Renter housing units (%)	16.46	27.33	24.29	29.56	32.92	29.57	26.88
College degree or higher (%)	22.34	22.21	30.56	29.34	33.10	30.12	31.00
Unemployment (%)	9.13	7.87	7.56	8.57	9.80	8.74	7.02
Poverty (%)	7.68	11.73	10.44	13.07	19.12	14.56	9.78
Mean household income ($)	80,515	68,452	85,014	73,582	75,545	75,854	81,038
Index	33	33	33	33	33	33	33

B.34 Oyster Creek Nuclear Generating Station, New Jersey

See Fig. B.34.
See Table B.34.

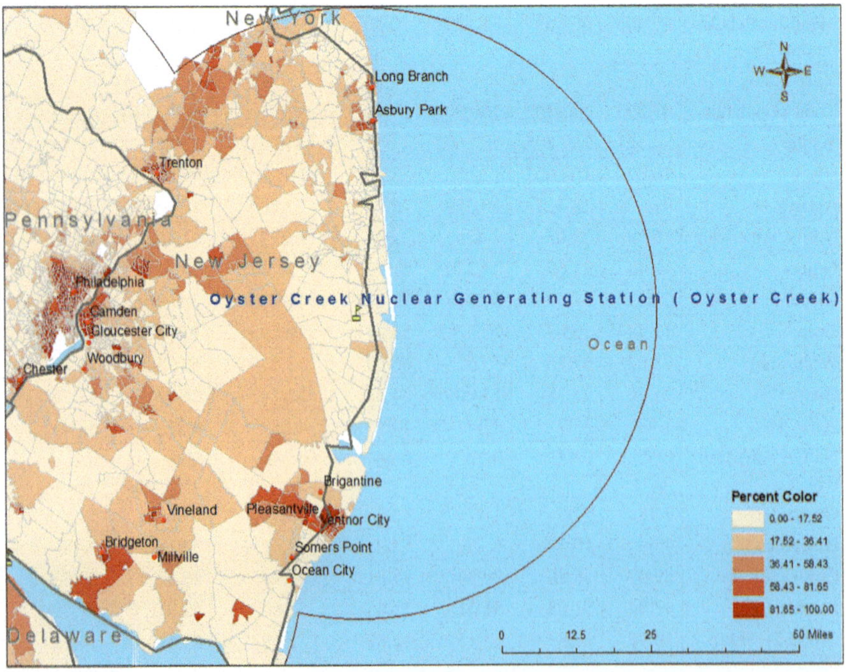

Fig. B.34 Spatial distribution of percent color among the populations surrounding Oyster Creek Nuclear Generating Station, New Jersey, in 2010

Table B.34 Demographic composition of population, as sorted by distance from Oyster Creek Nuclear Generating Station, New Jersey, in 1990, 2000, and 2010 *Source* (Kyne 2014)

	0–10	11–20	21–30	31–40	41–50	0–50	Outside
Year 1990							
Tract	29	70	86	237	161	583	9,564
Tract area (sq. mile)	287	760	787	1,008	570	3,412	105,920
Total population	91,665	237,253	335,439	843,941	624,342	2,132,640	35,469,420
White	89,171	227,599	299,980	684,923	523,233	1,824,906	28,231,784
Black	1,310	5,490	23,686	120,560	64,692	215,738	4,767,806
Asian	445	2,221	5,296	21,250	25,462	54,674	1,039,475
Native American	284	304	659	1,703	918	3,868	85,823
Others	457	1,635	5,821	15,501	10,020	33,434	1,344,511
Hispanic	2,092	5,849	16,672	37,725	29,157	91,495	3,001,051
Color	4,095	13,689	45,483	177,475	118,250	358,992	8,587,946
White (%)	97.28	95.93	89.43	81.16	83.81	85.57	79.59
Black (%)	1.43	2.31	7.06	14.29	10.36	10.12	13.44
Asian (%)	0.49	0.94	1.58	2.52	4.08	2.56	2.93
Native American (%)	0.31	0.13	0.20	0.20	0.15	0.18	0.24
Others (%)	0.50	0.69	1.74	1.84	1.60	1.57	3.79
Hispanic (%)	2.28	2.47	4.97	4.47	4.67	4.29	8.46
Color (%)	4.47	5.77	13.56	21.03	18.94	16.83	24.21
Female (%)	51.88	53.44	50.94	51.63	51.24	51.62	52.04
Old (65+ years) (%)	16.81	29.28	12.64	13.18	11.48	14.54	13.85
Kid (<5 years) (%)	8.67	6.41	9.40	8.30	8.48	8.33	8.17
Native-born (%)	95.46	93.69	94.47	93.14	90.67	92.78	88.63
Renter housing units (%)	11.29	10.34	23.21	26.06	27.70	23.05	36.72
Education (%)	13.55	15.09	20.12	24.41	27.90	23.18	21.76
Unemployment (%)	6.25	5.84	5.12	4.90	5.08	5.13	6.43
Poverty (%)	6.02	5.03	5.90	6.13	5.57	5.80	11.63
Mean household income ($)	65,149	63,752	76,608	82,080	89,603	80,406	70,261
Year 2000							
Total population	114,293	263,512	378,078	954,033	677,090	2,387,006	37,284,856
White	109,595	245,123	332,045	718,669	518,696	1,924,128	27,552,606
Black	1,773	7,186	22,581	137,758	72,571	241,869	5,083,308
Asian	799	4,201	7,574	41,148	54,148	107,870	1,649,311
Native American	198	562	603	1,874	821	4,058	112,754
Others	1,935	6,437	15,286	54,589	30,827	109,074	2,886,883
Hispanic	3,834	11,064	25,065	77,961	45,460	163,384	4,209,902

(continued)

Table B.34 (continued)

	0–10	11–20	21–30	31–40	41–50	0–50	Outside
Color	7,287	25,266	60,142	271,090	183,804	547,589	11,480,117
White (%)	95.89	93.02	87.82	75.33	76.61	80.61	73.90
Black (%)	1.55	2.73	5.97	14.44	10.72	10.13	13.63
Asian (%)	0.70	1.59	2.00	4.31	8.00	4.52	4.42
Native American (%)	0.17	0.21	0.16	0.20	0.12	0.17	0.30
Others (%)	1.69	2.44	4.04	5.72	4.55	4.57	7.74
Hispanic (%)	3.35	4.20	6.63	8.17	6.71	6.84	11.29
Color (%)	6.38	9.59	15.91	28.42	27.15	22.94	30.79
Female (%)	51.27	53.71	50.56	51.39	51.49	51.54	51.77
Old (65+ years) (%)	17.84	29.32	12.12	13.49	12.05	14.82	13.76
Kid (<5 years) (%)	7.42	5.74	9.20	8.06	7.97	7.93	7.68
Native-born (%)	96.16	93.33	93.13	88.92	84.73	89.23	84.99
Renter housing units (%)	10.74	11.18	21.58	24.98	27.76	22.45	36.20
College degree or higher (%)	17.29	18.83	25.99	28.76	33.98	28.04	26.21
Unemployment (%)	5.13	5.12	5.23	4.94	5.99	5.33	6.43
Poverty (%)	5.85	5.65	7.07	7.29	6.37	6.74	12.53
Mean household income ($)	73,388	70,711	85,188	89,543	97,777	88,022	75,580
Year 2010							
Total population	128,893	280,778	414,445	1,011,867	702,567	2,538,550	38,025,484
White	119,722	255,716	361,170	731,620	496,127	1,964,355	27,334,152
Black	2,072	9,461	23,916	147,076	77,463	259,988	5,253,896
Asian	1,616	5,631	9,479	65,377	93,761	175,864	2,257,388
Native American	101	250	439	1,771	2,358	4,919	99,845
Others	5,382	9,720	19,441	66,023	32,858	133,424	3,080,202
Hispanic	8,519	19,856	34,215	120,065	64,939	247,594	5,165,970
Color	13,354	37,989	72,166	343,540	245,981	713,030	13,069,155
White (%)	92.88	91.07	87.15	72.30	70.62	77.38	71.88
Black (%)	1.61	3.37	5.77	14.54	11.03	10.24	13.82
Asian (%)	1.25	2.01	2.29	6.46	13.35	6.93	5.94
Native American (%)	0.08	0.09	0.11	0.18	0.34	0.19	0.26
Others (%)	4.18	3.46	4.69	6.52	4.68	5.26	8.10
Hispanic (%)	6.61	7.07	8.26	11.87	9.24	9.75	13.59
Color (%)	10.36	13.53	17.41	33.95	35.01	28.09	34.37
Female (%)	51.07	52.90	51.05	51.07	51.57	51.41	51.48
Old (65+ years) (%)	18.23	27.87	12.57	13.80	12.35	14.98	13.83
Kid (<5 years) (%)	5.70	5.19	7.15	5.93	6.04	6.07	5.99

(continued)

Table B.34 (continued)

	0–10	11–20	21–30	31–40	41–50	0–50	Outside
Native-born (%)	93.94	92.04	91.74	84.86	79.72	85.81	83.44
Renter housing units (%)	9.74	11.86	20.42	22.46	25.72	20.79	34.21
College degree or higher (%)	22.60	23.19	30.67	33.36	39.75	32.90	30.72
Unemployment (%)	8.07	8.41	7.57	7.23	7.71	7.58	7.50
Poverty (%)	6.32	7.43	9.31	7.67	6.91	7.62	12.85
Mean household income ($)	79,599	71,633	89,548	93,284	99,085	90,835	78,076
Index	34	34	34	34	34	34	34

B.35 Salem Nuclear Generating Station, New Jersey

See Fig. B.35.
See Table B.35.

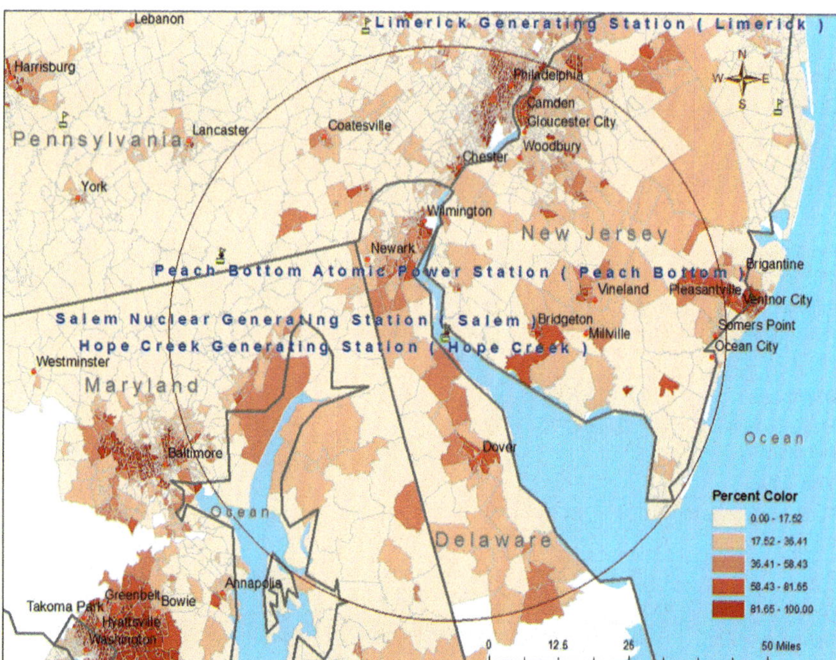

Fig. B.35 Spatial distribution of percent color among the populations surrounding Salem Nuclear Generating Station, New Jersey, in 2010

Table B.35 Demographic composition of population, as sorted by distance from Salem Nuclear Generating Station, New Jersey, in 1990, 2000, and 2010 *Source* (Kyne 2014)

	0–10	11–20	21–30	31–40	41–50	0–50	Outside
Year 1990							
Tract	15	93	189	386	405	1,088	5,764
Tract area (sq. mile)	332	845	1,857	1,673	1,008	5,715	63,957
Total population	34,738	307,326	653,350	1,471,049	1,545,323	4,011,786	21,047,592
White	26,049	254,777	519,515	1,137,890	903,351	2,841,582	17,748,828
Black	7,987	43,400	108,578	289,691	525,204	974,860	2,449,128
Asian	84	3,827	7,438	30,921	37,360	79,630	471,690
Native American	195	1,568	1,685	2,720	3,433	9,601	37,468
Others	436	3,746	16,128	9,828	75,967	106,105	340,414
Hispanic	745	7,385	29,586	26,039	105,050	168,805	907,142
Color	8,919	55,876	145,738	346,588	664,996	1,222,117	3,812,643
White (%)	74.99	82.90	79.52	77.35	58.46	70.83	84.33
Black (%)	22.99	14.12	16.62	19.69	33.99	24.30	11.64
Asian (%)	0.24	1.25	1.14	2.10	2.42	1.98	2.24
Native American (%)	0.56	0.51	0.26	0.18	0.22	0.24	0.18
Others (%)	1.26	1.22	2.47	0.67	4.92	2.64	1.62
Hispanic (%)	2.14	2.40	4.53	1.77	6.80	4.21	4.31
Color (%)	25.68	18.18	22.31	23.56	43.03	30.46	18.11
Female (%)	49.36	51.29	51.83	51.85	53.26	52.33	51.74
Old (65+ years) (%)	11.86	10.38	12.47	13.42	13.68	13.12	13.93
Kid (<5 years) (%)	8.96	8.87	8.90	8.60	8.96	8.82	8.14
Native-born (%)	98.26	96.88	96.27	95.43	94.59	95.38	92.94
Renter housing units (%)	30.02	28.72	28.10	28.48	32.31	29.98	29.16
Education (%)	10.36	15.61	22.02	19.62	24.22	21.40	21.86
Unemployment (%)	5.87	4.37	4.90	5.96	7.81	6.34	5.35
Poverty (%)	12.17	8.83	9.27	10.90	17.16	12.91	8.75
Mean household income ($)	58,742	64,718	72,905	65,050	69,536	67,979	71,701
Year 2000							
Total population	40,224	364,904	710,302	1,492,038	1,540,413	4,147,881	22,627,608
White	29,785	271,716	539,851	1,064,490	831,336	2,737,178	17,824,144
Black	8,705	66,465	121,708	328,162	531,243	1,056,283	2,899,718
Asian	485	7,095	13,450	51,940	55,473	128,443	804,543
Native American	196	1,639	2,535	3,059	4,446	11,875	44,385
Others	1,067	17,987	32,735	44,397	117,908	214,094	1,054,828
Hispanic	1,085	22,527	46,285	43,337	150,064	263,298	1,509,398
Color	10,870	102,797	190,551	446,616	754,453	1,505,287	5,532,300
White (%)	74.05	74.46	76.00	71.34	53.97	65.99	78.77

(continued)

Table B.35 (continued)

	0–10	11–20	21–30	31–40	41–50	0–50	Outside
Black (%)	21.64	18.21	17.13	21.99	34.49	25.47	12.81
Asian (%)	1.21	1.94	1.89	3.48	3.60	3.10	3.56
Native American (%)	0.49	0.45	0.36	0.21	0.29	0.29	0.20
Others (%)	2.65	4.93	4.61	2.98	7.65	5.16	4.66
Hispanic (%)	2.70	6.17	6.52	2.90	9.74	6.35	6.67
Color (%)	27.02	28.17	26.83	29.93	48.98	36.29	24.45
Female (%)	49.04	50.86	51.56	52.09	53.11	52.24	51.57
Old (65+ years) (%)	10.23	10.41	12.71	13.41	13.16	12.90	14.13
Kid (<5 years) (%)	8.76	8.11	7.87	7.74	8.05	7.92	7.60
Native-born (%)	96.97	94.23	94.74	93.64	92.73	93.58	89.92
Renter housing units (%)	24.46	27.99	26.58	28.85	33.17	30.01	28.20
College degree or higher (%)	16.76	19.63	26.79	23.51	28.99	25.69	26.70
Unemployment (%)	5.18	5.40	6.40	6.42	8.79	7.15	5.20
Poverty (%)	10.07	9.25	9.47	11.89	18.25	13.58	8.93
Mean household income ($)	69,448	69,565	80,882	69,965	73,920	73,254	78,004
Year 2010							
Total population	47,275	410,475	764,726	1,552,122	1,553,871	4,328,469	23,583,514
White	34,531	280,389	572,330	1,048,911	823,884	2,760,045	17,796,704
Black	10,020	93,492	134,705	366,094	517,310	1,121,621	3,252,160
Asian	914	14,162	20,355	73,125	72,830	181,386	1,190,074
Native American	167	1,858	2,448	2,724	3,875	11,072	46,105
Others	1,643	20,574	34,888	61,268	135,972	254,345	1,298,472
Hispanic	2,217	41,275	68,759	79,979	195,236	387,466	2,234,079
Color	13,744	155,999	235,078	543,170	799,515	1,747,506	6,992,336
White (%)	73.04	68.31	74.84	67.58	53.02	63.76	75.46
Black (%)	21.20	22.78	17.61	23.59	33.29	25.91	13.79
Asian (%)	1.93	3.45	2.66	4.71	4.69	4.19	5.05
Native American (%)	0.35	0.45	0.32	0.18	0.25	0.26	0.20
Others (%)	3.48	5.01	4.56	3.95	8.75	5.88	5.51
Hispanic (%)	4.69	10.06	8.99	5.15	12.56	8.95	9.47
Color (%)	29.07	38.00	30.74	35.00	51.45	40.37	29.65
Female (%)	47.94	51.07	51.55	51.72	52.85	52.00	51.29
Old (65+ years) (%)	10.70	10.79	13.20	12.63	12.80	12.60	14.15
Kid (<5 years) (%)	6.41	6.82	6.37	6.28	6.53	6.44	6.01
Native-born (%)	95.99	90.42	92.72	91.53	90.63	91.36	87.59

(continued)

Table B.35 (continued)

	0–10	11–20	21–30	31–40	41–50	0–50	Outside
Renter housing units (%)	16.46	26.81	24.89	28.84	33.46	29.60	26.88
College degree or higher (%)	22.34	22.74	30.14	28.50	33.80	30.08	31.01
Unemployment (%)	9.13	7.72	7.72	8.49	9.83	8.75	7.03
Poverty (%)	7.68	11.19	10.85	12.76	19.43	14.62	9.78
Mean household income ($)	80,515	69,550	83,926	73,432	75,908	75,893	81,020
Index	35	35	35	35	35	35	35

B.36 Indian Point Nuclear Generating, New York

See Fig. B.36.
See Table B.36.

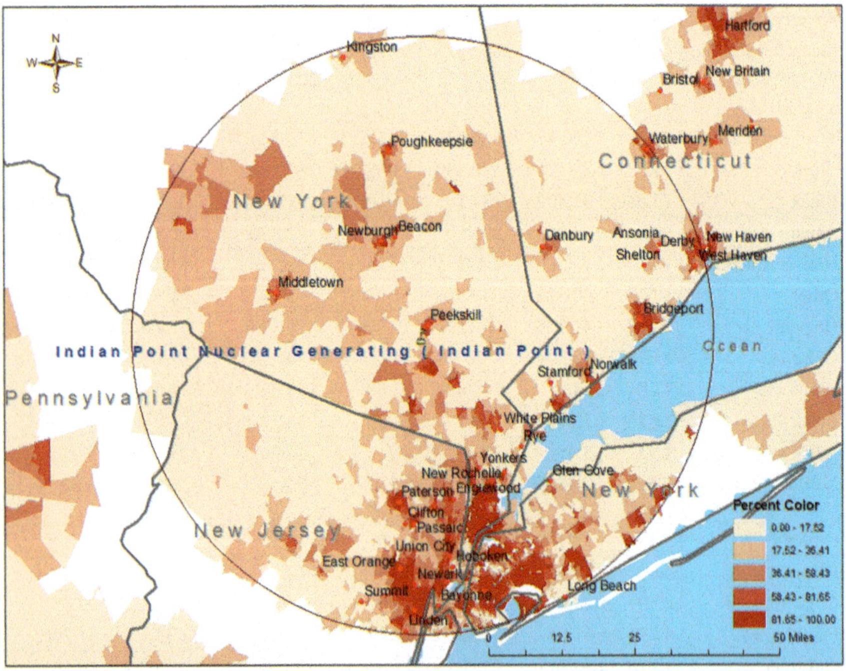

Fig. B.36 Spatial distribution of percent color among the populations surrounding Indian Point Nuclear Generating, New York, in 2010

Table B.36 Demographic composition of population, as sorted by distance from Indian Point Nuclear Generating, New York, in 1990, 2000, and 2010 *Source* (Kyne 2014)

	0–10	11–20	21–30	31–40	41–50	0–50	Outside
Year 1990							
Tract	55	187	596	1,252	1,670	3,760	4,002
Tract area (sq. mile)	292	1,005	1,487	1,919	1,107	5,810	63,011
Total population	229,882	749,366	2,453,297	4,911,005	5,442,748	13,786,298	15,221,250
White	201,501	647,849	1,855,732	3,162,924	3,096,381	8,964,387	13,431,768
Black	16,429	64,693	368,460	861,202	1,657,783	2,968,567	1,200,971
Asian	5,847	23,174	91,373	317,468	303,274	741,136	266,893
Native American	425	1,661	4,420	12,869	13,867	33,242	47,283
Others	5,669	11,994	133,340	556,475	371,457	1,078,935	274,366
Hispanic	18,731	43,014	324,710	1,214,297	825,975	2,426,727	648,859
Color	39,637	129,807	757,725	2,266,527	2,700,268	5,893,964	2,127,483
White (%)	87.65	86.45	75.64	64.40	56.89	65.02	88.24
Black (%)	7.15	8.63	15.02	17.54	30.46	21.53	7.89
Asian (%)	2.54	3.09	3.72	6.46	5.57	5.38	1.75
Native American (%)	0.18	0.22	0.18	0.26	0.25	0.24	0.31
Others (%)	2.47	1.60	5.44	11.33	6.82	7.83	1.80
Hispanic (%)	8.15	5.74	13.24	24.73	15.18	17.60	4.26
Color (%)	17.24	17.32	30.89	46.15	49.61	42.75	13.98
Female (%)	49.70	51.09	52.43	52.56	52.72	52.48	51.44
Old (65+ years) (%)	10.37	11.00	13.76	13.15	13.02	13.04	13.40
Kid (<5 years) (%)	8.35	8.53	8.11	7.60	8.36	8.05	8.44
Native-born (%)	89.09	87.42	81.67	74.57	76.13	77.39	93.56
Renter housing units (%)	25.75	26.01	41.22	60.66	53.40	52.34	28.04
Education (%)	32.51	34.60	29.01	28.94	20.61	26.10	22.16
Unemployment (%)	3.96	4.35	5.61	7.72	8.14	7.23	5.62
Poverty (%)	5.02	6.38	7.89	16.16	15.07	13.55	8.41
Mean household income ($)	103,352	113,938	97,399	82,528	70,724	82,538	72,323
Year 2000							
Total population	242,165	826,902	2,627,286	5,352,294	5,845,359	14,894,006	15,902,366
White	196,235	673,908	1,759,588	3,040,855	2,825,575	8,496,161	13,272,190
Black	17,902	70,385	410,274	800,710	1,750,064	3,049,335	1,370,075
Asian	8,780	34,826	130,263	500,998	486,427	1,161,294	459,169
Native American	431	2,370	7,084	23,711	18,208	51,804	54,916
Others	18,804	45,422	320,086	985,956	765,109	2,135,377	746,051
Hispanic	30,788	77,568	495,739	1,592,978	1,054,119	3,251,192	1,048,920
Color	61,284	195,385	1,073,881	2,944,270	3,443,913	7,718,733	3,123,957
White (%)	81.03	81.50	66.97	56.81	48.34	57.04	83.46

(continued)

Table B.36 (continued)

	0–10	11–20	21–30	31–40	41–50	0–50	Outside
Black (%)	7.39	8.51	15.62	14.96	29.94	20.47	8.62
Asian (%)	3.63	4.21	4.96	9.36	8.32	7.80	2.89
Native American (%)	0.18	0.29	0.27	0.44	0.31	0.35	0.35
Others (%)	7.76	5.49	12.18	18.42	13.09	14.34	4.69
Hispanic (%)	12.71	9.38	18.87	29.76	18.03	21.83	6.60
Color (%)	25.31	23.63	40.87	55.01	58.92	51.82	19.64
Female (%)	50.78	50.92	52.31	52.01	52.58	52.20	51.32
Old (65+ years) (%)	11.52	11.93	13.10	12.21	12.12	12.31	13.84
Kid (<5 years) (%)	8.73	8.92	8.43	7.62	8.39	8.16	7.68
Native-born (%)	85.06	84.71	76.46	68.28	68.34	70.93	91.29
Renter housing units (%)	25.56	26.47	41.05	60.63	53.30	52.23	27.18
College degree or higher (%)	38.04	39.90	33.57	33.57	25.04	30.72	26.40
Unemployment (%)	3.56	4.01	5.68	8.10	8.42	7.47	5.67
Poverty (%)	6.01	7.92	10.23	17.31	17.28	15.35	9.17
Mean household income ($)	109,993	123,688	103,111	89,616	74,011	88,163	77,552
Year 2010							
Total population	253,977	866,039	2,685,808	5,416,986	5,895,371	15,118,181	16,378,985
White	196,989	679,057	1,697,802	2,981,212	2,799,672	8,354,732	13,279,606
Black	21,031	75,492	428,486	765,664	1,716,532	3,007,205	1,502,391
Asian	12,594	45,390	166,660	646,636	662,158	1,533,438	695,603
Native American	261	1,937	5,135	17,716	16,853	41,902	52,616
Others	23,102	64,163	387,725	1,005,758	700,156	2,180,904	848,769
Hispanic	48,788	114,365	622,117	1,732,926	1,182,057	3,700,253	1,503,914
Color	83,856	244,517	1,238,828	3,138,954	3,625,473	8,331,628	3,940,985
White (%)	77.56	78.41	63.21	55.03	47.49	55.26	81.08
Black (%)	8.28	8.72	15.95	14.13	29.12	19.89	9.17
Asian (%)	4.96	5.24	6.21	11.94	11.23	10.14	4.25
Native American (%)	0.10	0.22	0.19	0.33	0.29	0.28	0.32
Others (%)	9.10	7.41	14.44	18.57	11.88	14.43	5.18
Hispanic (%)	19.21	13.21	23.16	31.99	20.05	24.48	9.18
Color (%)	33.02	28.23	46.12	57.95	61.50	55.11	24.06
Female (%)	49.24	51.22	51.94	51.90	52.29	51.98	51.10
Old (65+ years) (%)	12.80	13.01	13.16	12.55	12.00	12.47	14.14
Kid (<5 years) (%)	5.96	6.82	6.38	6.07	6.62	6.38	5.78
Native-born (%)	80.58	82.49	74.88	66.85	66.66	69.33	89.09

(continued)

Table B.36 (continued)

	0–10	11–20	21–30	31–40	41–50	0–50	Outside
Renter housing units (%)	22.67	24.79	38.10	56.15	49.51	48.41	25.33
College degree or higher (%)	40.49	43.98	37.58	39.18	30.93	36.01	30.52
Unemployment (%)	6.00	5.55	7.45	8.29	8.35	7.97	7.26
Poverty (%)	7.08	8.73	10.30	16.31	15.63	14.39	10.14
Mean household income ($)	111,702	124,939	105,091	93,828	78,343	91,901	79,841
Index	36	36	36	36	36	36	36

B.37 James A. FitzPatrick Nuclear Power Plant, New York

See Fig. B.37.
See Table B.37.

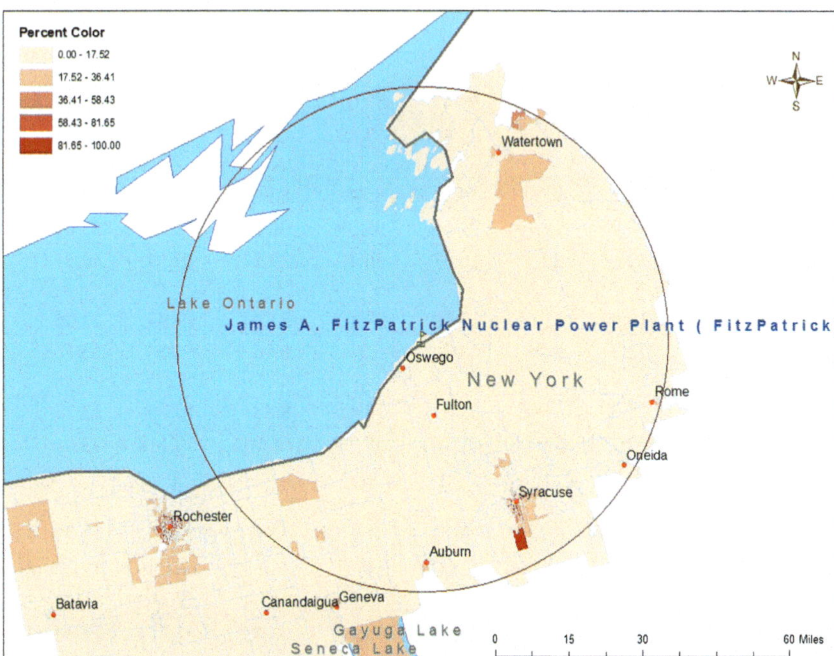

Fig. B.37 Spatial distribution of percent color among the populations surrounding James A. FitzPatrick Nuclear Power Plant, New York, in 2010

Table B.37 Demographic composition of population, as sorted by distance from James A. FitzPatrick Nuclear Power Plant, New York, in 1990, 2000, and 2010 *Source* (Kyne 2014)

	0–10	11–20	21–30	31–40	41–50	0–50	Outside
Year 1990							
Tract	9	16	31	115	41	212	4,707
Tract area (sq. mile)	374	658	1,451	1,844	611	4,938	49,616
Total population	30,260	66,188	121,603	397,569	141,586	757,206	17,233,108
White	29,760	65,244	118,190	349,634	135,091	697,919	12,699,958
Black	182	324	1,734	36,763	4,012	43,015	2,817,577
Asian	173	284	727	5,582	1,156	7,922	681,344
Native American	98	223	697	2,909	811	4,738	54,341
Others	47	116	267	2,675	516	3,621	979,897
Hispanic	320	437	1,092	6,065	1,427	9,341	2,142,401
Color	752	1,244	4,254	50,977	7,298	64,525	5,436,296
White (%)	98.35	98.57	97.19	87.94	95.41	92.17	73.70
Black (%)	0.60	0.49	1.43	9.25	2.83	5.68	16.35
Asian (%)	0.57	0.43	0.60	1.40	0.82	1.05	3.95
Native American (%)	0.32	0.34	0.57	0.73	0.57	0.63	0.32
Others (%)	0.16	0.18	0.22	0.67	0.36	0.48	5.69
Hispanic (%)	1.06	0.66	0.90	1.53	1.01	1.23	12.43
Color (%)	2.49	1.88	3.50	12.82	5.15	8.52	31.55
Female (%)	51.89	50.95	50.41	52.01	51.72	51.60	52.13
Old (65+ years) (%)	13.44	10.16	8.18	13.63	15.21	12.74	13.15
Kid (<5 years) (%)	9.02	9.31	10.34	8.74	8.96	9.10	8.27
Native-born (%)	98.06	98.16	97.66	95.39	96.53	96.32	83.61
Renter housing units (%)	32.65	22.43	16.31	36.40	29.72	30.59	44.51
Education (%)	18.41	11.65	19.08	20.18	23.84	19.97	23.26
Unemployment (%)	7.87	8.75	4.90	6.31	6.16	6.30	6.91
Poverty (%)	15.38	11.02	6.25	12.47	8.88	10.76	13.13
Mean household income ($)	53,774	55,410	65,701	55,782	65,934	59,114	73,010
Year 2000							
Total population	29,878	66,826	130,812	374,975	143,031	745,522	18,230,936
White	28,683	64,949	124,952	311,921	130,787	661,292	12,229,826
Black	113	435	2,081	39,869	4,994	47,492	2,938,750
Asian	304	157	965	7,671	2,275	11,372	1,040,954
Native American	102	216	626	2,601	1,666	5,211	74,103
Others	676	1,076	2,202	12,907	3,309	20,170	1,947,287
Hispanic	620	655	1,378	10,565	2,278	15,496	2,849,520
Color	1,461	2,211	6,599	67,432	13,319	91,022	7,123,756

(continued)

Table B.37 (continued)

	0–10	11–20	21–30	31–40	41–50	0–50	Outside
White (%)	96.00	97.19	95.52	83.18	91.44	88.70	67.08
Black (%)	0.38	0.65	1.59	10.63	3.49	6.37	16.12
Asian (%)	1.02	0.23	0.74	2.05	1.59	1.53	5.71
Native American (%)	0.34	0.32	0.48	0.69	1.16	0.70	0.41
Others (%)	2.26	1.61	1.68	3.44	2.31	2.71	10.68
Hispanic (%)	2.08	0.98	1.05	2.82	1.59	2.08	15.63
Color (%)	4.89	3.31	5.04	17.98	9.31	12.21	39.08
Female (%)	51.69	50.72	50.64	52.02	51.46	51.54	51.88
Old (65+ years) (%)	13.77	10.75	9.83	14.45	15.84	13.55	12.89
Kid (<5 years) (%)	6.95	7.80	8.21	7.69	7.54	7.73	7.87
Native-born (%)	97.77	98.69	97.27	94.51	95.75	95.74	78.96
Renter housing units (%)	33.73	21.30	16.77	35.22	28.94	29.56	43.81
College degree or higher (%)	18.46	13.14	23.36	22.55	28.79	22.98	27.55
Unemployment (%)	8.35	10.96	4.62	6.22	5.12	6.22	7.13
Poverty (%)	18.77	12.97	7.24	14.90	9.98	12.57	14.67
Mean household income ($)	53,246	56,785	69,001	58,249	69,951	61,978	77,711
Year 2010							
Total population	29,580	66,955	132,151	373,620	145,161	747,467	18,482,284
White	28,357	64,615	125,152	301,879	131,131	651,134	12,113,268
Black	388	648	2,561	45,876	6,166	55,639	2,934,952
Asian	422	328	1,823	10,353	2,967	15,893	1,381,821
Native American	69	109	452	2,412	1,063	4,105	62,771
Others	344	1,255	2,163	13,100	3,834	20,696	1,989,473
Hispanic	943	1,259	2,084	16,013	3,450	23,749	3,265,131
Color	1,925	3,040	8,339	81,394	15,808	110,506	7,744,309
White (%)	95.87	96.51	94.70	80.80	90.33	87.11	65.54
Black (%)	1.31	0.97	1.94	12.28	4.25	7.44	15.88
Asian (%)	1.43	0.49	1.38	2.77	2.04	2.13	7.48
Native American (%)	0.23	0.16	0.34	0.65	0.73	0.55	0.34
Others (%)	1.16	1.87	1.64	3.51	2.64	2.77	10.76
Hispanic (%)	3.19	1.88	1.58	4.29	2.38	3.18	17.67
Color (%)	6.51	4.54	6.31	21.79	10.89	14.78	41.90
Female (%)	51.59	49.71	50.61	51.56	50.77	51.07	51.66
Old (65+ years) (%)	13.17	11.83	11.48	14.18	15.70	13.75	13.27
Kid (<5 years) (%)	5.31	5.77	5.78	6.02	5.63	5.85	6.03

(continued)

Table B.37 (continued)

	0–10	11–20	21–30	31–40	41–50	0–50	Outside
Native-born (%)	97.69	97.92	96.69	93.30	95.70	94.95	77.59
Renter housing units (%)	31.23	21.20	17.05	33.76	27.97	28.56	40.60
College degree or higher (%)	20.37	14.46	26.79	25.86	31.07	25.90	32.38
Unemployment (%)	10.42	10.16	6.12	7.49	6.17	7.32	7.53
Poverty (%)	18.87	15.70	7.50	16.67	11.47	13.99	14.17
Mean household income ($)	56,009	56,240	71,278	57,132	70,169	62,051	81,148
Index	37	37	37	37	37	37	37

B.38 Nine Mile Point Nuclear Station, New York

See Fig. B.38.
See Table B.38.

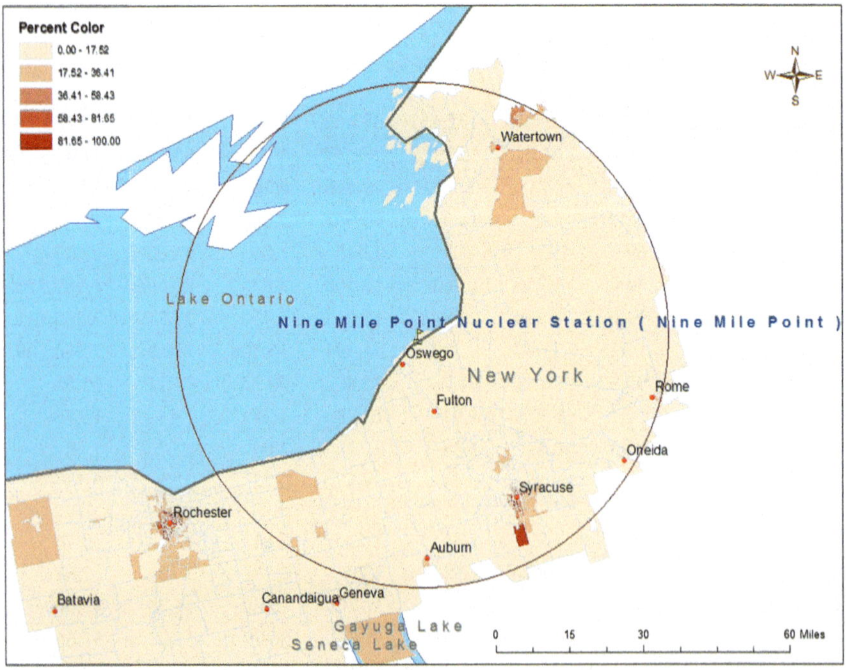

Fig. B.38 Spatial distribution of percent color among the populations surrounding Nine Mile Point Nuclear Station, New York, in 2010

Table B.38 Demographic composition of population, as sorted by distance from Nine Mile Point Nuclear Station, New York, in 1990, 2000, and 2010 *Source* (Kyne 2014)

	0–10	11–20	21–30	31–40	41–50	0–50	Outside
Year 1990							
Tract	9	16	32	112	42	211	4,708
Tract area (sq. mile)	374	658	1,480	1,810	585	4,908	49,647
Total population	30,260	66,188	125,097	383,940	148,776	754,261	17,236,052
White	29,760	65,244	121,632	336,899	141,450	694,985	12,702,892
Black	182	324	1,758	36,190	4,561	43,015	2,817,577
Asian	173	284	745	5,433	1,287	7,922	681,344
Native American	98	223	697	2,852	857	4,727	54,352
Others	47	116	274	2,563	621	3,621	979,897
Hispanic	320	437	1,138	5,810	1,636	9,341	2,142,401
Color	752	1,244	4,353	49,941	8,224	64,514	5,436,307
White (%)	98.35	98.57	97.23	87.75	95.08	92.14	73.70
Black (%)	0.60	0.49	1.41	9.43	3.07	5.70	16.35
Asian (%)	0.57	0.43	0.60	1.42	0.87	1.05	3.95
Native American (%)	0.32	0.34	0.56	0.74	0.58	0.63	0.32
Others (%)	0.16	0.18	0.22	0.67	0.42	0.48	5.69
Hispanic (%)	1.06	0.66	0.91	1.51	1.10	1.24	12.43
Color (%)	2.49	1.88	3.48	13.01	5.53	8.55	31.54
Female (%)	51.89	50.95	50.37	52.06	51.75	51.61	52.13
Old (65+ years) (%)	13.44	10.16	8.16	13.72	15.16	12.76	13.15
Kid (<5 years) (%)	9.02	9.31	10.32	8.67	9.08	9.10	8.27
Native-born (%)	98.06	98.16	97.68	95.34	96.48	96.31	83.62
Renter housing units (%)	32.65	22.43	16.15	36.40	30.91	30.64	44.51
Education (%)	18.41	11.65	19.06	20.20	23.75	19.99	23.26
Unemployment (%)	7.87	8.75	4.94	6.24	6.37	6.30	6.91
Poverty (%)	15.38	11.02	6.15	12.68	8.91	10.78	13.12
Mean household income ($)	53,774	55,410	65,848	55,736	65,074	59,090	73,009
Year 2000							
Total population	29,878	66,826	134,215	362,647	148,811	742,377	18,234,080
White	28,683	64,949	128,308	300,729	135,610	658,279	12,232,839
Black	113	435	2,081	39,364	5,499	47,492	2,938,750
Asian	304	157	965	7,501	2,426	11,353	1,040,973
Native American	102	216	657	2,536	1,661	5,172	74,142
Others	676	1,076	2,217	12,512	3,615	20,096	1,947,361
Hispanic	620	655	1,460	10,141	2,620	15,496	2,849,520
Color	1,461	2,211	6,714	66,060	14,444	90,890	7,123,888

(continued)

Table B.38 (continued)

	0–10	11–20	21–30	31–40	41–50	0–50	Outside
White (%)	96.00	97.19	95.60	82.93	91.13	88.67	67.09
Black (%)	0.38	0.65	1.55	10.85	3.70	6.40	16.12
Asian (%)	1.02	0.23	0.72	2.07	1.63	1.53	5.71
Native American (%)	0.34	0.32	0.49	0.70	1.12	0.70	0.41
Others (%)	2.26	1.61	1.65	3.45	2.43	2.71	10.68
Hispanic (%)	2.08	0.98	1.09	2.80	1.76	2.09	15.63
Color (%)	4.89	3.31	5.00	18.22	9.71	12.24	39.07
Female (%)	51.69	50.72	50.59	52.07	51.48	51.55	51.88
Old (65+ years) (%)	13.77	10.75	9.85	14.52	15.83	13.57	12.89
Kid (<5 years) (%)	6.95	7.80	8.16	7.67	7.62	7.73	7.87
Native-born (%)	97.77	98.69	97.29	94.45	95.68	95.73	78.96
Renter housing units (%)	33.73	21.30	16.62	35.19	30.10	29.60	43.80
College degree or higher (%)	18.46	13.14	23.27	22.65	28.37	22.97	27.55
Unemployment (%)	8.35	10.96	4.57	6.21	5.23	6.22	7.13
Poverty (%)	18.77	12.97	7.13	15.06	10.23	12.59	14.67
Mean household income ($)	53,246	56,785	69,176	58,246	68,871	61,943	77,710
Year 2010							
Total population	29,580	66,955	135,763	361,573	150,396	744,267	18,485,484
White	28,357	64,615	128,594	291,435	134,984	647,985	12,116,417
Black	388	648	2,571	45,191	6,841	55,639	2,934,952
Asian	422	328	1,906	10,037	3,192	15,885	1,381,829
Native American	69	109	460	2,398	1,069	4,105	62,771
Others	344	1,255	2,232	12,512	4,310	20,653	1,989,516
Hispanic	943	1,259	2,216	15,455	3,825	23,698	3,265,182
Color	1,925	3,040	8,621	79,584	17,249	110,419	7,744,396
White (%)	95.87	96.51	94.72	80.60	89.75	87.06	65.55
Black (%)	1.31	0.97	1.89	12.50	4.55	7.48	15.88
Asian (%)	1.43	0.49	1.40	2.78	2.12	2.13	7.48
Native American (%)	0.23	0.16	0.34	0.66	0.71	0.55	0.34
Others (%)	1.16	1.87	1.64	3.46	2.87	2.77	10.76
Hispanic (%)	3.19	1.88	1.63	4.27	2.54	3.18	17.66
Color (%)	6.51	4.54	6.35	22.01	11.47	14.84	41.89
Female (%)	51.59	49.71	50.57	51.64	50.72	51.08	51.66
Old (65+ years) (%)	13.17	11.83	11.49	14.31	15.44	13.76	13.27
Kid (<5 years) (%)	5.31	5.77	5.78	5.96	5.85	5.86	6.03

(continued)

Table B.38 (continued)

	0–10	11–20	21–30	31–40	41–50	0–50	Outside
Native-born (%)	97.69	97.92	96.67	93.26	95.52	94.93	77.59
Renter housing units (%)	31.23	21.20	16.92	33.81	28.86	28.61	40.60
College degree or higher (%)	20.37	14.46	26.67	25.89	31.17	25.94	32.37
Unemployment (%)	10.42	10.16	6.03	7.54	6.32	7.34	7.53
Poverty (%)	18.87	15.70	7.46	16.90	11.66	14.04	14.17
Mean household income ($)	56,009	56,240	71,360	56,980	69,449	62,017	81,146
Index	38	38	38	38	38	38	38

B.39 R.E. Ginna Nuclear Power Plant, New York

See Fig. B.39.
See Table B.39.

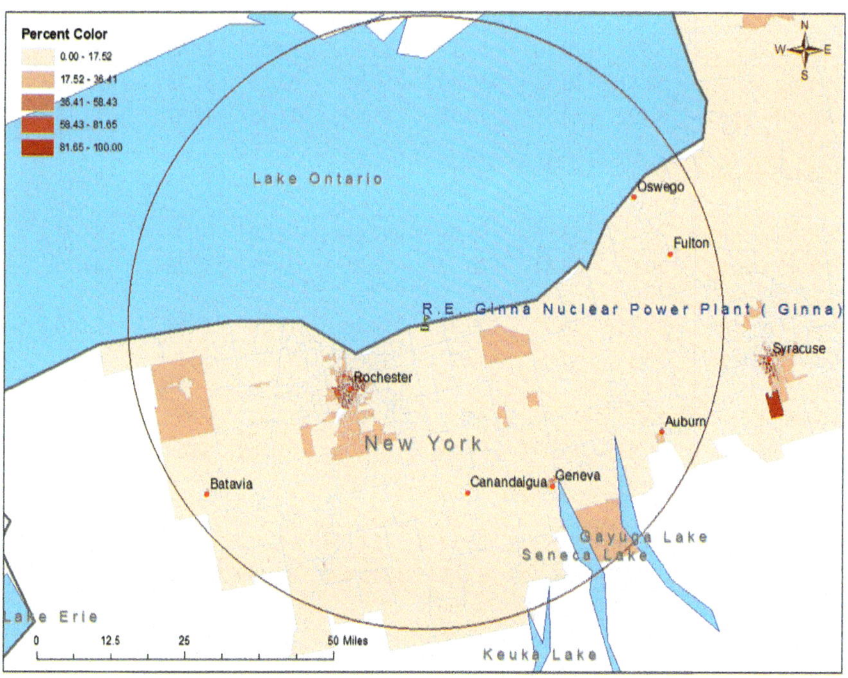

Fig. B.39 Spatial distribution of percent color among the populations surrounding R.E. Ginna Nuclear Power Plant, New York, in 2010

Table B.39 Demographic composition of population, as sorted by distance from R.E. Ginna Nuclear Power Plant, New York, in 1990, 2000, and 2010 *Source* (Kyne 2014)

	0–10	11–20	21–30	31–40	41–50	0–50	Outside
Year 1990							
Tract	13	155	56	32	8	264	4,655
Tract area (sq. mile)	139	1,264	1,527	1,098	487	4,515	50,040
Total population	49,871	556,467	228,991	118,552	33,211	987,092	17,003,222
White	48,507	453,285	217,993	114,729	32,176	866,690	12,531,187
Black	775	79,475	6,728	2,424	580	89,982	2,770,610
Asian	427	9,263	2,686	545	233	13,154	676,112
Native American	58	1,487	559	333	100	2,537	56,542
Others	100	12,965	1,019	528	125	14,737	968,781
Hispanic	495	22,563	3,434	1,245	380	28,117	2,123,625
Color	1,761	111,830	13,313	4,547	1,281	132,732	5,368,089
White (%)	97.26	81.46	95.20	96.78	96.88	87.80	73.70
Black (%)	1.55	14.28	2.94	2.04	1.75	9.12	16.29
Asian (%)	0.86	1.66	1.17	0.46	0.70	1.33	3.98
Native American (%)	0.12	0.27	0.24	0.28	0.30	0.26	0.33
Others (%)	0.20	2.33	0.44	0.45	0.38	1.49	5.70
Hispanic (%)	0.99	4.05	1.50	1.05	1.14	2.85	12.49
Color (%)	3.53	20.10	5.81	3.84	3.86	13.45	31.57
Female (%)	50.81	52.53	50.40	50.91	52.35	51.75	52.13
Old (65+ years) (%)	9.49	13.58	10.38	11.99	11.45	12.37	13.18
Kid (<5 years) (%)	9.31	9.32	8.71	9.03	8.11	9.11	8.26
Native-born (%)	96.40	93.56	95.59	97.64	98.22	94.82	83.53
Renter housing units (%)	19.90	35.84	22.77	22.21	23.47	30.26	44.70
Education (%)	25.81	26.05	20.54	18.58	18.29	23.68	23.10
Unemployment (%)	3.98	5.60	4.02	5.28	5.91	5.11	6.99
Poverty (%)	3.47	12.09	6.03	8.52	13.64	9.86	13.21
Mean household income ($)	79,631	65,307	71,841	61,174	54,589	66,637	72,765
Year 2000							
Total population	59,508	555,306	250,746	121,854	32,772	1,020,186	17,956,272
White	56,473	416,290	231,935	115,490	31,269	851,457	12,039,661
Black	1,037	92,335	8,160	2,772	475	104,779	2,881,463
Asian	823	13,036	4,799	758	468	19,884	1,032,442
Native American	98	1,940	632	361	20	3,051	76,263
Others	1,068	31,716	5,215	2,462	545	41,006	1,926,451
Hispanic	849	35,111	5,051	2,534	507	44,052	2,820,964
Color	3,449	151,307	21,474	7,684	1,751	185,665	7,029,113
White (%)	94.90	74.97	92.50	94.78	95.41	83.46	67.05

(continued)

Table B.39 (continued)

	0–10	11–20	21–30	31–40	41–50	0–50	Outside
Black (%)	1.74	16.63	3.25	2.27	1.45	10.27	16.05
Asian (%)	1.38	2.35	1.91	0.62	1.43	1.95	5.75
Native American (%)	0.16	0.35	0.25	0.30	0.06	0.30	0.42
Others (%)	1.79	5.71	2.08	2.02	1.66	4.02	10.73
Hispanic (%)	1.43	6.32	2.01	2.08	1.55	4.32	15.71
Color (%)	5.80	27.25	8.56	6.31	5.34	18.20	39.15
Female (%)	50.75	52.25	50.38	51.26	52.10	51.58	51.88
Old (65+ years) (%)	11.45	13.77	11.66	12.62	11.86	12.92	12.91
Kid (<5 years) (%)	8.11	7.90	7.25	7.19	6.33	7.62	7.88
Native-born (%)	94.77	92.55	95.02	97.67	97.43	94.05	78.80
Renter housing units (%)	19.43	35.73	22.67	21.65	22.63	29.80	43.98
College degree or higher (%)	31.72	30.25	26.07	22.58	20.08	28.13	27.33
Unemployment (%)	3.62	6.13	5.26	6.00	6.26	5.74	7.18
Poverty (%)	4.05	13.35	6.20	8.48	14.11	10.49	14.82
Mean household income ($)	82,529	67,914	75,301	64,752	60,079	69,874	77,488
Year 2010							
Total population	65,466	547,508	262,996	123,563	35,681	1,035,214	18,194,538
White	61,862	401,996	237,087	116,892	33,023	850,860	11,913,542
Black	1,433	99,888	12,574	2,595	1,507	117,997	2,872,594
Asian	1,179	16,782	6,640	1,196	406	26,203	1,371,511
Native American	92	1,520	470	342	144	2,568	64,308
Others	900	27,322	6,225	2,538	601	37,586	1,972,583
Hispanic	1,293	44,871	8,011	3,615	1,118	58,908	3,229,972
Color	4,529	169,549	31,091	8,863	3,396	217,428	7,637,387
White (%)	94.49	73.42	90.15	94.60	92.55	82.19	65.48
Black (%)	2.19	18.24	4.78	2.10	4.22	11.40	15.79
Asian (%)	1.80	3.07	2.52	0.97	1.14	2.53	7.54
Native American (%)	0.14	0.28	0.18	0.28	0.40	0.25	0.35
Others (%)	1.37	4.99	2.37	2.05	1.68	3.63	10.84
Hispanic (%)	1.98	8.20	3.05	2.93	3.13	5.69	17.75
Color (%)	6.92	30.97	11.82	7.17	9.52	21.00	41.98
Female (%)	50.51	52.31	50.36	51.12	49.08	51.45	51.64
Old (65+ years) (%)	14.00	13.97	13.22	14.46	11.81	13.77	13.26
Kid (<5 years) (%)	5.36	5.97	5.67	5.39	4.68	5.74	6.04
Native-born (%)	93.10	91.85	93.69	97.33	97.50	93.24	77.41

(continued)

Table B.39 (continued)

	0–10	11–20	21–30	31–40	41–50	0–50	Outside
Renter housing units (%)	19.54	33.34	21.37	20.20	19.09	27.68	40.83
College degree or higher (%)	35.17	33.70	31.19	26.37	22.13	31.92	32.14
Unemployment (%)	4.07	7.74	6.47	5.76	6.23	6.87	7.56
Poverty (%)	4.86	16.57	7.72	10.54	15.99	12.84	14.24
Mean household income ($)	78,783	62,564	73,457	64,045	57,958	66,241	81,212
Index	39	39	39	39	39	39	39

B.40 Brunswick Steam Electric Plant, North Carolina

See Fig. B.40.
See Table B.40.

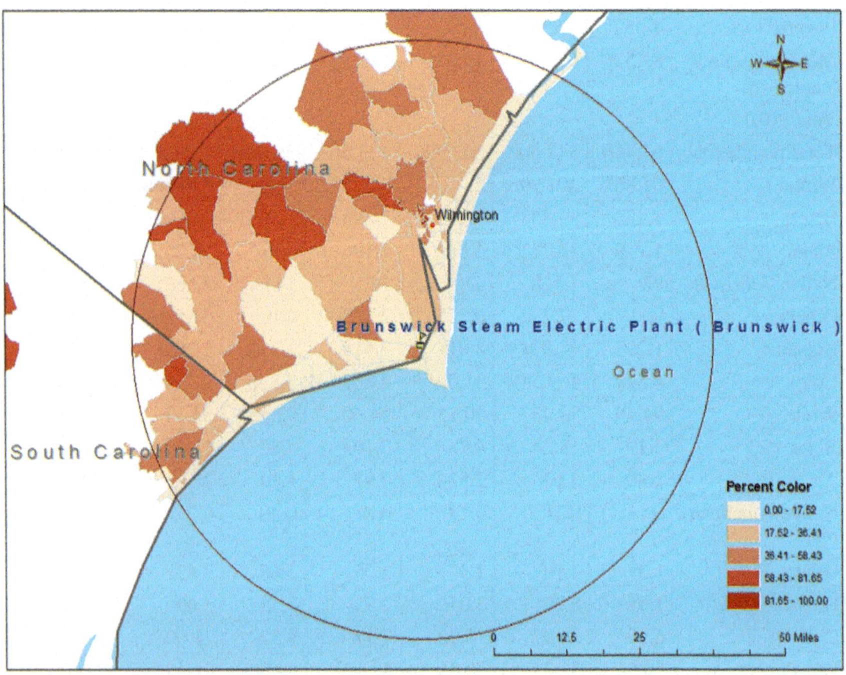

Fig. B.40 Spatial distribution of percent color among the populations surrounding Brunswick Steam Electric Plant, North Carolina, in 2010

Table B.40 Demographic composition of population, as sorted by distance from Brunswick Steam Electric Plant, North Carolina, in 1990, 2000, and 2010 *Source* (Kyne 2014)

	0–10	11–20	21–30	31–40	41–50	0–50	Outside
Year 1990							
Tract	11	31	34	24	8	108	3,190
Tract area (sq. mile)	160	579	625	912	764	3,040	82,799
Total population	13,649	71,422	83,925	41,652	21,689	232,337	9,882,965
White	12,374	59,685	62,338	30,490	11,682	176,569	7,242,354
Black	1,181	10,966	20,729	9,920	9,356	52,152	2,443,188
Asian	13	351	289	60	9	722	70,966
Native American	51	310	443	1,037	599	2,440	89,099
Others	29	109	125	145	44	452	37,369
Hispanic	58	450	545	268	94	1,415	95,933
Color	1,315	12,067	21,895	11,254	10,037	56,568	2,689,528
White (%)	90.66	83.57	74.28	73.20	53.86	76.00	73.28
Black (%)	8.65	15.35	24.70	23.82	43.14	22.45	24.72
Asian (%)	0.10	0.49	0.34	0.14	0.04	0.31	0.72
Native American (%)	0.37	0.43	0.53	2.49	2.76	1.05	0.90
Others (%)	0.21	0.15	0.15	0.35	0.20	0.19	0.38
Hispanic (%)	0.42	0.63	0.65	0.64	0.43	0.61	0.97
Color (%)	9.63	16.90	26.09	27.02	46.28	24.35	27.21
Female (%)	51.27	52.68	51.95	51.62	52.47	52.13	51.57
Old (65+ years) (%)	17.20	12.66	12.45	15.54	14.72	13.56	11.80
Kid (<5 years) (%)	6.49	7.77	7.40	7.41	7.36	7.46	8.44
Native-born (%)	98.35	98.34	98.97	98.68	99.67	98.75	98.36
Renter housing units (%)	13.29	22.94	26.29	12.04	14.54	20.28	28.16
Education (%)	13.63	19.51	17.46	14.26	8.75	16.47	17.15
Unemployment (%)	7.04	5.45	5.62	6.33	5.94	5.79	5.04
Poverty (%)	12.29	13.54	15.71	16.58	22.30	15.61	13.76
Mean household income ($)	52,717	58,299	50,771	50,726	40,276	52,330	54,044
Year 2000							
Total population	18,639	98,916	112,072	60,380	26,766	316,773	11,744,552
White	17,173	83,477	84,383	46,394	14,776	246,203	8,251,640
Black	1,113	12,313	23,379	10,865	10,265	57,935	2,858,946
Asian	69	778	772	181	78	1,878	151,002
Native American	31	498	722	1,244	808	3,303	112,341
Others	232	1,858	2,826	1,683	848	7,447	370,630
Hispanic	187	1,928	3,272	2,003	709	8,099	457,693
Color	1,561	16,548	29,063	14,610	12,129	73,911	3,684,060
White (%)	92.13	84.39	75.29	76.84	55.20	77.72	70.26

(continued)

Table B.40 (continued)

	0–10	11–20	21–30	31–40	41–50	0–50	Outside
Black (%)	5.97	12.45	20.86	17.99	38.35	18.29	24.34
Asian (%)	0.37	0.79	0.69	0.30	0.29	0.59	1.29
Native American (%)	0.17	0.50	0.64	2.06	3.02	1.04	0.96
Others (%)	1.24	1.88	2.52	2.79	3.17	2.35	3.16
Hispanic (%)	1.00	1.95	2.92	3.32	2.65	2.56	3.90
Color (%)	8.37	16.73	25.93	24.20	45.31	23.33	31.37
Female (%)	51.54	51.84	51.08	51.40	50.95	51.40	51.18
Old (65+ years) (%)	18.03	14.24	12.30	17.65	13.94	14.40	12.01
Kid (<5 years) (%)	4.59	7.33	7.06	6.49	7.76	6.95	8.00
Native-born (%)	98.01	97.01	96.58	96.15	97.95	96.83	95.44
Renter housing units (%)	12.78	21.30	26.63	12.24	15.03	20.04	26.48
College degree or higher (%)	22.17	27.64	25.68	16.26	10.09	22.90	21.76
Unemployment (%)	3.33	4.74	6.23	4.91	6.89	5.40	5.50
Poverty (%)	8.95	11.14	15.32	12.67	19.91	13.49	12.87
Mean household income ($)	64,696	67,610	61,352	59,324	45,142	61,885	62,663
Year 2010							
Total population	28,812	125,531	138,154	80,437	29,461	402,395	13,380,211
White	26,509	102,598	106,478	64,888	17,535	318,008	9,169,719
Black	1,555	15,090	24,132	11,315	10,110	62,202	3,188,224
Asian	217	1,059	1,798	364	21	3,459	253,045
Native American	58	463	603	1,476	985	3,585	116,536
Others	473	6,321	5,143	2,394	810	15,141	652,687
Hispanic	569	6,224	7,265	3,220	1,472	18,750	913,412
Color	2,654	24,976	35,797	17,101	12,667	93,195	4,652,672
White (%)	92.01	81.73	77.07	80.67	59.52	79.03	68.53
Black (%)	5.40	12.02	17.47	14.07	34.32	15.46	23.83
Asian (%)	0.75	0.84	1.30	0.45	0.07	0.86	1.89
Native American (%)	0.20	0.37	0.44	1.83	3.34	0.89	0.87
Others (%)	1.64	5.04	3.72	2.98	2.75	3.76	4.88
Hispanic (%)	1.97	4.96	5.26	4.00	5.00	4.66	6.83
Color (%)	9.21	19.90	25.91	21.26	43.00	23.16	34.77
Female (%)	51.18	52.03	51.05	51.58	50.18	51.41	51.28
Old (65+ years) (%)	22.83	14.59	13.85	20.45	15.15	16.14	12.69
Kid (<5 years) (%)	3.56	6.64	5.28	5.43	4.93	5.58	6.69
Native-born (%)	96.52	95.44	94.66	96.38	96.79	95.54	93.44

(continued)

Table B.40 (continued)

	0–10	11–20	21–30	31–40	41–50	0–50	Outside
Renter housing units (%)	13.37	24.92	28.60	12.31	15.96	21.30	26.84
College degree or higher (%)	33.08	32.83	30.26	21.31	13.64	28.17	25.31
Unemployment (%)	8.14	7.91	8.66	7.98	10.14	8.35	8.93
Poverty (%)	10.50	13.02	17.63	11.55	20.53	14.64	15.85
Mean household income ($)	68,524	70,118	60,215	59,789	48,173	63,017	60,785
Index	40	40	40	40	40	40	40

B.41 McGuire Nuclear Station, North Carolina

See Fig. B.41.
See Table B.41.

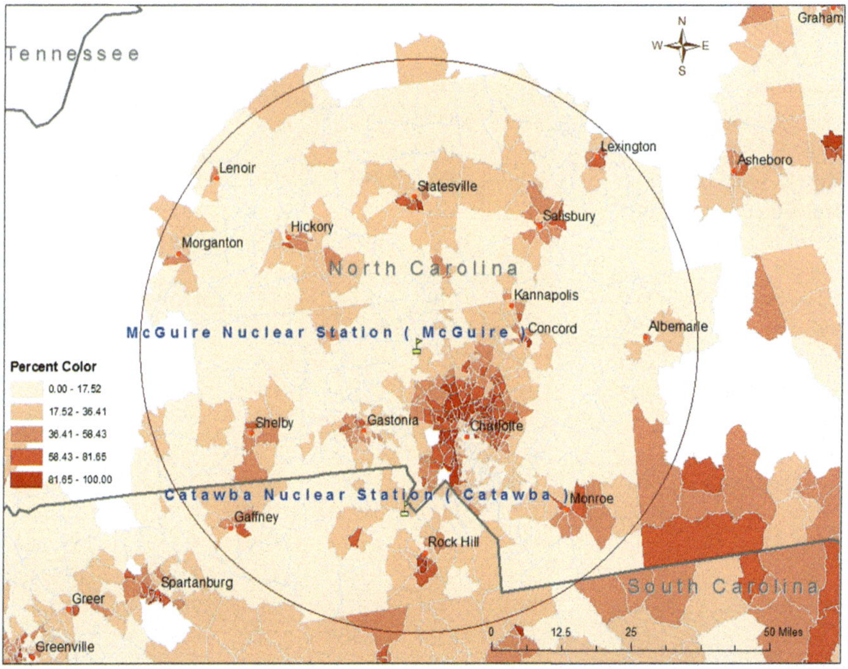

Fig. B.41 Spatial distribution of percent color among the populations surrounding McGuire Nuclear Station, North Carolina, in 2010

Table B.41 Demographic composition of population, as sorted by distance from McGuire Nuclear Station, North Carolina, in 1990, 2000, and 2010 *Source* (Kyne 2014)

	0–10	11–20	21–30	31–40	41–50	0–50	Outside
Year 1990							
Tract	47	219	186	125	36	613	2,685
Tract area (sq. mile)	322	923	1,573	2,057	1,491	6,366	79,474
Total population	61,535	546,470	508,186	391,233	128,988	1,636,412	8,478,890
White	54,050	391,448	442,955	327,287	113,863	1,329,603	6,089,320
Black	6,764	146,100	56,574	60,356	14,070	283,864	2,211,476
Asian	304	5,103	5,588	1,570	278	12,843	58,845
Native American	319	1,984	1,626	1,319	489	5,737	85,802
Others	93	1,832	1,463	696	292	4,376	33,445
Hispanic	352	4,844	4,341	2,107	747	12,391	84,957
Color	7,756	157,703	67,988	65,125	15,503	314,075	2,432,021
White (%)	87.84	71.63	87.16	83.66	88.27	81.25	71.82
Black (%)	10.99	26.74	11.13	15.43	10.91	17.35	26.08
Asian (%)	0.49	0.93	1.10	0.40	0.22	0.78	0.69
Native American (%)	0.52	0.36	0.32	0.34	0.38	0.35	1.01
Others (%)	0.15	0.34	0.29	0.18	0.23	0.27	0.39
Hispanic (%)	0.57	0.89	0.85	0.54	0.58	0.76	1.00
Color (%)	12.60	28.86	13.38	16.65	12.02	19.19	28.68
Female (%)	49.99	52.44	51.38	51.63	50.49	51.67	51.57
Old (65+ years) (%)	10.97	12.06	9.92	12.09	11.31	11.30	11.95
Kid (<5 years) (%)	8.05	8.56	8.52	8.42	7.76	8.43	8.41
Native-born (%)	98.92	98.03	97.55	99.09	99.33	98.27	98.39
Renter housing units (%)	16.31	36.65	25.01	25.94	19.16	28.43	27.82
Education (%)	18.96	15.63	22.79	13.64	11.27	17.19	17.12
Unemployment (%)	3.52	4.60	3.30	4.27	3.66	3.99	5.28
Poverty (%)	6.45	12.36	6.40	10.11	9.26	9.49	14.65
Mean household income ($)	70,219	54,885	69,750	55,420	54,957	60,197	52,782
Year 2000							
Total population	117,264	661,218	659,306	478,606	156,105	2,072,499	9,988,826
White	102,010	432,402	534,841	389,554	136,304	1,595,111	6,902,732
Black	11,676	183,934	85,412	69,361	15,194	365,577	2,551,304
Asian	1,348	13,363	13,960	5,169	542	34,382	118,498
Native American	507	2,884	2,781	2,671	723	9,566	106,078
Others	1,717	28,642	22,314	11,857	3,342	67,872	310,205
Hispanic	2,503	40,194	30,988	16,073	5,154	94,912	370,880
Color	16,770	246,061	139,643	96,433	22,545	521,452	3,236,519
White (%)	86.99	65.39	81.12	81.39	87.32	76.97	69.10

(continued)

Table B.41 (continued)

	0–10	11–20	21–30	31–40	41–50	0–50	Outside
Black (%)	9.96	27.82	12.95	14.49	9.73	17.64	25.54
Asian (%)	1.15	2.02	2.12	1.08	0.35	1.66	1.19
Native American (%)	0.43	0.44	0.42	0.56	0.46	0.46	1.06
Others (%)	1.46	4.33	3.38	2.48	2.14	3.27	3.11
Hispanic (%)	2.13	6.08	4.70	3.36	3.30	4.58	3.71
Color (%)	14.30	37.21	21.18	20.15	14.44	25.16	32.40
Female (%)	50.20	51.00	50.92	51.12	50.45	50.92	51.24
Old (65+ years) (%)	8.66	10.70	10.37	11.86	11.74	10.82	12.33
Kid (<5 years) (%)	9.01	8.30	8.39	8.35	7.98	8.36	7.90
Native-born (%)	96.60	92.40	93.26	96.41	96.99	94.18	95.74
Renter housing units (%)	15.74	35.67	25.32	24.94	18.28	27.53	26.02
College degree or higher (%)	35.59	21.72	28.59	17.32	13.10	23.07	21.52
Unemployment (%)	3.50	5.96	3.90	5.20	4.51	4.87	5.64
Poverty (%)	4.39	11.67	7.54	10.34	9.46	9.46	13.61
Mean household income ($)	97,183	64,982	80,572	63,591	61,406	71,160	60,881
Year 2010							
Total population	188,937	782,274	784,185	585,783	176,671	2,517,850	11,264,756
White	145,604	479,941	601,362	467,712	151,042	1,845,661	7,642,066
Black	31,770	226,337	112,315	86,475	16,981	473,878	2,776,548
Asian	4,590	25,249	25,830	8,565	1,088	65,322	191,182
Native American	809	2,556	3,198	2,258	580	9,401	110,720
Others	6,164	48,191	41,480	20,773	6,980	123,588	544,240
Hispanic	9,622	79,836	70,926	33,962	9,135	203,481	728,681
Color	48,726	343,042	223,401	138,425	29,987	783,581	3,962,286
White (%)	77.06	61.35	76.69	79.84	85.49	73.30	67.84
Black (%)	16.82	28.93	14.32	14.76	9.61	18.82	24.65
Asian (%)	2.43	3.23	3.29	1.46	0.62	2.59	1.70
Native American (%)	0.43	0.33	0.41	0.39	0.33	0.37	0.98
Others (%)	3.26	6.16	5.29	3.55	3.95	4.91	4.83
Hispanic (%)	5.09	10.21	9.04	5.80	5.17	8.08	6.47
Color (%)	25.79	43.85	28.49	23.63	16.97	31.12	35.17
Female (%)	50.98	51.39	51.31	50.98	51.63	51.26	51.28
Old (65+ years) (%)	8.43	10.36	11.16	12.09	14.15	11.13	13.16
Kid (<5 years) (%)	7.43	7.24	6.86	7.02	6.27	7.01	6.58
Native-born (%)	93.27	89.71	90.39	94.87	96.37	91.86	93.87

(continued)

Table B.41 (continued)

	0–10	11–20	21–30	31–40	41–50	0–50	Outside
Renter housing units (%)	18.22	33.98	26.16	23.46	18.13	26.90	26.57
College degree or higher (%)	41.75	25.94	31.32	22.69	16.84	27.42	24.95
Unemployment (%)	6.40	10.78	8.05	9.15	8.82	9.08	8.87
Poverty (%)	7.00	15.92	11.07	13.90	14.03	13.13	16.43
Mean household income ($)	93,371	62,194	74,370	64,032	57,552	68,424	59,192
Index	41	41	41	41	41	41	41

B.42 Shearon Harris Nuclear Power Plant, North Carolina

See Fig. B.42.
See Table B.42.

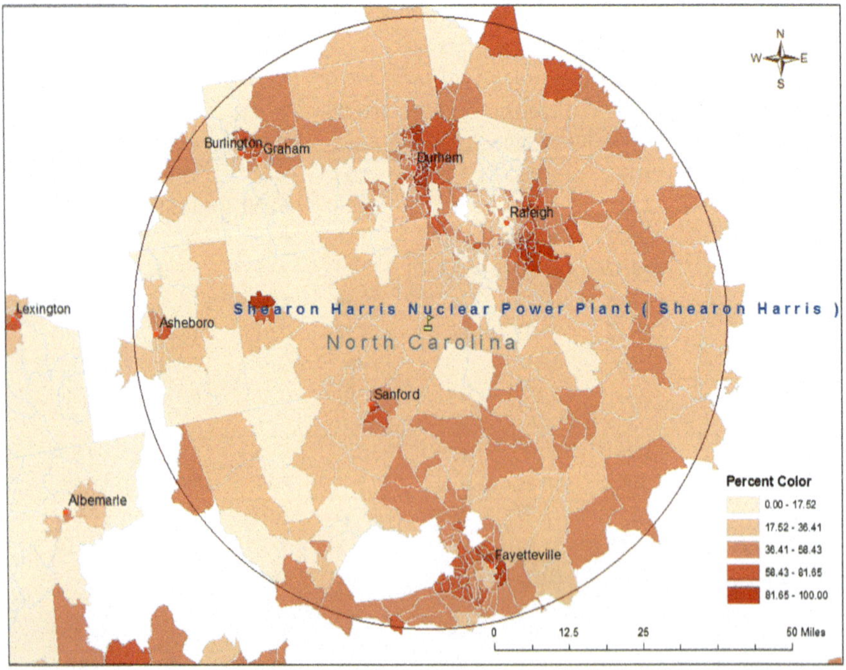

Fig. B.42 Spatial distribution of percent color among the populations surrounding Shearon Harris Nuclear Power Plant, North Carolina, in 2010

Table B.42 Demographic composition of population, as sorted by distance from Shearon Harris Nuclear Power Plant, North Carolina, in 1990, 2000, and 2010 *Source* (Kyne 2014)

	0–10	11–20	21–30	31–40	41–50	0–50	Outside
Year 1990							
Tract	19	111	176	102	78	486	1,709
Tract area (sq. mile)	272	954	1,727	1,866	1,399	6,217	47,602
Total population	24,724	249,080	499,206	346,954	273,388	1,393,352	5,235,264
White	18,076	195,649	357,131	244,455	188,777	1,004,088	4,007,155
Black	6,338	46,154	129,542	91,404	74,580	348,018	1,107,321
Asian	137	4,971	8,639	3,861	2,972	20,580	29,811
Native American	64	871	1,691	2,114	4,317	9,057	73,549
Others	116	1,394	2,228	5,122	2,743	11,603	17,442
Hispanic	344	3,538	5,872	9,573	5,928	25,255	43,763
Color	6,832	55,329	145,225	105,925	87,170	400,481	1,250,842
White (%)	73.11	78.55	71.54	70.46	69.05	72.06	76.54
Black (%)	25.64	18.53	25.95	26.34	27.28	24.98	21.15
Asian (%)	0.55	2.00	1.73	1.11	1.09	1.48	0.57
Native American (%)	0.26	0.35	0.34	0.61	1.58	0.65	1.40
Others (%)	0.47	0.56	0.45	1.48	1.00	0.83	0.33
Hispanic (%)	1.39	1.42	1.18	2.76	2.17	1.81	0.84
Color (%)	27.63	22.21	29.09	30.53	31.89	28.74	23.89
Female (%)	51.80	50.51	52.42	49.17	51.82	51.14	51.66
Old (65+ years) (%)	10.86	8.60	9.71	9.54	11.24	9.79	12.72
Kid (<5 years) (%)	9.54	7.98	8.30	9.19	9.14	8.65	8.11
Native-born (%)	98.41	95.86	96.36	97.74	97.45	96.86	98.64
Renter housing units (%)	24.64	36.58	38.68	31.79	30.25	34.82	26.99
Education (%)	20.10	33.01	33.55	16.53	16.11	25.66	15.24
Unemployment (%)	3.26	3.74	3.94	5.42	5.82	4.55	4.86
Poverty (%)	9.70	11.29	11.18	11.46	13.11	11.63	13.32
Mean household income ($)	60,506	63,876	64,641	54,434	53,679	59,938	53,138
Year 2000							
Total population	54,745	364,328	656,183	421,566	333,000	1,829,822	6,219,491
White	43,286	274,941	426,252	285,718	210,313	1,240,510	4,561,655
Black	8,362	56,852	172,471	106,660	95,130	439,475	1,294,679
Asian	1,113	14,102	17,636	4,849	5,512	43,212	71,779
Native American	258	1,532	3,128	2,988	4,718	12,624	88,332
Others	1,742	16,830	36,745	21,364	17,311	93,992	203,055
Hispanic	2,056	21,836	46,593	25,944	20,476	116,905	256,059
Color	12,396	98,885	248,860	146,147	130,529	636,817	1,763,543
White (%)	79.07	75.47	64.96	67.78	63.16	67.79	73.34

(continued)

Table B.42 (continued)

	0–10	11–20	21–30	31–40	41–50	0–50	Outside
Black (%)	15.27	15.60	26.28	25.30	28.57	24.02	20.82
Asian (%)	2.03	3.87	2.69	1.15	1.66	2.36	1.15
Native American (%)	0.47	0.42	0.48	0.71	1.42	0.69	1.42
Others (%)	3.18	4.62	5.60	5.07	5.20	5.14	3.26
Hispanic (%)	3.76	5.99	7.10	6.15	6.15	6.39	4.12
Color (%)	22.64	27.14	37.93	34.67	39.20	34.80	28.36
Female (%)	50.96	49.81	51.45	49.59	51.41	50.67	51.15
Old (65+ years) (%)	7.41	8.26	8.76	10.26	11.55	9.47	12.81
Kid (<5 years) (%)	11.05	7.78	8.30	8.95	8.69	8.50	7.88
Native-born (%)	93.99	90.01	90.01	94.80	94.34	92.02	95.43
Renter housing units (%)	20.04	33.06	35.91	28.83	29.64	32.15	25.89
College degree or higher (%)	40.46	42.91	37.96	21.84	19.89	32.03	19.76
Unemployment (%)	2.30	5.04	4.31	5.61	5.95	4.94	5.44
Poverty (%)	5.35	9.65	11.58	11.15	12.00	10.99	12.66
Mean household income ($)	85,239	81,543	75,356	63,473	61,973	71,827	61,520
Year 2010							
Total population	84,598	469,143	806,279	487,213	374,805	2,222,038	7,049,140
White	64,958	336,291	493,610	324,251	221,779	1,440,889	5,012,838
Black	10,978	69,906	215,284	116,846	108,922	521,936	1,459,006
Asian	3,410	30,026	29,999	8,170	7,002	78,607	121,556
Native American	181	2,072	2,170	2,065	4,280	10,768	95,258
Others	5,071	30,848	65,216	35,881	32,822	169,838	360,482
Hispanic	4,979	42,502	91,906	45,294	35,270	219,951	503,457
Color	21,980	151,942	351,651	182,026	165,707	873,306	2,268,808
White (%)	76.78	71.68	61.22	66.55	59.17	64.85	71.11
Black (%)	12.98	14.90	26.70	23.98	29.06	23.49	20.70
Asian (%)	4.03	6.40	3.72	1.68	1.87	3.54	1.72
Native American (%)	0.21	0.44	0.27	0.42	1.14	0.48	1.35
Others (%)	5.99	6.58	8.09	7.36	8.76	7.64	5.11
Hispanic (%)	5.89	9.06	11.40	9.30	9.41	9.90	7.14
Color (%)	25.98	32.39	43.61	37.36	44.21	39.30	32.19
Female (%)	51.05	50.40	52.12	50.59	51.83	51.33	51.23
Old (65+ years) (%)	7.11	9.26	9.03	11.36	12.31	10.07	13.39
Kid (<5 years) (%)	9.72	6.73	7.47	7.67	6.96	7.36	6.50
Native-born (%)	91.73	86.75	87.26	93.20	93.41	89.67	93.57
Renter housing units (%)	18.55	32.27	33.66	28.48	29.89	31.10	26.29

(continued)

Table B.42 (continued)

	0–10	11–20	21–30	31–40	41–50	0–50	Outside
College degree or higher (%)	46.34	47.19	40.91	25.98	22.03	35.96	23.11
Unemployment (%)	5.54	6.29	7.21	8.99	9.45	7.64	9.11
Poverty (%)	5.57	10.90	14.03	13.54	16.20	13.30	16.23
Mean household income ($)	85,138	81,927	72,288	63,019	57,944	70,362	59,138
Index	42	42	42	42	42	42	42

B.43 Davis-Besse Nuclear Power Station, Ohio

See Fig. B.43.
See Table B.43.

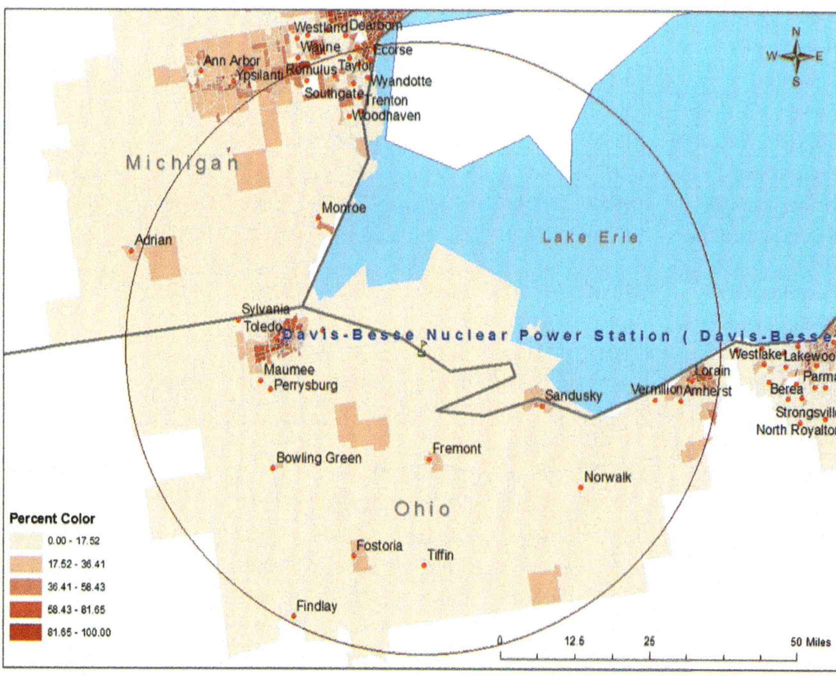

Fig. B.43 Spatial distribution of percent color among the populations surrounding Davis-Besse Nuclear Power Station, Ohio, in 2010

Table B.43 Demographic composition of population, as sorted by distance from Davis-Besse Nuclear Power Station, Ohio, in 1990, 2000, and 2010 *Source* (Kyne 2014)

	0–10	11–20	21–30	31–40	41–50	0–50	Outside
Year 1990							
Tract	6	27	163	65	102	363	5,402
Tract area (sq. mile)	416	758	1,397	1,146	758	4,476	137,063
Total population	22,999	94,762	586,871	264,486	374,460	1,343,578	18,798,744
White	22,333	90,228	498,548	254,720	339,036	1,204,865	16,079,312
Black	167	2,171	73,677	4,422	27,526	107,963	2,333,273
Asian	64	249	4,603	2,568	3,137	10,621	181,491
Native American	64	132	1,558	711	2,161	4,626	76,625
Others	375	1,974	8,491	2,064	2,600	15,504	128,028
Hispanic	940	4,149	16,610	4,880	9,851	36,430	285,469
Color	1,235	6,727	96,344	12,447	42,568	159,321	2,870,556
White (%)	97.10	95.22	84.95	96.31	90.54	89.68	85.53
Black (%)	0.73	2.29	12.55	1.67	7.35	8.04	12.41
Asian (%)	0.28	0.26	0.78	0.97	0.84	0.79	0.97
Native American (%)	0.28	0.14	0.27	0.27	0.58	0.34	0.41
Others (%)	1.63	2.08	1.45	0.78	0.69	1.15	0.68
Hispanic (%)	4.09	4.38	2.83	1.85	2.63	2.71	1.52
Color (%)	5.37	7.10	16.42	4.71	11.37	11.86	15.27
Female (%)	50.32	51.45	51.98	51.56	51.52	51.70	51.65
Old (65+ years) (%)	13.82	13.77	13.03	10.48	12.07	12.33	12.49
Kid (<5 years) (%)	7.67	8.37	9.29	8.40	8.77	8.88	8.87
Native-born (%)	98.87	98.43	97.71	97.85	96.32	97.42	96.91
Renter housing units (%)	17.36	20.40	31.01	26.13	25.38	27.49	28.29
Education (%)	10.54	11.30	15.38	17.42	9.97	13.84	17.38
Unemployment (%)	8.49	6.99	8.25	6.24	8.09	7.71	7.34
Poverty (%)	7.21	8.36	14.14	9.06	9.90	11.45	12.90
Mean household income ($)	58,521	58,550	57,630	66,489	61,610	60,453	60,011
Year 2000							
Total population	23,501	96,100	582,186	284,237	367,992	1,354,016	19,937,568
White	22,792	89,248	470,175	267,834	318,925	1,168,974	16,431,891
Black	69	2,497	82,714	5,523	32,483	123,286	2,566,796
Asian	51	435	5,192	3,268	3,747	12,693	299,572
Native American	80	246	1,687	511	1,827	4,351	83,490
Others	515	3,668	22,421	7,097	11,003	44,704	555,827
Hispanic	951	5,318	23,783	7,336	13,614	51,002	485,047
Color	1,263	9,300	122,576	20,013	57,217	210,369	3,738,808
White (%)	96.98	92.87	80.76	94.23	86.67	86.33	82.42

(continued)

Table B.43 (continued)

	0–10	11–20	21–30	31–40	41–50	0–50	Outside
Black (%)	0.29	2.60	14.21	1.94	8.83	9.11	12.87
Asian (%)	0.22	0.45	0.89	1.15	1.02	0.94	1.50
Native American (%)	0.34	0.26	0.29	0.18	0.50	0.32	0.42
Others (%)	2.19	3.82	3.85	2.50	2.99	3.30	2.79
Hispanic (%)	4.05	5.53	4.09	2.58	3.70	3.77	2.43
Color (%)	5.37	9.68	21.05	7.04	15.55	15.54	18.75
Female (%)	50.77	51.37	51.65	51.07	51.49	51.45	51.22
Old (65+ years) (%)	14.64	15.20	13.44	11.29	13.35	13.11	12.79
Kid (<5 years) (%)	6.74	7.27	8.21	7.53	8.15	7.96	8.09
Native-born (%)	98.96	98.36	97.39	97.21	96.40	97.18	95.86
Renter housing units (%)	14.76	19.92	30.25	24.63	22.79	26.03	26.25
College degree or higher (%)	13.58	13.90	18.64	21.94	12.33	17.12	21.69
Unemployment (%)	3.99	4.75	5.71	5.15	5.68	5.48	5.34
Poverty (%)	6.70	7.30	12.73	8.01	8.50	10.10	10.60
Mean household income ($)	65,801	63,897	62,473	73,446	68,645	66,499	68,575
Year 2010							
Total population	23,312	96,554	562,982	303,934	362,811	1,349,593	20,115,524
White	22,134	89,114	444,314	280,843	300,132	1,136,537	16,357,529
Black	262	2,837	88,499	8,709	40,783	141,090	2,651,766
Asian	127	403	6,293	4,604	5,395	16,822	417,412
Native American	65	200	1,910	666	1,462	4,303	72,984
Others	724	4,000	21,966	9,112	15,039	50,841	615,834
Hispanic	1,146	6,689	28,863	12,381	21,487	70,566	685,865
Color	1,825	11,077	134,953	30,332	74,527	252,714	4,159,412
White (%)	94.95	92.29	78.92	92.40	82.72	84.21	81.32
Black (%)	1.12	2.94	15.72	2.87	11.24	10.45	13.18
Asian (%)	0.54	0.42	1.12	1.51	1.49	1.25	2.08
Native American (%)	0.28	0.21	0.34	0.22	0.40	0.32	0.36
Others (%)	3.11	4.14	3.90	3.00	4.15	3.77	3.06
Hispanic (%)	4.92	6.93	5.13	4.07	5.92	5.23	3.41
Color (%)	7.83	11.47	23.97	9.98	20.54	18.73	20.68
Female (%)	51.08	50.85	51.59	50.72	51.38	51.28	51.07
Old (65+ years) (%)	16.22	16.28	13.28	12.82	13.19	13.42	13.52
Kid (<5 years) (%)	5.10	5.95	6.54	5.80	6.15	6.20	6.24
Native-born (%)	98.79	97.71	97.09	96.65	95.63	96.67	95.09

(continued)

Table B.43 (continued)

	0–10	11–20	21–30	31–40	41–50	0–50	Outside
Renter housing units (%)	12.77	18.34	29.35	23.14	22.55	25.08	24.84
College degree or higher (%)	15.75	15.67	20.68	24.81	15.14	19.60	24.88
Unemployment (%)	8.32	7.89	11.49	9.49	12.38	10.94	9.89
Poverty (%)	9.87	10.16	16.78	10.60	13.40	13.90	14.52
Mean household income ($)	59,922	60,610	56,287	69,453	58,802	60,178	63,067
Index	43	43	43	43	43	43	43

B.44 Perry Nuclear Power Plant, Ohio

See Fig. B.44.
See Table B.44.

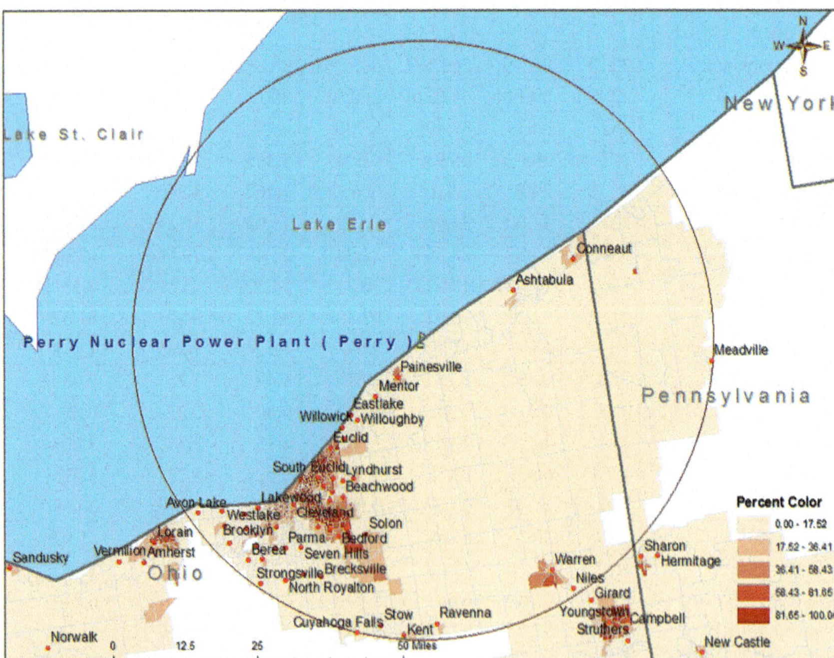

Fig. B.44 Spatial distribution of percent color among the populations surrounding Perry Nuclear Power Plant, Ohio, in 2010

Table B.44 Demographic composition of population, as sorted by distance from Perry Nuclear Power Plant, Ohio, in 1990, 2000, and 2010 *Source* (Kyne 2014)

	0–10	11–20	21–30	31–40	41–50	0–50	Outside
Year 1990							
Tract	22	51	120	254	84	531	5,639
Tract area (sq. mile)	141	1,212	2,142	931	479	4,905	85,975
Total population	68,920	199,578	407,025	761,604	328,910	1,766,037	20,962,608
White	66,129	194,305	304,537	487,990	307,495	1,360,456	18,687,648
Black	2,212	3,015	96,726	250,079	16,988	369,020	1,870,760
Asian	143	1,339	4,312	8,066	3,183	17,043	207,277
Native American	198	356	544	1,762	556	3,416	34,879
Others	242	563	910	13,696	693	16,104	162,001
Hispanic	962	1,724	2,813	23,082	2,374	30,955	321,504
Color	3,471	6,511	104,206	282,413	23,219	419,820	2,425,832
White (%)	95.95	97.36	74.82	64.07	93.49	77.03	89.15
Black (%)	3.21	1.51	23.76	32.84	5.16	20.90	8.92
Asian (%)	0.21	0.67	1.06	1.06	0.97	0.97	0.99
Native American (%)	0.29	0.18	0.13	0.23	0.17	0.19	0.17
Others (%)	0.35	0.28	0.22	1.80	0.21	0.91	0.77
Hispanic (%)	1.40	0.86	0.69	3.03	0.72	1.75	1.53
Color (%)	5.04	3.26	25.60	37.08	7.06	23.77	11.57
Female (%)	50.92	51.52	53.46	52.74	52.01	52.56	51.91
Old (65+ years) (%)	11.25	12.35	16.91	14.04	13.22	14.25	14.24
Kid (<5 years) (%)	9.48	8.54	8.13	9.16	7.96	8.64	8.32
Native-born (%)	98.02	97.35	94.02	95.72	95.95	95.64	97.37
Renter housing units (%)	25.14	23.46	32.05	38.98	25.85	32.90	28.07
Education (%)	13.03	17.24	23.04	15.23	20.23	18.16	17.43
Unemployment (%)	5.80	5.42	6.67	9.75	5.95	7.61	6.16
Poverty (%)	8.58	7.18	10.74	19.59	9.48	13.84	11.63
Mean household income ($)	58,729	65,223	67,983	51,882	64,978	59,780	59,105
Year 2000							
Total population	75,998	209,426	400,393	759,545	343,210	1,788,572	21,845,622
White	70,583	201,024	275,267	450,578	313,076	1,310,528	18,816,172
Black	2,332	3,524	112,698	262,807	19,756	401,117	2,098,911
Asian	384	1,570	5,609	12,817	4,551	24,931	330,193
Native American	123	241	420	1,754	546	3,084	43,426
Others	2,583	3,068	6,397	31,578	5,287	48,913	556,919
Hispanic	3,116	2,979	3,923	35,423	3,735	49,176	556,834
Color	6,970	10,121	127,497	322,959	32,580	500,127	3,268,987
White (%)	92.87	95.99	68.75	59.32	91.22	73.27	86.13

(continued)

Table B.44 (continued)

	0–10	11–20	21–30	31–40	41–50	0–50	Outside
Black (%)	3.07	1.68	28.15	34.60	5.76	22.43	9.61
Asian (%)	0.51	0.75	1.40	1.69	1.33	1.39	1.51
Native American (%)	0.16	0.12	0.10	0.23	0.16	0.17	0.20
Others (%)	3.40	1.46	1.60	4.16	1.54	2.73	2.55
Hispanic (%)	4.10	1.42	0.98	4.66	1.09	2.75	2.55
Color (%)	9.17	4.83	31.84	42.52	9.49	27.96	14.96
Female (%)	50.59	51.70	53.17	52.34	51.69	52.25	51.55
Old (65+ years) (%)	11.44	14.11	16.89	13.61	14.40	14.47	14.51
Kid (<5 years) (%)	8.50	7.29	7.65	8.50	7.39	7.96	7.55
Native-born (%)	96.27	96.88	93.58	95.23	95.84	95.22	96.51
Renter housing units (%)	22.40	21.86	30.32	37.04	24.71	30.94	27.05
College degree or higher (%)	16.04	21.34	27.70	20.31	24.95	22.84	21.67
Unemployment (%)	4.30	3.65	5.34	7.63	4.67	5.87	5.28
Poverty (%)	7.50	6.13	10.94	17.57	8.22	12.52	10.65
Mean household income ($)	65,424	71,288	72,616	59,867	72,744	66,774	65,640
Year 2010							
Total population	80,167	207,252	374,819	695,322	341,899	1,699,459	22,425,676
White	73,646	196,845	241,739	399,049	305,993	1,217,272	18,841,550
Black	3,453	3,854	119,345	254,470	22,446	403,568	2,334,724
Asian	631	2,218	6,877	15,941	5,675	31,342	492,731
Native American	101	439	462	1,695	524	3,221	37,769
Others	2,336	3,896	6,396	24,167	7,261	44,056	718,903
Hispanic	4,966	4,812	5,677	40,579	5,247	61,281	928,246
Color	10,256	13,741	136,703	320,734	39,218	520,652	4,082,172
White (%)	91.87	94.98	64.49	57.39	89.50	71.63	84.02
Black (%)	4.31	1.86	31.84	36.60	6.57	23.75	10.41
Asian (%)	0.79	1.07	1.83	2.29	1.66	1.84	2.20
Native American (%)	0.13	0.21	0.12	0.24	0.15	0.19	0.17
Others (%)	2.91	1.88	1.71	3.48	2.12	2.59	3.21
Hispanic (%)	6.19	2.32	1.51	5.84	1.53	3.61	4.14
Color (%)	12.79	6.63	36.47	46.13	11.47	30.64	18.20
Female (%)	49.73	51.71	52.78	52.46	51.21	52.06	51.21
Old (65+ years) (%)	12.00	16.05	16.81	13.87	14.73	14.87	14.53
Kid (<5 years) (%)	6.51	5.76	5.70	6.12	5.66	5.91	6.03
Native-born (%)	95.22	96.24	93.96	94.73	95.17	94.86	95.28
Renter housing units (%)	22.95	21.81	29.08	34.67	25.11	29.63	26.39

(continued)

Table B.44 (continued)

	0–10	11–20	21–30	31–40	41–50	0–50	Outside
College degree or higher (%)	19.23	24.07	29.83	23.26	28.97	25.78	25.29
Unemployment (%)	7.49	6.53	9.03	12.54	8.00	9.82	7.80
Poverty (%)	12.12	9.53	13.72	20.89	12.63	15.85	13.03
Mean household income ($)	62,082	66,964	66,200	55,466	66,754	61,796	65,084
Index	44	44	44	44	44	44	44

B.45 Beaver Valley Power Station, Pennsylvania

See Fig. B.45.
See Table B.45.

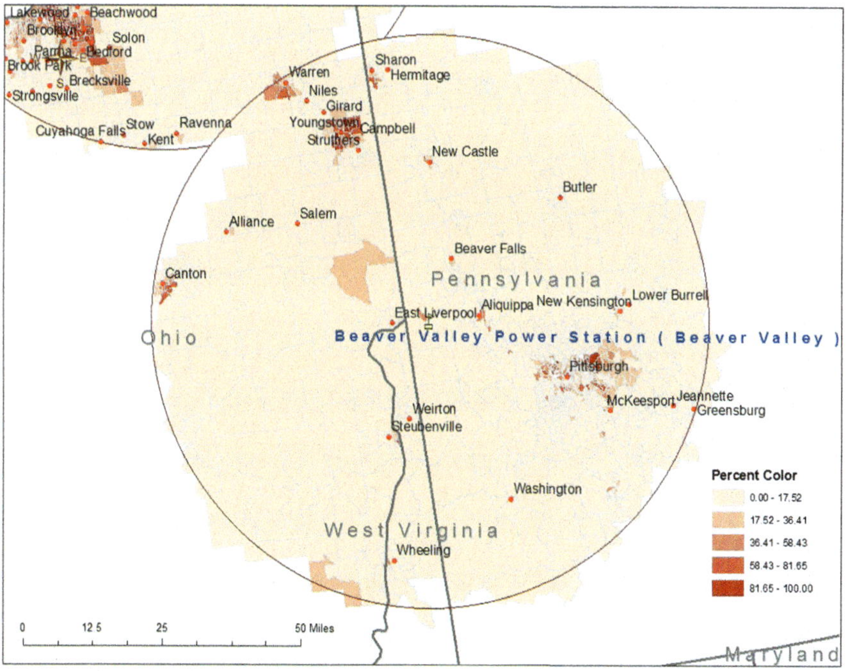

Fig. B.45 Spatial distribution of percent color among the populations surrounding Beaver Valley Power Station, Pennsylvania, in 2010

Table B.45 Demographic composition of population, as sorted by distance from Beaver Valley Power Station, Pennsylvania, in 1990, 2000, and 2010 *Source* (Kyne 2014)

	0–10	11–20	21–30	31–40	41–50	0–50	Outside
Year 1990							
Tract	33	97	313	269	96	808	5,846
Tract area (sq. mile)	279	963	1,639	1,747	1,019	5,647	109,463
Total population	119,111	341,379	1,060,141	954,382	349,524	2,824,537	21,697,584
White	111,021	330,728	935,988	853,255	328,368	2,559,360	19,214,768
Black	7,439	8,428	110,586	92,163	18,478	237,094	2,058,081
Asian	226	1,523	10,299	3,687	1,492	17,227	214,602
Native American	203	383	1,490	1,171	484	3,731	37,662
Others	234	279	1,782	4,133	696	7,124	172,432
Hispanic	604	1,485	6,729	8,240	1,658	18,716	341,636
Color	8,506	11,898	128,961	105,646	22,322	277,333	2,641,997
White (%)	93.21	96.88	88.29	89.40	93.95	90.61	88.56
Black (%)	6.25	2.47	10.43	9.66	5.29	8.39	9.49
Asian (%)	0.19	0.45	0.97	0.39	0.43	0.61	0.99
Native American (%)	0.17	0.11	0.14	0.12	0.14	0.13	0.17
Others (%)	0.20	0.08	0.17	0.43	0.20	0.25	0.79
Hispanic (%)	0.51	0.44	0.63	0.86	0.47	0.66	1.57
Color (%)	7.14	3.49	12.16	11.07	6.39	9.82	12.18
Female (%)	52.76	51.82	52.86	52.94	52.38	52.70	51.86
Old (65+ years) (%)	16.96	15.06	17.16	17.59	17.65	17.10	13.93
Kid (<5 years) (%)	7.82	7.76	7.45	7.54	7.45	7.53	8.36
Native-born (%)	97.90	97.95	96.89	97.84	98.08	97.53	97.35
Renter housing units (%)	26.42	22.30	31.20	27.65	25.13	28.01	28.02
Education (%)	10.95	18.01	21.14	15.52	13.11	17.44	17.07
Unemployment (%)	8.83	6.38	6.50	7.56	7.79	7.08	6.40
Poverty (%)	16.37	9.55	12.72	13.08	13.58	12.72	12.33
Mean household income ($)	48,114	60,829	59,438	51,531	50,431	55,351	58,440
Year 2000							
Total population	115,882	360,292	1,022,100	923,557	336,094	2,757,925	22,684,612
White	106,422	342,247	881,207	797,390	309,290	2,436,556	19,407,626
Black	7,671	9,982	108,182	102,859	19,560	248,254	2,307,773
Asian	292	3,036	16,336	5,864	1,996	27,524	337,450
Native American	139	450	1,294	1,517	541	3,941	46,339
Others	1,369	4,529	15,105	15,928	4,703	41,634	585,441
Hispanic	1,014	3,300	9,403	11,710	2,758	28,185	589,599
Color	10,179	20,125	146,362	132,175	28,564	337,405	3,530,736
White (%)	91.84	94.99	86.22	86.34	92.02	88.35	85.55

(continued)

Table B.45 (continued)

	0–10	11–20	21–30	31–40	41–50	0–50	Outside
Black (%)	6.62	2.77	10.58	11.14	5.82	9.00	10.17
Asian (%)	0.25	0.84	1.60	0.63	0.59	1.00	1.49
Native American (%)	0.12	0.12	0.13	0.16	0.16	0.14	0.20
Others (%)	1.18	1.26	1.48	1.72	1.40	1.51	2.58
Hispanic (%)	0.88	0.92	0.92	1.27	0.82	1.02	2.60
Color (%)	8.78	5.59	14.32	14.31	8.50	12.23	15.56
Female (%)	52.39	51.30	52.36	52.65	52.17	52.30	51.50
Old (65+ years) (%)	18.20	15.79	17.38	18.49	18.24	17.68	14.18
Kid (<5 years) (%)	7.09	7.22	6.69	6.95	6.64	6.86	7.61
Native-born (%)	98.46	97.54	96.36	97.95	98.12	97.35	96.50
Renter housing units (%)	25.65	20.54	30.00	26.31	24.40	26.73	26.93
College degree or higher (%)	14.47	24.76	26.89	18.94	16.78	22.18	21.13
Unemployment (%)	5.41	4.40	6.34	5.87	6.26	5.88	5.40
Poverty (%)	11.31	7.56	11.79	11.86	12.51	11.32	11.30
Mean household income ($)	53,823	71,293	64,496	56,723	56,077	61,272	64,972
Year 2010							
Total population	108,656	363,062	988,565	876,230	319,082	2,655,595	23,310,344
White	98,914	338,398	842,323	740,456	291,560	2,311,651	19,481,786
Black	7,166	11,120	100,457	107,701	19,285	245,729	2,551,531
Asian	374	6,372	24,466	8,219	2,050	41,481	494,603
Native American	110	389	1,092	1,142	446	3,179	40,950
Others	2,092	6,783	20,227	18,712	5,741	53,555	741,473
Hispanic	1,322	4,767	14,187	16,600	4,313	41,189	969,293
Color	10,672	27,561	155,220	146,119	30,482	370,054	4,353,806
White (%)	91.03	93.21	85.21	84.50	91.37	87.05	83.58
Black (%)	6.60	3.06	10.16	12.29	6.04	9.25	10.95
Asian (%)	0.34	1.76	2.47	0.94	0.64	1.56	2.12
Native American (%)	0.10	0.11	0.11	0.13	0.14	0.12	0.18
Others (%)	1.93	1.87	2.05	2.14	1.80	2.02	3.18
Hispanic (%)	1.22	1.31	1.44	1.89	1.35	1.55	4.16
Color (%)	9.82	7.59	15.70	16.68	9.55	13.93	18.68
Female (%)	51.95	50.85	51.96	52.11	52.03	51.87	51.17
Old (65+ years) (%)	18.33	15.89	16.87	17.70	18.10	17.22	14.35
Kid (<5 years) (%)	5.47	5.55	5.10	5.33	5.19	5.26	6.07
Native-born (%)	97.95	96.96	95.54	97.75	98.40	96.90	95.34

(continued)

Table B.45 (continued)

	0–10	11–20	21–30	31–40	41–50	0–50	Outside
Renter housing units (%)	24.30	20.17	28.59	25.88	24.29	25.92	26.27
College degree or higher (%)	18.03	30.45	32.11	22.56	19.19	26.58	24.51
Unemployment (%)	7.23	6.26	7.04	8.15	8.18	7.43	7.94
Poverty (%)	12.58	9.41	12.81	13.74	15.43	12.96	13.59
Mean household income ($)	56,082	75,637	65,853	55,540	52,271	61,673	64,111
Index	45	45	45	45	45	45	45

B.46 Limerick Generating Station, Pennsylvania

See Fig. B.46.
See Table B.46.

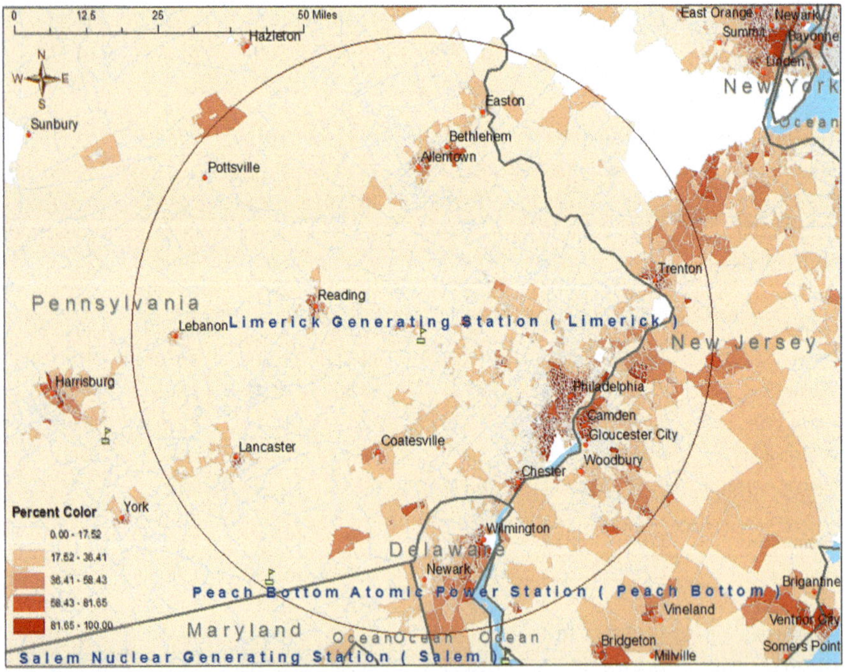

Fig. B.46 Spatial distribution of percent color among the populations surrounding Limerick Generating Station, Pennsylvania, in 2010

Table B.46 Demographic composition of population, as sorted by distance from Limerick Generating Station, Pennsylvania, in 1990, 2000, and 2010 *Source* (Kyne 2014)

	0–10	11–20	21–30	31–40	41–50	0–50	Outside
Year 1990							
Tract	62	231	704	499	113	1,609	3,837
Tract area (sq. mile)	350	913	1,512	2,035	950	5,759	51,506
Total population	184,712	790,709	2,732,462	1,859,942	418,565	5,986,390	14,291,528
White	174,314	727,752	1,895,372	1,602,011	383,381	4,782,830	12,411,323
Black	8,109	37,191	703,293	186,321	25,974	960,888	1,274,186
Asian	1,247	14,697	60,158	28,854	6,822	111,778	301,876
Native American	267	1,068	4,374	2,951	686	9,346	23,466
Others	767	9,993	69,256	39,790	1,715	121,521	280,645
Hispanic	2,376	22,040	107,438	68,080	5,389	205,323	750,634
Color	11,691	73,896	869,378	282,711	38,662	1,276,338	2,306,925
White (%)	94.37	92.04	69.36	86.13	91.59	79.90	86.84
Black (%)	4.39	4.70	25.74	10.02	6.21	16.05	8.92
Asian (%)	0.68	1.86	2.20	1.55	1.63	1.87	2.11
Native American (%)	0.14	0.14	0.16	0.16	0.16	0.16	0.16
Others (%)	0.42	1.26	2.53	2.14	0.41	2.03	1.96
Hispanic (%)	1.29	2.79	3.93	3.66	1.29	3.43	5.25
Color (%)	6.33	9.35	31.82	15.20	9.24	21.32	16.14
Female (%)	50.10	51.79	52.94	51.55	51.52	52.17	51.80
Old (65+ years) (%)	12.26	13.21	15.09	13.65	11.33	14.04	14.69
Kid (<5 years) (%)	8.62	8.40	8.51	8.51	8.64	8.51	7.91
Native-born (%)	97.67	95.62	94.37	95.87	96.79	95.27	92.48
Renter housing units (%)	23.84	27.19	30.51	26.78	26.80	28.50	28.63
Education (%)	20.18	30.16	21.30	19.54	19.06	21.75	20.27
Unemployment (%)	3.47	3.51	6.79	4.79	4.36	5.41	5.98
Poverty (%)	3.98	5.11	13.81	8.09	7.20	10.13	9.50
Mean household income ($)	74,664	86,960	65,923	70,297	68,310	70,490	68,899
Year 2000							
Total population	222,564	876,175	2,774,273	1,930,389	479,320	6,282,721	15,196,283
White	204,582	762,985	1,778,519	1,577,481	408,245	4,731,812	12,438,488
Black	10,623	47,060	751,754	208,857	47,267	1,065,561	1,422,197
Asian	3,314	28,755	94,131	52,555	11,318	190,073	531,170
Native American	286	1,514	6,131	3,770	935	12,636	27,973
Others	3,751	35,877	143,727	87,723	11,580	282,658	776,436
Hispanic	3,593	45,929	172,828	115,833	13,017	351,200	1,194,391
Color	19,782	131,299	1,053,980	399,603	77,541	1,682,205	3,345,967
White (%)	91.92	87.08	64.11	81.72	85.17	75.31	81.85

(continued)

Table B.46 (continued)

	0–10	11–20	21–30	31–40	41–50	0–50	Outside
Black (%)	4.77	5.37	27.10	10.82	9.86	16.96	9.36
Asian (%)	1.49	3.28	3.39	2.72	2.36	3.03	3.50
Native American (%)	0.13	0.17	0.22	0.20	0.20	0.20	0.18
Others (%)	1.69	4.09	5.18	4.54	2.42	4.50	5.11
Hispanic (%)	1.61	5.24	6.23	6.00	2.72	5.59	7.86
Color (%)	8.89	14.99	37.99	20.70	16.18	26.78	22.02
Female (%)	50.23	51.45	52.84	51.44	51.62	52.03	51.49
Old (65+ years) (%)	11.72	13.53	14.67	14.27	11.71	14.06	14.82
Kid (<5 years) (%)	8.64	7.82	7.77	7.53	7.64	7.72	7.48
Native-born (%)	96.63	93.55	92.49	93.73	95.38	93.38	89.38
Renter housing units (%)	22.29	26.23	31.26	26.69	25.74	28.47	28.03
College degree or higher (%)	28.85	36.19	25.84	23.93	23.27	26.62	24.85
Unemployment (%)	4.39	3.94	7.50	4.94	5.64	5.91	5.61
Poverty (%)	4.37	6.49	14.88	8.90	7.34	10.92	9.53
Mean household income ($)	87,022	94,814	69,524	75,574	73,711	75,826	75,547
Year 2010							
Total population	257,462	938,523	2,844,249	2,009,040	509,935	6,559,209	15,656,351
White	229,340	787,408	1,736,334	1,581,055	403,574	4,737,711	12,422,821
Black	14,839	53,033	797,752	236,469	67,295	1,169,388	1,539,158
Asian	7,365	48,477	132,483	78,191	19,640	286,156	777,753
Native American	197	963	5,002	4,497	853	11,512	29,452
Others	5,721	48,642	172,678	108,828	18,573	354,442	887,167
Hispanic	6,414	79,425	258,163	179,807	27,973	551,782	1,639,817
Color	32,397	192,254	1,216,711	519,083	121,554	2,081,999	4,140,991
White (%)	89.08	83.90	61.05	78.70	79.14	72.23	79.35
Black (%)	5.76	5.65	28.05	11.77	13.20	17.83	9.83
Asian (%)	2.86	5.17	4.66	3.89	3.85	4.36	4.97
Native American (%)	0.08	0.10	0.18	0.22	0.17	0.18	0.19
Others (%)	2.22	5.18	6.07	5.42	3.64	5.40	5.67
Hispanic (%)	2.49	8.46	9.08	8.95	5.49	8.41	10.47
Color (%)	12.58	20.48	42.78	25.84	23.84	31.74	26.45
Female (%)	49.99	51.45	52.53	51.30	51.37	51.81	51.14
Old (65+ years) (%)	11.65	13.86	13.69	14.17	12.23	13.67	14.75
Kid (<5 years) (%)	6.75	6.17	6.34	6.07	6.27	6.24	5.88
Native-born (%)	95.33	90.73	90.30	91.51	92.30	91.08	87.44

(continued)

Table B.46 (continued)

	0–10	11–20	21–30	31–40	41–50	0–50	Outside
Renter housing units (%)	19.91	25.96	31.64	26.37	23.71	28.20	26.72
College degree or higher (%)	35.90	41.46	30.11	27.73	27.26	31.01	29.09
Unemployment (%)	5.41	5.88	9.24	7.63	7.37	7.93	7.29
Poverty (%)	5.36	8.28	16.44	10.08	8.34	12.26	10.51
Mean household income ($)	93,455	96,613	70,631	77,758	75,337	77,828	77,247
Index	46	46	46	46	46	46	46

B.47 Peach Bottom Atomic Power Station, Pennsylvania

See Fig. B.47.
See Table B.47.

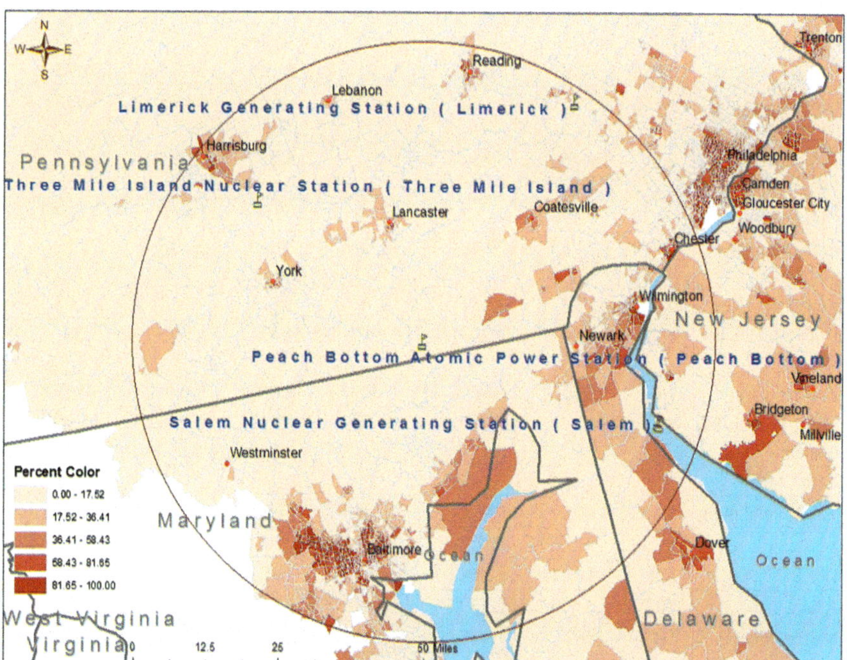

Fig. B.47 Spatial distribution of percent color among the populations surrounding Peach Bottom Atomic Power Station, Pennsylvania, in 2010

Table B.47 Demographic composition of population, as sorted by distance from Peach Bottom Atomic Power Station, Pennsylvania, in 1990, 2000, and 2010 *Source* (Kyne 2014)

	0–10	11–20	21–30	31–40	41–50	0–50	Outside
Year 1990							
Tract	10	81	228	485	204	1,008	3,834
Tract area (sq. mile)	317	966	1,579	1,909	1,023	5,794	55,155
Total population	35,760	291,208	821,840	1,702,769	718,048	3,569,625	13,759,639
White	34,789	269,850	747,917	1,171,446	610,100	2,834,102	11,620,667
Black	693	14,051	46,719	499,072	85,484	646,019	1,742,575
Asian	171	2,703	14,232	18,641	10,491	46,238	235,281
Native American	56	462	1,526	4,181	1,517	7,742	24,682
Others	61	4,126	11,433	9,443	10,447	35,510	136,391
Hispanic	226	7,205	21,130	23,351	21,951	73,863	281,754
Color	1,108	23,956	82,927	542,986	118,188	769,165	2,269,976
White (%)	97.28	92.67	91.01	68.80	84.97	79.39	84.45
Black (%)	1.94	4.83	5.68	29.31	11.91	18.10	12.66
Asian (%)	0.48	0.93	1.73	1.09	1.46	1.30	1.71
Native American (%)	0.16	0.16	0.19	0.25	0.21	0.22	0.18
Others (%)	0.17	1.42	1.39	0.55	1.45	0.99	0.99
Hispanic (%)	0.63	2.47	2.57	1.37	3.06	2.07	2.05
Color (%)	3.10	8.23	10.09	31.89	16.46	21.55	16.50
Female (%)	49.56	50.81	51.12	52.31	51.99	51.82	51.93
Old (65+ years) (%)	10.65	11.11	11.60	13.24	14.25	12.87	14.30
Kid (<5 years) (%)	9.69	9.45	8.81	8.77	8.45	8.78	8.17
Native-born (%)	99.32	97.96	96.67	96.69	97.00	96.88	95.69
Renter housing units (%)	16.57	24.15	29.20	35.88	31.19	32.36	27.12
Education (%)	9.30	17.33	23.32	19.70	16.21	19.52	20.63
Unemployment (%)	4.22	3.50	3.29	5.59	4.47	4.63	5.60
Poverty (%)	7.41	6.55	6.86	12.08	8.56	9.68	10.39
Mean household income ($)	63,748	69,715	73,801	66,101	63,128	67,489	65,008
Year 2000							
Total population	40,396	344,798	931,445	1,759,116	769,821	3,845,576	14,515,564
White	39,286	314,594	807,708	1,138,072	596,443	2,896,103	11,565,779
Black	531	16,111	62,687	542,850	114,133	736,312	2,092,423
Asian	202	3,514	22,303	30,882	17,572	74,473	374,010
Native American	30	563	1,915	4,504	2,250	9,262	29,011
Others	354	10,020	36,825	42,800	39,428	129,427	454,340
Hispanic	297	12,121	44,816	44,060	48,682	149,976	506,571
Color	1,310	35,048	143,047	642,036	190,731	1,012,172	3,165,543

(continued)

Table B.47 (continued)

	0–10	11–20	21–30	31–40	41–50	0–50	Outside
White (%)	97.25	91.24	86.72	64.70	77.48	75.31	79.68
Black (%)	1.31	4.67	6.73	30.86	14.83	19.15	14.42
Asian (%)	0.50	1.02	2.39	1.76	2.28	1.94	2.58
Native American (%)	0.07	0.16	0.21	0.26	0.29	0.24	0.20
Others (%)	0.88	2.91	3.95	2.43	5.12	3.37	3.13
Hispanic (%)	0.74	3.52	4.81	2.50	6.32	3.90	3.49
Color (%)	3.24	10.16	15.36	36.50	24.78	26.32	21.81
Female (%)	49.66	50.98	51.29	52.34	51.89	51.84	51.71
Old (65+ years) (%)	11.10	11.85	12.70	13.15	14.48	13.17	14.56
Kid (<5 years) (%)	8.46	8.08	7.94	7.64	7.83	7.80	7.38
Native-born (%)	98.60	97.25	94.95	94.85	95.14	95.19	93.89
Renter housing units (%)	14.92	21.42	27.57	32.48	30.37	29.83	26.34
College degree or higher (%)	13.70	22.60	28.06	25.01	20.18	24.44	25.21
Unemployment (%)	3.23	3.16	4.08	5.75	5.00	4.92	5.48
Poverty (%)	5.82	6.53	7.20	11.71	10.04	9.75	10.30
Mean household income ($)	69,195	75,538	78,651	71,599	67,536	72,773	70,781
Year 2010							
Total population	43,032	387,349	1,024,312	1,847,088	822,243	4,124,024	15,066,382
White	41,506	347,498	851,839	1,144,429	595,702	2,980,974	11,501,764
Black	763	22,544	92,530	584,976	137,182	837,995	2,359,545
Asian	119	5,046	37,446	55,118	25,725	123,454	547,915
Native American	56	749	2,136	4,879	1,509	9,329	28,165
Others	588	11,512	40,361	57,686	62,125	172,272	628,993
Hispanic	752	21,957	74,840	87,726	85,660	270,935	882,434
Color	1,927	54,631	219,035	755,927	262,768	1,294,288	4,006,588
White (%)	96.45	89.71	83.16	61.96	72.45	72.28	76.34
Black (%)	1.77	5.82	9.03	31.67	16.68	20.32	15.66
Asian (%)	0.28	1.30	3.66	2.98	3.13	2.99	3.64
Native American (%)	0.13	0.19	0.21	0.26	0.18	0.23	0.19
Others (%)	1.37	2.97	3.94	3.12	7.56	4.18	4.17
Hispanic (%)	1.75	5.67	7.31	4.75	10.42	6.57	5.86
Color (%)	4.48	14.10	21.38	40.93	31.96	31.38	26.59
Female (%)	49.04	50.67	51.35	52.34	51.63	51.76	51.34
Old (65+ years) (%)	12.65	12.87	13.41	12.81	13.71	13.14	14.50
Kid (<5 years) (%)	5.47	6.22	6.48	6.35	6.48	6.39	5.87

(continued)

Table B.47 (continued)

	0–10	11–20	21–30	31–40	41–50	0–50	Outside
Native-born (%)	98.44	96.31	92.52	92.19	92.50	92.78	91.84
Renter housing units (%)	13.08	20.85	26.66	31.04	29.70	28.67	25.62
College degree or higher (%)	14.75	26.10	31.65	30.15	24.02	28.77	29.28
Unemployment (%)	7.06	5.54	5.88	7.57	7.61	6.96	7.09
Poverty (%)	8.30	7.87	8.99	11.81	12.35	10.81	11.28
Mean household income ($)	74,946	80,493	80,633	75,452	69,182	75,930	74,175
Index	47	47	47	47	47	47	47

B.48 Susquehanna Steam Electric Station, Pennsylvania

See Fig. B.48.
See Table B.48.

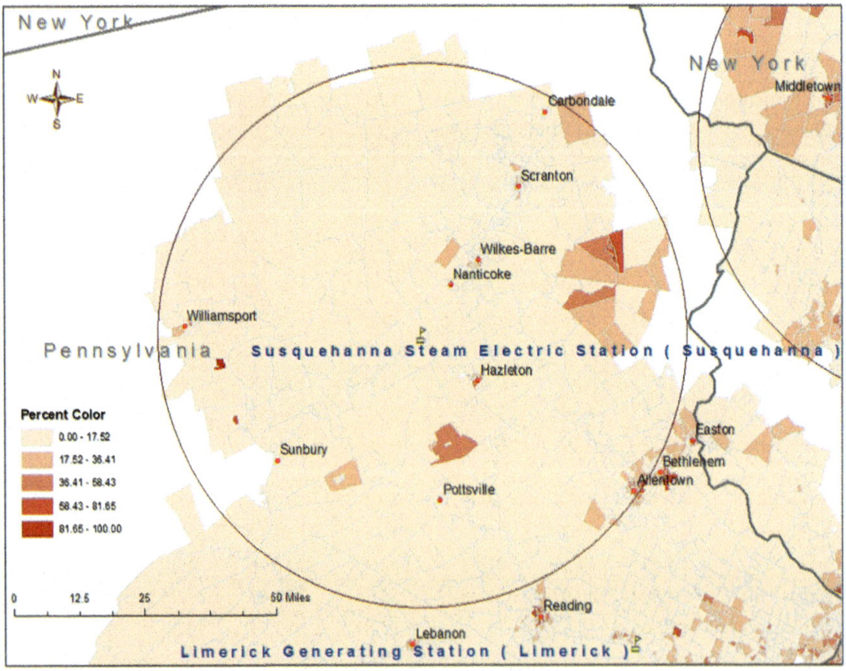

Fig. B.48 Spatial distribution of percent color among the populations surrounding Susquehanna Steam Electric Station, Pennsylvania, in 2010

Table B.48 Demographic composition of population, as sorted by distance from Susquehanna Steam Electric Station, Pennsylvania, in 1990, 2000, and 2010 *Source* (Kyne 2014)

	0–10	11–20	21–30	31–40	41–50	0–50	Outside
Year 1990							
Tract	14	84	77	97	31	303	2,915
Tract area (sq. mile)	287	1,075	1,627	1,746	1,128	5,863	40,192
Total population	51,798	271,019	291,624	369,871	118,025	1,102,337	10,779,295
White	51,037	265,342	289,016	364,679	116,600	1,086,674	9,436,499
Black	495	3,600	1,121	2,235	616	8,067	1,079,491
Asian	211	1,161	892	1,784	386	4,434	130,648
Native American	27	159	219	408	103	916	15,051
Others	29	746	382	778	300	2,235	117,574
Hispanic	289	1,794	1,143	2,412	880	6,518	213,956
Color	996	6,714	3,410	6,852	1,951	19,923	1,430,575
White (%)	98.53	97.91	99.11	98.60	98.79	98.58	87.54
Black (%)	0.96	1.33	0.38	0.60	0.52	0.73	10.01
Asian (%)	0.41	0.43	0.31	0.48	0.33	0.40	1.21
Native American (%)	0.05	0.06	0.08	0.11	0.09	0.08	0.14
Others (%)	0.06	0.28	0.13	0.21	0.25	0.20	1.09
Hispanic (%)	0.56	0.66	0.39	0.65	0.75	0.59	1.98
Color (%)	1.92	2.48	1.17	1.85	1.65	1.81	13.27
Female (%)	51.51	52.69	52.55	52.20	50.28	52.17	52.07
Old (65+ years) (%)	17.26	19.27	20.22	17.37	12.58	18.07	15.13
Kid (<5 years) (%)	7.54	6.97	7.07	7.71	8.07	7.39	8.08
Native-born (%)	98.69	98.37	98.57	98.39	98.69	98.48	96.73
Renter housing units (%)	20.01	30.27	20.33	25.36	17.94	24.19	26.98
Education (%)	11.14	12.48	10.32	13.33	11.85	12.05	18.56
Unemployment (%)	5.92	6.00	6.35	5.41	4.21	5.67	6.00
Poverty (%)	9.11	11.70	11.14	9.78	8.44	10.43	11.20
Mean household income ($)	53,076	47,121	47,544	52,612	58,235	50,447	61,232
Year 2000							
Total population	53,318	262,272	290,229	377,812	132,857	1,116,488	11,164,566
White	51,573	252,219	282,224	365,141	129,902	1,081,059	9,405,118
Black	989	5,003	4,265	4,725	805	15,787	1,195,882
Asian	271	1,320	1,194	2,601	546	5,932	214,420
Native American	99	372	280	529	181	1,461	18,050
Others	389	3,346	2,268	4,829	1,391	12,223	331,122
Hispanic	569	3,410	2,799	5,862	1,800	14,440	377,681
Color	2,140	11,743	9,578	15,644	3,950	43,055	1,910,001
White (%)	96.73	96.17	97.24	96.65	97.78	96.83	84.24

(continued)

Table B.48 (continued)

	0–10	11–20	21–30	31–40	41–50	0–50	Outside
Black (%)	1.85	1.91	1.47	1.25	0.61	1.41	10.71
Asian (%)	0.51	0.50	0.41	0.69	0.41	0.53	1.92
Native American (%)	0.19	0.14	0.10	0.14	0.14	0.13	0.16
Others (%)	0.73	1.28	0.78	1.28	1.05	1.09	2.97
Hispanic (%)	1.07	1.30	0.96	1.55	1.35	1.29	3.38
Color (%)	4.01	4.48	3.30	4.14	2.97	3.86	17.11
Female (%)	50.52	51.93	51.12	51.73	50.20	51.38	51.77
Old (65+ years) (%)	17.22	19.18	20.01	17.31	13.20	17.96	15.40
Kid (<5 years) (%)	6.21	6.09	5.96	6.50	6.85	6.29	7.28
Native-born (%)	98.36	98.08	98.53	97.97	98.29	98.20	95.63
Renter housing units (%)	19.62	28.64	19.95	24.04	17.44	23.11	26.43
College degree or higher (%)	13.78	15.25	14.28	16.86	16.09	15.56	23.06
Unemployment (%)	6.32	6.23	5.69	4.85	5.52	5.53	5.68
Poverty (%)	9.88	12.60	10.25	9.49	8.16	10.27	11.05
Mean household income ($)	56,513	51,664	52,833	57,830	66,608	55,919	66,631
Year 2010							
Total population	55,117	264,433	292,379	385,363	139,000	1,136,292	11,476,413
White	53,031	242,828	279,934	362,588	133,583	1,071,964	9,388,132
Black	797	10,266	6,394	9,364	1,883	28,704	1,318,348
Asian	623	2,368	1,551	4,658	881	10,081	325,366
Native American	122	276	513	462	99	1,472	16,733
Others	544	8,695	3,987	8,291	2,554	24,071	427,834
Hispanic	1,214	16,572	6,865	14,718	3,641	43,010	613,498
Color	3,088	31,046	16,898	31,980	7,636	90,648	2,397,756
White (%)	96.22	91.83	95.74	94.09	96.10	94.34	81.80
Black (%)	1.45	3.88	2.19	2.43	1.35	2.53	11.49
Asian (%)	1.13	0.90	0.53	1.21	0.63	0.89	2.84
Native American (%)	0.22	0.10	0.18	0.12	0.07	0.13	0.15
Others (%)	0.99	3.29	1.36	2.15	1.84	2.12	3.73
Hispanic (%)	2.20	6.27	2.35	3.82	2.62	3.79	5.35
Color (%)	5.60	11.74	5.78	8.30	5.49	7.98	20.89
Female (%)	50.59	51.49	50.03	51.19	50.13	50.80	51.38
Old (65+ years) (%)	16.87	17.39	18.60	16.98	14.10	17.13	15.09
Kid (<5 years) (%)	4.61	5.19	5.11	5.35	5.27	5.21	5.82
Native-born (%)	97.76	95.45	97.86	96.80	97.88	96.94	94.17

(continued)

Table B.48 (continued)

	0–10	11–20	21–30	31–40	41–50	0–50	Outside
Renter housing units (%)	18.66	28.44	20.32	23.32	18.17	22.91	26.17
College degree or higher (%)	17.19	18.20	18.16	19.96	19.56	18.89	27.15
Unemployment (%)	5.64	7.00	8.04	6.45	6.37	6.92	7.33
Poverty (%)	10.54	15.86	11.66	11.79	8.73	12.26	12.39
Mean household income ($)	56,670	52,042	55,312	58,621	66,313	57,003	68,322
Index	48	48	48	48	48	48	48

B.49 Three Mile Island Nuclear Station, Pennsylvania

See Fig. B.49.
See Table B.49.

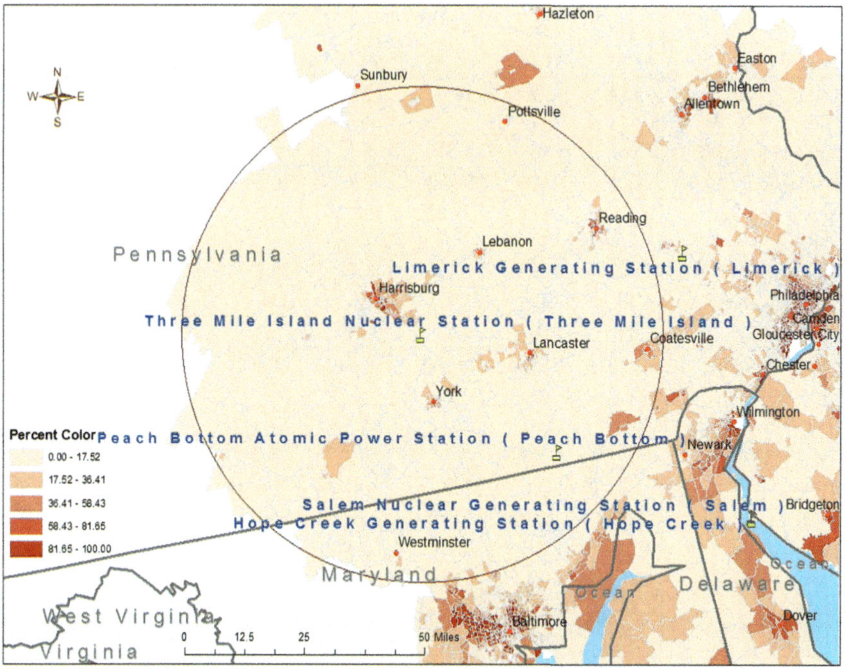

Fig. B.49 Spatial distribution of percent color among the populations surrounding Three Mile Island Nuclear Station, Pennsylvania, in 2010

Table B.49 Demographic composition of population, as sorted by distance from Three Mile Island Nuclear Station, Pennsylvania, in 1990, 2000, and 2010 *Source* (Kyne 2014)

	0–10	11–20	21–30	31–40	41–50	0–50	Outside
Year 1990							
Tract	41	150	118	74	65	448	4,176
Tract area (sq. mile)	297	933	1,520	2,098	1,093	5,941	52,519
Total population	156,537	568,626	476,325	293,340	223,969	1,718,797	14,944,296
White	145,102	516,219	450,334	289,009	218,524	1,619,188	12,300,242
Black	8,529	40,264	10,337	2,013	3,254	64,397	2,212,075
Asian	1,367	6,012	4,606	1,081	1,448	14,514	258,234
Native American	313	794	571	220	314	2,212	28,012
Others	1,238	5,309	10,466	1,022	425	18,460	145,702
Hispanic	2,409	9,714	16,221	2,384	1,198	31,926	308,538
Color	12,590	56,304	31,016	5,666	6,257	111,833	2,790,162
White (%)	92.70	90.78	94.54	98.52	97.57	94.20	82.31
Black (%)	5.45	7.08	2.17	0.69	1.45	3.75	14.80
Asian (%)	0.87	1.06	0.97	0.37	0.65	0.84	1.73
Native American (%)	0.20	0.14	0.12	0.07	0.14	0.13	0.19
Others (%)	0.79	0.93	2.20	0.35	0.19	1.07	0.97
Hispanic (%)	1.54	1.71	3.41	0.81	0.53	1.86	2.06
Color (%)	8.04	9.90	6.51	1.93	2.79	6.51	18.67
Female (%)	51.73	51.68	51.36	50.42	51.51	51.36	51.99
Old (65+ years) (%)	13.52	13.87	13.67	12.21	13.68	13.47	14.15
Kid (<5 years) (%)	8.05	8.08	8.54	9.07	8.57	8.44	8.26
Native-born (%)	98.01	97.95	98.02	98.93	98.29	98.18	95.64
Renter housing units (%)	27.73	29.62	28.82	19.57	20.02	26.37	28.48
Education (%)	17.02	19.25	14.93	11.85	19.73	16.70	20.78
Unemployment (%)	3.76	3.82	3.42	3.16	3.88	3.60	5.68
Poverty (%)	7.03	7.04	7.60	6.74	7.02	7.14	10.67
Mean household income ($)	62,567	64,844	60,486	61,943	71,398	63,803	65,556
Year 2000							
Total population	177,279	618,682	521,514	329,596	258,966	1,906,037	15,671,503
White	159,236	539,121	479,171	319,472	247,414	1,744,414	12,132,784
Black	10,592	48,334	13,859	2,961	5,955	81,701	2,598,211
Asian	2,777	9,539	6,197	2,432	2,630	23,575	408,520
Native American	235	917	675	442	472	2,741	32,421
Others	4,454	20,735	21,585	4,313	2,509	53,596	499,577
Hispanic	4,874	21,767	30,131	5,181	2,444	64,397	554,829
Color	20,286	87,669	54,181	12,808	13,151	188,095	3,774,376
White (%)	89.82	87.14	91.88	96.93	95.54	91.52	77.42

(continued)

Table B.49 (continued)

	0–10	11–20	21–30	31–40	41–50	0–50	Outside
Black (%)	5.97	7.81	2.66	0.90	2.30	4.29	16.58
Asian (%)	1.57	1.54	1.19	0.74	1.02	1.24	2.61
Native American (%)	0.13	0.15	0.13	0.13	0.18	0.14	0.21
Others (%)	2.51	3.35	4.14	1.31	0.97	2.81	3.19
Hispanic (%)	2.75	3.52	5.78	1.57	0.94	3.38	3.54
Color (%)	11.44	14.17	10.39	3.89	5.08	9.87	24.08
Female (%)	51.62	51.41	51.14	50.29	50.78	51.08	51.83
Old (65+ years) (%)	13.68	14.63	14.40	13.26	13.45	14.08	14.36
Kid (<5 years) (%)	7.35	7.20	7.70	7.85	7.35	7.48	7.45
Native-born (%)	96.81	96.85	96.89	97.88	97.69	97.15	93.80
Renter housing units (%)	26.00	27.73	26.88	18.88	17.79	24.58	27.48
College degree or higher (%)	23.05	23.78	18.70	16.06	25.13	21.21	25.52
Unemployment (%)	3.29	3.72	3.59	3.37	3.44	3.55	5.60
Poverty (%)	6.54	7.40	7.72	6.59	6.26	7.11	10.60
Mean household income ($)	68,924	69,167	65,922	66,958	78,348	69,104	71,261
Year 2010							
Total population	191,325	682,479	561,898	357,731	275,252	2,068,685	16,240,443
White	163,766	577,673	504,793	342,331	258,225	1,846,788	12,009,524
Black	14,572	59,853	19,086	5,373	8,488	107,372	2,904,915
Asian	4,489	15,871	9,683	2,463	4,143	36,649	606,349
Native American	176	1,352	688	532	507	3,255	31,163
Others	8,322	27,730	27,648	7,032	3,889	74,621	688,492
Hispanic	9,836	39,676	46,888	9,213	5,959	111,572	974,882
Color	32,310	126,631	81,083	20,474	21,185	281,683	4,723,573
White (%)	85.60	84.64	89.84	95.70	93.81	89.27	73.95
Black (%)	7.62	8.77	3.40	1.50	3.08	5.19	17.89
Asian (%)	2.35	2.33	1.72	0.69	1.51	1.77	3.73
Native American (%)	0.09	0.20	0.12	0.15	0.18	0.16	0.19
Others (%)	4.35	4.06	4.92	1.97	1.41	3.61	4.24
Hispanic (%)	5.14	5.81	8.34	2.58	2.16	5.39	6.00
Color (%)	16.89	18.55	14.43	5.72	7.70	13.62	29.09
Female (%)	51.12	51.22	51.07	50.66	50.74	51.01	51.48
Old (65+ years) (%)	13.08	14.70	14.92	14.49	14.11	14.50	14.19
Kid (<5 years) (%)	6.71	5.98	6.19	6.19	5.67	6.10	5.95
Native-born (%)	95.58	95.07	95.75	97.51	96.53	95.92	91.56
Renter housing units (%)	25.62	27.52	26.58	18.58	16.57	24.20	26.74

(continued)

Table B.49 (continued)

	0–10	11–20	21–30	31–40	41–50	0–50	Outside
College degree or higher (%)	25.91	27.64	21.40	19.27	28.98	24.54	29.84
Unemployment (%)	5.30	5.89	6.17	4.98	5.40	5.69	7.24
Poverty (%)	9.12	9.36	9.54	7.02	7.36	8.71	11.50
Mean household income ($)	69,255	70,611	65,528	70,953	85,147	71,013	74,991
Index	49	49	49	49	49	49	49

B.50 Catawba Nuclear Station, South Carolina

See Fig. B.50.
See Table B.50.

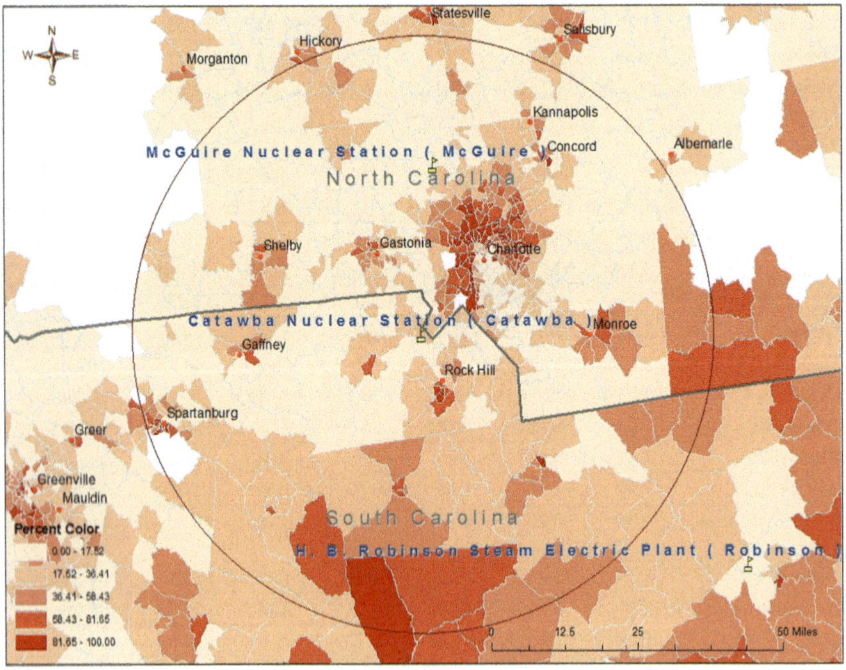

Fig. B.50 Spatial distribution of percent color among the populations surrounding Catawba Nuclear Station, South Carolina, in 2010

Table B.50 Demographic composition of population, as sorted by distance from Catawba Nuclear Station, South Carolina, in 1990, 2000, and 2010 *Source* (Kyne 2014)

	0–10	11–20	21–30	31–40	41–50	0–50	Outside
Year 1990							
Tract	46	192	165	108	45	556	2,742
Tract area (sq. mile)	309	824	1,695	2,276	1,300	6,403	79,437
Total population	101,221	506,174	375,954	306,069	148,143	1,437,561	8,677,741
White	85,352	374,242	287,011	250,088	124,291	1,120,984	6,297,939
Black	14,449	122,466	83,012	53,605	22,835	296,367	2,198,973
Asian	798	5,916	3,473	985	463	11,635	60,053
Native American	419	1,868	1,536	784	306	4,913	86,626
Others	214	1,676	916	612	253	3,671	34,150
Hispanic	627	5,375	2,422	1,670	550	10,644	86,704
Color	16,263	135,152	90,326	56,959	24,141	322,841	2,423,255
White (%)	84.32	73.94	76.34	81.71	83.90	77.98	72.58
Black (%)	14.27	24.19	22.08	17.51	15.41	20.62	25.34
Asian (%)	0.79	1.17	0.92	0.32	0.31	0.81	0.69
Native American (%)	0.41	0.37	0.41	0.26	0.21	0.34	1.00
Others (%)	0.21	0.33	0.24	0.20	0.17	0.26	0.39
Hispanic (%)	0.62	1.06	0.64	0.55	0.37	0.74	1.00
Color (%)	16.07	26.70	24.03	18.61	16.30	22.46	27.92
Female (%)	51.99	52.36	51.47	51.89	51.30	51.89	51.53
Old (65+ years) (%)	10.57	10.72	9.45	12.87	13.87	11.16	11.96
Kid (<5 years) (%)	8.43	8.84	9.07	8.08	7.89	8.61	8.39
Native-born (%)	98.38	97.27	98.20	99.21	99.36	98.22	98.39
Renter housing units (%)	27.25	36.73	27.35	22.96	20.29	29.08	27.73
Education (%)	20.31	24.62	15.15	12.06	9.25	17.59	17.05
Unemployment (%)	4.34	4.39	4.36	4.60	3.78	4.36	5.18
Poverty (%)	8.76	11.06	9.75	10.77	9.29	10.31	14.39
Mean household income ($)	65,223	66,912	57,991	54,520	50,828	60,268	52,954
Year 2000							
Total population	137,392	596,426	521,467	385,237	179,167	1,819,689	10,241,636
White	110,288	416,181	360,546	312,168	147,562	1,346,745	7,151,098
Black	21,441	143,626	127,229	60,596	25,027	377,919	2,538,962
Asian	1,950	12,971	10,344	1,921	1,770	28,956	123,924
Native American	862	3,129	2,264	1,165	657	8,077	107,567
Others	2,864	20,508	21,073	9,407	4,147	57,999	320,078
Hispanic	3,459	29,849	26,838	14,721	5,608	80,475	385,317
Color	28,947	194,624	172,093	80,278	34,100	510,042	3,247,929
White (%)	80.27	69.78	69.14	81.03	82.36	74.01	69.82

(continued)

Table B.50 (continued)

	0–10	11–20	21–30	31–40	41–50	0–50	Outside
Black (%)	15.61	24.08	24.40	15.73	13.97	20.77	24.79
Asian (%)	1.42	2.17	1.98	0.50	0.99	1.59	1.21
Native American (%)	0.63	0.52	0.43	0.30	0.37	0.44	1.05
Others (%)	2.08	3.44	4.04	2.44	2.31	3.19	3.13
Hispanic (%)	2.52	5.00	5.15	3.82	3.13	4.42	3.76
Color (%)	21.07	32.63	33.00	20.84	19.03	28.03	31.71
Female (%)	51.84	51.23	50.84	50.99	50.30	51.02	51.21
Old (65+ years) (%)	10.10	10.43	8.83	12.19	12.70	10.54	12.34
Kid (<5 years) (%)	8.35	8.30	8.88	8.29	7.84	8.42	7.90
Native-born (%)	96.29	92.64	93.15	96.24	96.87	94.24	95.69
Renter housing units (%)	26.50	34.97	26.73	22.44	20.37	27.95	25.98
College degree or higher (%)	26.82	31.48	22.52	18.49	12.77	23.98	21.40
Unemployment (%)	5.94	5.63	4.78	4.99	4.11	5.13	5.57
Poverty (%)	8.57	10.44	9.73	9.89	9.31	9.87	13.43
Mean household income ($)	73,800	81,761	67,243	67,343	59,438	71,896	61,009
Year 2010							
Total population	200,869	693,864	670,767	464,075	205,922	2,235,497	11,547,109
White	152,165	471,548	418,438	372,209	167,638	1,581,998	7,905,729
Black	35,104	161,031	190,158	69,123	27,425	482,841	2,767,585
Asian	5,771	23,823	18,911	6,456	3,427	58,388	198,116
Native American	556	3,374	2,981	1,405	440	8,756	111,365
Others	7,273	34,088	40,279	14,882	6,992	103,514	564,314
Hispanic	14,464	58,967	63,030	31,482	12,116	180,059	752,103
Color	58,960	256,111	284,248	113,063	45,505	757,887	3,987,980
White (%)	75.75	67.96	62.38	80.20	81.41	70.77	68.47
Black (%)	17.48	23.21	28.35	14.89	13.32	21.60	23.97
Asian (%)	2.87	3.43	2.82	1.39	1.66	2.61	1.72
Native American (%)	0.28	0.49	0.44	0.30	0.21	0.39	0.96
Others (%)	3.62	4.91	6.00	3.21	3.40	4.63	4.89
Hispanic (%)	7.20	8.50	9.40	6.78	5.88	8.05	6.51
Color (%)	29.35	36.91	42.38	24.36	22.10	33.90	34.54
Female (%)	51.82	51.64	51.28	51.12	50.60	51.34	51.27
Old (65+ years) (%)	10.03	10.52	9.27	12.45	13.22	10.75	13.19
Kid (<5 years) (%)	7.45	7.17	7.41	6.88	6.38	7.14	6.57
Native-born (%)	92.75	89.74	90.29	94.34	95.55	91.67	93.86

(continued)

Table B.50 (continued)

	0–10	11–20	21–30	31–40	41–50	0–50	Outside
Renter housing units (%)	25.73	32.68	27.48	23.41	21.19	27.57	26.46
College degree or higher (%)	32.92	36.63	25.99	22.89	16.67	28.42	24.82
Unemployment (%)	8.22	9.49	9.53	9.76	9.16	9.41	8.81
Poverty (%)	10.07	14.15	13.50	13.39	12.46	13.28	16.31
Mean household income ($)	72,266	79,641	62,770	65,891	57,326	69,185	59,269
Index	50	50	50	50	50	50	50

B.51 H.B. Robinson Steam Electric Plant, South Carolina

See Fig. B.51.
See Table B.51.

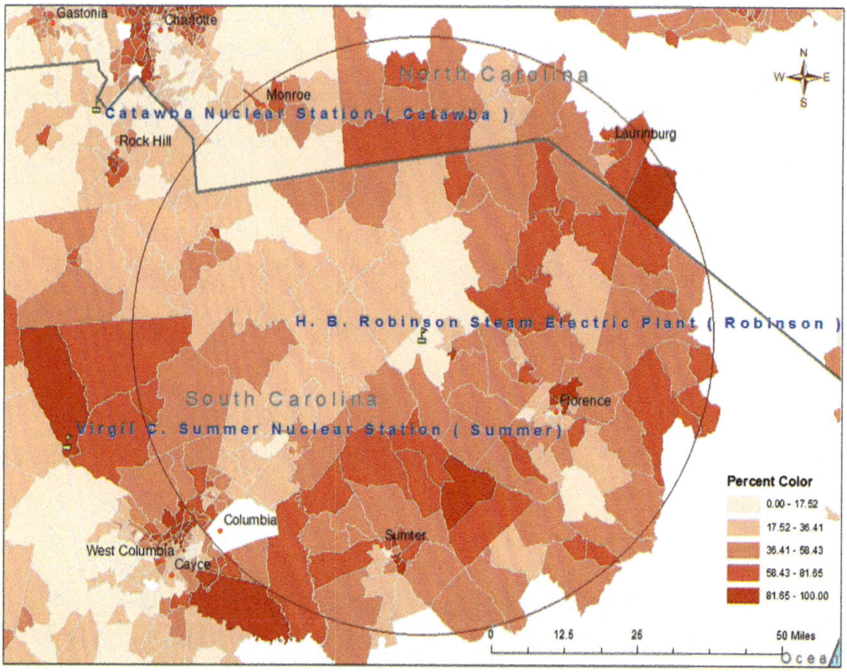

Fig. B.51 Spatial distribution of percent color among the populations surrounding H.B. Robinson Steam Electric Plant, South Carolina, in 2010

Table B.51 Demographic composition of population, as sorted by distance from H.B. Robinson Steam Electric Plant, South Carolina, in 1990, 2000, and 2010 *Source* (Kyne 2014)

	0–10	11–20	21–30	31–40	41–50	0–50	Outside
Year 1990							
Tract	8	16	38	55	35	152	3,146
Tract area (sq. mile)	309	1,060	1,346	2,225	1,423	6,363	79,477
Total population	30,342	53,517	142,372	207,559	134,251	568,041	9,547,261
White	20,650	30,268	82,578	128,638	85,691	347,825	7,071,098
Black	9,607	23,051	58,902	76,615	46,752	214,927	2,280,413
Asian	19	32	475	1,058	379	1,963	69,725
Native American	50	74	301	926	1,101	2,452	89,087
Others	16	93	114	308	341	872	36,949
Hispanic	77	239	459	1,126	971	2,872	94,476
Color	9,751	23,269	60,035	79,568	49,014	221,637	2,524,459
White (%)	68.06	56.56	58.00	61.98	63.83	61.23	74.06
Black (%)	31.66	43.07	41.37	36.91	34.82	37.84	23.89
Asian (%)	0.06	0.06	0.33	0.51	0.28	0.35	0.73
Native American (%)	0.16	0.14	0.21	0.45	0.82	0.43	0.93
Others (%)	0.05	0.17	0.08	0.15	0.25	0.15	0.39
Hispanic (%)	0.25	0.45	0.32	0.54	0.72	0.51	0.99
Color (%)	32.14	43.48	42.17	38.34	36.51	39.02	26.44
Female (%)	53.37	52.44	53.06	51.15	52.22	52.12	51.55
Old (65+ years) (%)	12.03	12.75	12.47	10.56	11.98	11.66	11.86
Kid (<5 years) (%)	8.30	8.33	8.61	9.21	9.17	8.92	8.39
Native-born (%)	99.70	99.66	99.03	99.03	99.38	99.21	98.32
Renter housing units (%)	23.34	21.68	26.60	26.65	26.46	25.94	28.02
Education (%)	13.58	8.13	15.01	11.40	9.84	11.76	17.44
Unemployment (%)	5.10	6.86	6.03	6.62	5.92	6.24	4.99
Poverty (%)	17.68	22.42	19.66	18.52	18.18	19.05	13.49
Mean household income ($)	52,591	40,920	50,421	46,014	45,461	46,892	54,396
Year 2000							
Total population	32,295	59,414	155,000	222,116	147,285	616,110	11,445,215
White	21,100	32,655	86,903	130,786	89,174	360,618	8,137,225
Black	10,516	25,967	64,492	84,828	51,884	237,687	2,679,194
Asian	134	220	1,207	1,231	704	3,496	149,384
Native American	142	129	286	1,665	1,661	3,883	111,761
Others	403	442	2,112	3,593	3,875	10,425	367,652
Hispanic	410	554	1,980	2,879	6,002	11,825	453,967
Color	11,273	26,985	68,670	92,437	61,184	260,549	3,497,422

(continued)

Table B.51 (continued)

	0–10	11–20	21–30	31–40	41–50	0–50	Outside
White (%)	65.34	54.96	56.07	58.88	60.55	58.53	71.10
Black (%)	32.56	43.71	41.61	38.19	35.23	38.58	23.41
Asian (%)	0.41	0.37	0.78	0.55	0.48	0.57	1.31
Native American (%)	0.44	0.22	0.18	0.75	1.13	0.63	0.98
Others (%)	1.25	0.74	1.36	1.62	2.63	1.69	3.21
Hispanic (%)	1.27	0.93	1.28	1.30	4.08	1.92	3.97
Color (%)	34.91	45.42	44.30	41.62	41.54	42.29	30.56
Female (%)	52.78	50.76	52.85	51.41	50.85	51.65	51.16
Old (65+ years) (%)	12.29	11.97	13.47	11.27	12.27	12.18	12.06
Kid (<5 years) (%)	8.22	7.75	7.72	8.32	8.49	8.15	7.97
Native-born (%)	98.90	99.05	98.11	98.50	96.53	98.01	95.34
Renter housing units (%)	20.92	18.64	24.19	23.42	24.85	23.34	26.41
College degree or higher (%)	16.22	9.66	18.29	12.81	11.66	13.82	22.21
Unemployment (%)	10.49	7.47	7.06	6.93	7.63	7.37	5.40
Poverty (%)	19.68	20.70	17.16	15.39	17.52	17.08	12.66
Mean household income ($)	55,802	47,818	56,811	52,563	52,119	53,277	63,124
Year 2010							
Total population	33,936	60,353	164,994	234,292	156,145	649,720	13,132,886
White	22,259	33,581	91,837	135,359	93,029	376,065	9,111,662
Black	10,546	25,967	68,854	90,478	54,262	250,107	3,000,319
Asian	18	256	1,244	1,705	1,198	4,421	252,083
Native American	74	56	219	1,769	1,902	4,020	116,101
Others	1,039	493	2,840	4,981	5,754	15,107	652,721
Hispanic	1,036	423	3,384	7,564	11,051	23,458	908,704
Color	12,193	27,057	75,417	104,227	70,019	288,913	4,456,954
White (%)	65.59	55.64	55.66	57.77	59.58	57.88	69.38
Black (%)	31.08	43.03	41.73	38.62	34.75	38.49	22.85
Asian (%)	0.05	0.42	0.75	0.73	0.77	0.68	1.92
Native American (%)	0.22	0.09	0.13	0.76	1.22	0.62	0.88
Others (%)	3.06	0.82	1.72	2.13	3.69	2.33	4.97
Hispanic (%)	3.05	0.70	2.05	3.23	7.08	3.61	6.92
Color (%)	35.93	44.83	45.71	44.49	44.84	44.47	33.94
Female (%)	52.35	50.88	52.11	50.50	51.69	51.33	51.28
Old (65+ years) (%)	14.05	13.17	14.11	12.80	12.75	13.22	12.77
Kid (<5 years) (%)	6.24	6.52	6.10	6.87	7.37	6.73	6.66

(continued)

Table B.51 (continued)

	0–10	11–20	21–30	31–40	41–50	0–50	Outside
Native-born (%)	97.37	98.96	97.58	97.07	95.09	96.92	93.34
Renter housing units (%)	21.34	21.77	26.64	25.22	27.42	25.58	26.68
College degree or higher (%)	20.22	10.72	19.64	14.23	12.77	15.27	25.90
Unemployment (%)	11.16	13.17	9.92	11.16	12.20	11.26	8.81
Poverty (%)	17.00	23.35	19.87	19.16	21.49	20.18	15.60
Mean household income ($)	55,522	43,121	53,567	48,269	48,010	49,510	61,393
Index	51	51	51	51	51	51	51

B.52 Oconee Nuclear Station, South Carolina

See Fig. B.52.
See Table B.52.

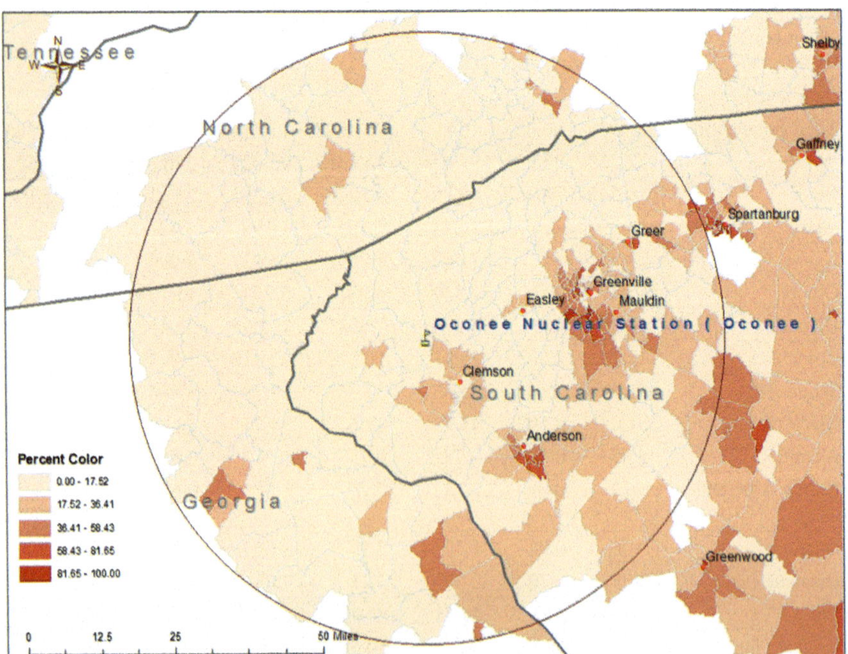

Fig. B.52 Spatial distribution of percent color among the populations surrounding Oconee Nuclear Station, South Carolina, in 2010

Table B.52 Demographic composition of population, as sorted by distance from Oconee Nuclear Station, South Carolina, in 1990, 2000, and 2010 *Source* (Kyne 2014)

	0–10	11–20	21–30	31–40	41–50	0–50	Outside
Year 1990							
Tract	15	29	103	91	38	276	4,991
Tract area (sq. mile)	281	1,020	1,522	2,107	1,246	6,175	139,090
Total population	51,644	108,245	320,112	285,328	107,983	873,312	15,720,199
White	46,636	97,024	254,169	255,012	100,817	753,658	11,268,635
Black	4,251	10,517	63,798	27,548	6,182	112,296	4,127,916
Asian	600	320	910	1,852	371	4,053	141,391
Native American	98	217	531	594	440	1,880	104,941
Others	59	166	702	328	177	1,432	77,286
Hispanic	429	553	2,271	2,087	780	6,120	192,619
Color	5,375	11,609	67,316	31,991	7,707	123,998	4,550,561
White (%)	90.30	89.63	79.40	89.38	93.36	86.30	71.68
Black (%)	8.23	9.72	19.93	9.65	5.72	12.86	26.26
Asian (%)	1.16	0.30	0.28	0.65	0.34	0.46	0.90
Native American (%)	0.19	0.20	0.17	0.21	0.41	0.22	0.67
Others (%)	0.11	0.15	0.22	0.11	0.16	0.16	0.49
Hispanic (%)	0.83	0.51	0.71	0.73	0.72	0.70	1.23
Color (%)	10.41	10.72	21.03	11.21	7.14	14.20	28.95
Female (%)	49.79	51.22	52.65	51.24	52.17	51.79	51.54
Old (65+ years) (%)	12.44	12.23	14.21	12.55	19.85	14.02	10.99
Kid (<5 years) (%)	6.85	7.42	8.12	7.96	6.60	7.72	8.76
Native-born (%)	97.51	99.26	98.95	98.17	98.53	98.59	97.93
Renter housing units (%)	28.97	19.39	27.90	22.67	18.98	24.08	29.58
Education (%)	22.72	12.49	14.31	20.70	16.20	16.92	18.03
Unemployment (%)	5.61	4.60	5.55	3.75	4.82	4.75	5.35
Poverty (%)	14.89	10.97	13.73	9.74	11.88	11.93	14.26
Mean household income ($)	50,519	51,741	50,654	58,351	49,849	53,158	56,668
Year 2000							
Total population	58,346	127,350	348,172	366,349	130,984	1,031,201	19,216,576
White	51,022	113,699	266,220	315,714	119,497	866,152	12,958,866
Black	4,744	11,143	70,882	37,340	7,169	131,278	5,127,713
Asian	1,107	492	2,382	4,898	419	9,298	318,911
Native American	105	238	738	696	710	2,487	136,845
Others	1,371	1,774	7,954	7,715	3,182	21,996	674,232
Hispanic	1,585	1,711	9,423	10,319	4,316	27,354	868,414
Color	8,081	14,385	86,578	55,909	13,588	178,541	6,636,156

(continued)

Table B.52 (continued)

	0–10	11–20	21–30	31–40	41–50	0–50	Outside
White (%)	87.45	89.28	76.46	86.18	91.23	83.99	67.44
Black (%)	8.13	8.75	20.36	10.19	5.47	12.73	26.68
Asian (%)	1.90	0.39	0.68	1.34	0.32	0.90	1.66
Native American (%)	0.18	0.19	0.21	0.19	0.54	0.24	0.71
Others (%)	2.35	1.39	2.28	2.11	2.43	2.13	3.51
Hispanic (%)	2.72	1.34	2.71	2.82	3.30	2.65	4.52
Color (%)	13.85	11.30	24.87	15.26	10.37	17.31	34.53
Female (%)	49.60	50.69	52.03	50.97	51.73	51.31	51.04
Old (65+ years) (%)	12.95	13.22	14.07	13.02	20.02	14.29	10.91
Kid (<5 years) (%)	6.71	7.36	7.82	7.76	6.65	7.53	8.30
Native-born (%)	95.46	98.28	96.94	95.80	96.35	96.54	94.34
Renter housing units (%)	28.12	18.37	26.02	22.09	18.42	22.85	27.89
College degree or higher (%)	25.55	16.81	17.60	25.26	21.04	21.12	22.88
Unemployment (%)	4.87	5.03	5.17	4.58	4.87	4.88	5.53
Poverty (%)	16.79	10.82	13.67	9.33	11.28	11.65	13.00
Mean household income ($)	55,769	56,897	57,906	67,407	57,481	60,951	66,003
Year 2010							
Total population	64,579	136,681	371,175	436,380	144,005	1,152,820	22,098,600
White	56,070	122,318	283,428	367,066	127,670	956,552	14,319,430
Black	5,311	10,358	72,758	46,137	7,615	142,179	5,981,316
Asian	1,410	668	2,775	8,699	811	14,363	543,977
Native American	131	290	597	842	1,100	2,960	140,395
Others	1,657	3,047	11,617	13,636	6,809	36,766	1,113,483
Hispanic	3,001	3,088	22,851	24,853	8,473	62,266	1,654,608
Color	10,590	16,167	104,261	86,084	19,392	236,494	8,596,418
White (%)	86.82	89.49	76.36	84.12	88.66	82.97	64.80
Black (%)	8.22	7.58	19.60	10.57	5.29	12.33	27.07
Asian (%)	2.18	0.49	0.75	1.99	0.56	1.25	2.46
Native American (%)	0.20	0.21	0.16	0.19	0.76	0.26	0.64
Others (%)	2.57	2.23	3.13	3.12	4.73	3.19	5.04
Hispanic (%)	4.65	2.26	6.16	5.70	5.88	5.40	7.49
Color (%)	16.40	11.83	28.09	19.73	13.47	20.51	38.90
Female (%)	49.62	50.71	51.91	51.33	51.29	51.34	51.21
Old (65+ years) (%)	14.53	15.07	14.77	13.97	21.05	15.28	11.58
Kid (<5 years) (%)	5.12	5.60	6.78	6.37	4.92	6.16	6.95

(continued)

Table B.52 (continued)

	0–10	11–20	21–30	31–40	41–50	0–50	Outside
Native-born (%)	93.72	97.83	94.53	93.46	94.47	94.46	92.12
Renter housing units (%)	27.44	19.21	26.63	22.84	19.05	23.42	27.48
College degree or higher (%)	32.65	19.03	20.31	28.55	23.32	24.33	26.22
Unemployment (%)	9.11	9.79	9.75	7.23	7.83	8.51	8.91
Poverty (%)	19.64	15.27	18.60	12.48	14.99	15.49	15.79
Mean household income ($)	56,152	56,421	55,156	62,997	55,109	58,313	63,381
Index	52	52	52	52	52	52	52

B.53 Virgil C. Summer Nuclear Station, South Carolina

See Fig. B.53.
See Table B.53.

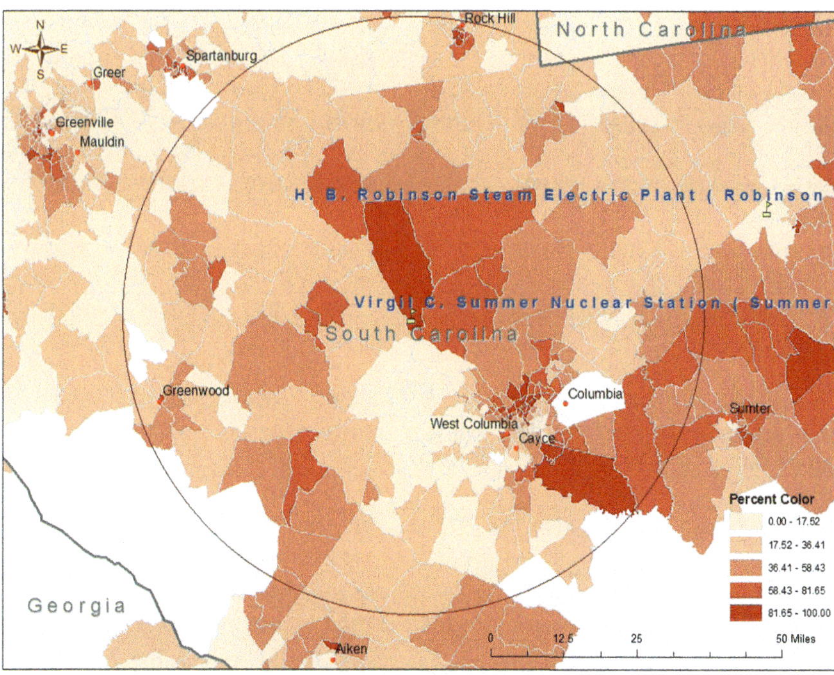

Fig. B.53 Spatial distribution of percent color among the populations surrounding Virgil C. Summer Nuclear Station, South Carolina, in 2010

Table B.53 Demographic composition of population, as sorted by distance from Virgil C. Summer Nuclear Station, South Carolina, in 1990, 2000, and 2010 *Source* (Kyne 2014)

	0–10	11–20	21–30	31–40	41–50	0–50	Outside
Year 1990							
Tract	5	29	116	64	37	251	852
Tract area (sq. mile)	457	827	1,380	2,295	1,477	6,436	25,584
Total population	14,256	93,431	326,936	195,017	119,756	749,396	2,737,290
White	7,775	71,777	212,349	132,565	76,637	501,103	1,906,577
Black	6,450	20,912	109,418	59,903	42,319	239,002	800,999
Asian	20	494	3,257	1,012	229	5,012	16,285
Native American	2	78	647	565	495	1,787	7,146
Others	11	166	1,257	988	66	2,488	6,288
Hispanic	38	500	3,443	2,260	421	6,662	21,668
Color	6,499	21,954	116,377	63,301	43,391	251,522	843,251
White (%)	54.54	76.82	64.95	67.98	63.99	66.87	69.65
Black (%)	45.24	22.38	33.47	30.72	35.34	31.89	29.26
Asian (%)	0.14	0.53	1.00	0.52	0.19	0.67	0.59
Native American (%)	0.01	0.08	0.20	0.29	0.41	0.24	0.26
Others (%)	0.08	0.18	0.38	0.51	0.06	0.33	0.23
Hispanic (%)	0.27	0.54	1.05	1.16	0.35	0.89	0.79
Color (%)	45.59	23.50	35.60	32.46	36.23	33.56	30.81
Female (%)	51.34	51.47	51.91	51.46	52.07	51.75	51.61
Old (65+ years) (%)	11.39	10.27	10.51	10.65	11.15	10.63	11.56
Kid (<5 years) (%)	7.93	8.90	8.13	8.52	9.07	8.47	8.88
Native-born (%)	99.71	98.87	97.62	98.66	99.46	98.38	98.62
Renter housing units (%)	11.61	18.67	34.93	22.68	24.67	27.80	26.32
Education (%)	13.47	26.78	24.72	12.99	9.87	19.45	15.88
Unemployment (%)	5.11	4.29	5.16	5.44	6.52	5.33	5.65
Poverty (%)	14.75	9.72	12.90	13.55	16.32	13.25	15.94
Mean household income ($)	52,998	66,688	56,992	50,938	48,823	55,365	52,202
Year 2000							
Total population	18,174	113,750	372,457	222,318	138,851	865,550	3,146,462
White	10,930	82,464	228,113	145,827	89,405	556,739	2,138,939
Black	7,087	28,061	129,336	68,664	46,769	279,917	902,810
Asian	58	1,078	5,618	1,669	415	8,838	29,051
Native American	22	341	1,357	778	518	3,016	11,672
Others	85	1,798	8,014	5,406	1,729	17,032	63,998
Hispanic	177	2,630	9,060	5,948	1,928	19,743	73,085
Color	7,324	32,759	148,317	78,743	50,306	317,449	1,040,162

(continued)

Table B.53 (continued)

	0–10	11–20	21–30	31–40	41–50	0–50	Outside
White (%)	60.14	72.50	61.25	65.59	64.39	64.32	67.98
Black (%)	39.00	24.67	34.73	30.89	33.68	32.34	28.69
Asian (%)	0.32	0.95	1.51	0.75	0.30	1.02	0.92
Native American (%)	0.12	0.30	0.36	0.35	0.37	0.35	0.37
Others (%)	0.47	1.58	2.15	2.43	1.25	1.97	2.03
Hispanic (%)	0.97	2.31	2.43	2.68	1.39	2.28	2.32
Color (%)	40.30	28.80	39.82	35.42	36.23	36.68	33.06
Female (%)	51.28	51.82	51.96	51.38	51.92	51.77	51.38
Old (65+ years) (%)	10.69	10.89	11.04	11.24	10.48	10.97	12.42
Kid (<5 years) (%)	8.23	7.95	7.51	8.04	8.32	7.85	7.90
Native-born (%)	99.03	96.81	96.37	97.52	98.70	97.15	97.10
Renter housing units (%)	8.77	18.17	32.90	21.69	23.00	26.10	23.84
College degree or higher (%)	21.91	31.82	28.82	14.65	11.92	22.83	19.76
Unemployment (%)	3.11	4.47	6.13	6.09	7.62	6.07	5.79
Poverty (%)	11.29	9.64	13.29	13.29	14.09	12.88	14.44
Mean household income ($)	65,454	74,481	64,127	55,401	55,276	61,971	59,765
Year 2010							
Total population	22,287	126,697	436,932	233,924	151,179	971,019	3,540,409
White	14,494	88,230	256,755	152,389	96,958	608,826	2,425,174
Black	7,260	32,158	156,161	71,933	48,484	315,996	953,488
Asian	115	1,155	9,088	2,249	811	13,418	42,923
Native American	37	118	808	1,211	495	2,669	11,426
Others	381	5,036	14,120	6,142	4,431	30,110	107,398
Hispanic	143	4,940	21,071	9,788	5,756	41,698	167,056
Color	7,888	40,699	192,459	87,465	57,621	386,132	1,217,621
White (%)	65.03	69.64	58.76	65.14	64.13	62.70	68.50
Black (%)	32.58	25.38	35.74	30.75	32.07	32.54	26.93
Asian (%)	0.52	0.91	2.08	0.96	0.54	1.38	1.21
Native American (%)	0.17	0.09	0.18	0.52	0.33	0.27	0.32
Others (%)	1.71	3.97	3.23	2.63	2.93	3.10	3.03
Hispanic (%)	0.64	3.90	4.82	4.18	3.81	4.29	4.72
Color (%)	35.39	32.12	44.05	37.39	38.11	39.77	34.39
Female (%)	52.11	51.75	51.25	50.92	50.98	51.22	51.36
Old (65+ years) (%)	12.46	12.33	10.90	12.79	11.60	11.69	13.61

(continued)

Table B.53 (continued)

	0–10	11–20	21–30	31–40	41–50	0–50	Outside
Kid (<5 years) (%)	5.96	6.61	6.24	6.99	6.98	6.58	6.55
Native-born (%)	98.58	96.06	94.52	96.37	97.07	95.66	95.20
Renter housing units (%)	12.15	20.04	32.53	23.80	25.19	27.17	24.59
College degree or higher (%)	30.39	34.70	32.25	18.44	14.99	26.49	23.28
Unemployment (%)	5.38	6.67	8.07	10.15	12.38	8.95	9.34
Poverty (%)	12.28	10.39	14.05	17.49	18.67	15.10	16.75
Mean household income ($)	68,719	73,422	61,786	53,274	50,067	59,643	58,749
Index	53	53	53	53	53	53	53

B.54 Sequoyah Nuclear Plant, Tennessee

See Fig. B.54.
See Table B.54.

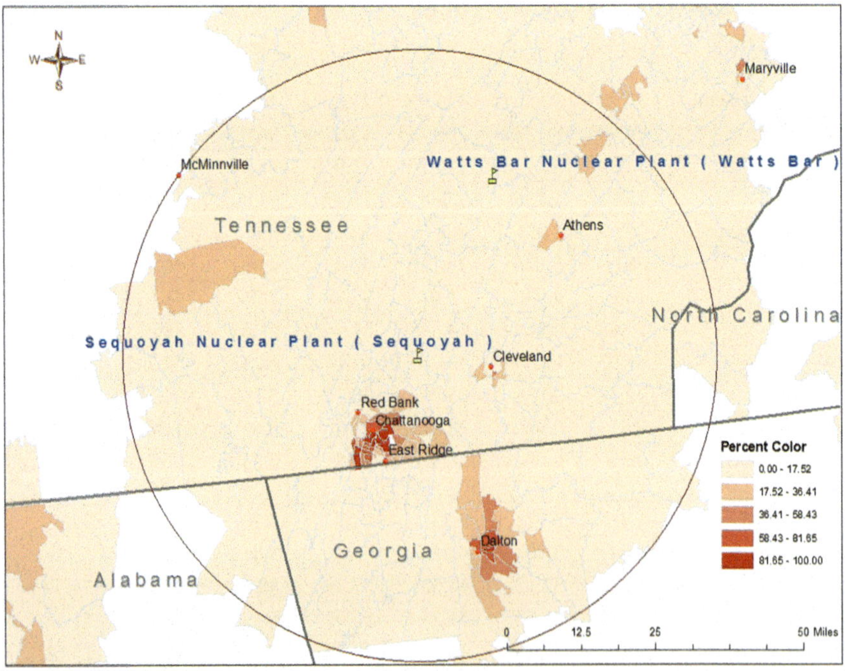

Fig. B.54 Spatial distribution of percent color among the populations surrounding Sequoyah Nuclear Plant, Tennessee, in 2010

Table B.54 Demographic composition of population, as sorted by distance from Sequoyah Nuclear Plant, Tennessee, in 1990, 2000, and 2010 *Source* (Kyne 2014)

	0–10	11–20	21–30	31–40	41–50	0–50	Outside
Year 1990							
Tract	20	84	41	42	18	205	4,442
Tract area (sq. mile)	312	830	1,704	2,070	1,311	6,227	147,763
Total population	76,278	292,438	165,308	155,698	57,971	747,693	14,648,283
White	74,365	239,110	154,592	147,359	56,638	672,064	10,954,868
Black	977	49,972	9,673	6,042	800	67,464	3,474,204
Asian	491	1,983	428	415	102	3,419	122,687
Native American	250	755	510	410	343	2,268	43,787
Others	181	634	96	1,469	93	2,473	52,731
Hispanic	442	2,011	713	2,481	255	5,902	150,147
Color	2,207	54,605	11,320	9,299	1,487	78,918	3,779,257
White (%)	97.49	81.76	93.52	94.64	97.70	89.89	74.79
Black (%)	1.28	17.09	5.85	3.88	1.38	9.02	23.72
Asian (%)	0.64	0.68	0.26	0.27	0.18	0.46	0.84
Native American (%)	0.33	0.26	0.31	0.26	0.59	0.30	0.30
Others (%)	0.24	0.22	0.06	0.94	0.16	0.33	0.36
Hispanic (%)	0.58	0.69	0.43	1.59	0.44	0.79	1.03
Color (%)	2.89	18.67	6.85	5.97	2.57	10.55	25.80
Female (%)	50.69	52.88	52.07	50.55	50.79	51.83	51.77
Old (65+ years) (%)	8.67	13.84	12.25	12.04	12.24	12.46	11.61
Kid (<5 years) (%)	7.71	7.82	8.13	7.86	7.74	7.88	8.69
Native-born (%)	98.85	98.49	99.54	98.60	99.53	98.86	98.18
Renter housing units (%)	18.23	34.24	23.17	24.80	15.65	26.95	29.60
Education (%)	18.19	17.53	8.70	9.58	7.66	13.26	17.49
Unemployment (%)	4.58	5.79	6.12	6.37	5.92	5.86	6.25
Poverty (%)	7.39	14.59	15.63	14.96	13.88	14.10	16.05
Mean household income ($)	65,651	52,404	46,665	48,602	46,180	51,188	55,264
Year 2000							
Total population	90,338	322,102	191,921	177,853	73,129	855,343	17,467,492
White	85,955	252,461	177,250	158,606	69,645	743,917	12,307,383
Black	2,147	57,830	9,594	6,063	974	76,608	4,348,410
Asian	789	3,779	869	997	284	6,718	255,997
Native American	288	860	565	488	228	2,429	59,697
Others	1,138	7,201	3,633	11,691	2,008	25,671	496,006
Hispanic	1,250	6,293	3,485	18,300	2,084	31,412	590,616
Color	5,121	72,418	16,199	26,746	4,112	124,596	5,432,851

(continued)

Table B.54 (continued)

	0–10	11–20	21–30	31–40	41–50	0–50	Outside
White (%)	95.15	78.38	92.36	89.18	95.24	86.97	70.46
Black (%)	2.38	17.95	5.00	3.41	1.33	8.96	24.89
Asian (%)	0.87	1.17	0.45	0.56	0.39	0.79	1.47
Native American (%)	0.32	0.27	0.29	0.27	0.31	0.28	0.34
Others (%)	1.26	2.24	1.89	6.57	2.75	3.00	2.84
Hispanic (%)	1.38	1.95	1.82	10.29	2.85	3.67	3.38
Color (%)	5.67	22.48	8.44	15.04	5.62	14.57	31.10
Female (%)	50.89	52.14	51.34	50.18	49.98	51.23	51.24
Old (65+ years) (%)	10.60	13.89	12.53	12.37	12.23	12.78	11.23
Kid (<5 years) (%)	7.50	7.41	8.06	8.41	7.87	7.81	8.29
Native-born (%)	97.65	97.06	98.29	92.23	97.67	96.45	95.46
Renter housing units (%)	17.20	33.66	21.95	24.24	15.37	25.92	27.76
College degree or higher (%)	22.05	21.29	11.39	10.58	8.55	15.87	21.80
Unemployment (%)	3.32	5.85	4.67	4.72	4.17	4.94	5.68
Poverty (%)	6.30	13.48	12.72	14.49	12.62	12.67	13.96
Mean household income ($)	74,061	59,963	56,218	52,671	54,467	58,659	64,327
Year 2010							
Total population	104,716	344,451	213,882	195,827	78,444	937,320	19,479,114
White	97,329	263,853	194,285	171,839	73,885	801,191	13,191,726
Black	4,396	62,369	10,116	5,896	1,375	84,152	5,058,618
Asian	1,640	5,411	1,590	1,299	70	10,010	435,058
Native American	250	1,000	692	660	295	2,897	61,388
Others	1,101	11,818	7,199	16,133	2,819	39,070	732,324
Hispanic	1,489	16,626	9,764	29,257	4,034	61,170	1,147,361
Color	8,510	90,122	25,491	40,313	7,284	171,720	6,923,635
White (%)	92.95	76.60	90.84	87.75	94.19	85.48	67.72
Black (%)	4.20	18.11	4.73	3.01	1.75	8.98	25.97
Asian (%)	1.57	1.57	0.74	0.66	0.09	1.07	2.23
Native American (%)	0.24	0.29	0.32	0.34	0.38	0.31	0.32
Others (%)	1.05	3.43	3.37	8.24	3.59	4.17	3.76
Hispanic (%)	1.42	4.83	4.57	14.94	5.14	6.53	5.89
Color (%)	8.13	26.16	11.92	20.59	9.29	18.32	35.54
Female (%)	51.47	51.77	51.16	50.31	49.15	51.07	51.26
Old (65+ years) (%)	13.67	14.40	13.89	13.23	14.30	13.95	11.76
Kid (<5 years) (%)	5.59	6.24	6.47	6.87	5.73	6.31	6.86
Native-born (%)	97.50	95.17	96.97	90.76	96.17	95.00	93.36

(continued)

Table B.54 (continued)

	0–10	11–20	21–30	31–40	41–50	0–50	Outside
Renter housing units (%)	18.13	32.94	23.28	23.30	16.56	25.80	26.98
College degree or higher (%)	26.17	23.80	14.61	13.03	9.30	18.55	24.82
Unemployment (%)	6.70	9.30	9.79	9.75	9.33	9.18	8.72
Poverty (%)	7.72	17.23	16.73	19.33	15.61	16.33	16.27
Mean household income ($)	74,655	56,716	52,512	50,189	48,170	55,772	62,483
Index	54	54	54	54	54	54	54

B.55 Watts Bar Nuclear Plant, Tennessee

See Fig. B.55.
See Table B.55.

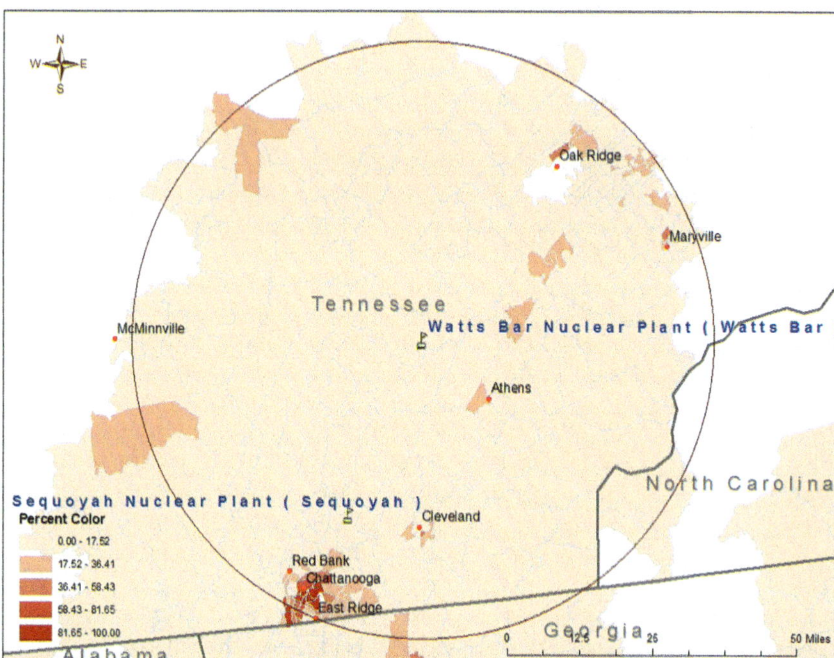

Fig. B.55 Spatial distribution of percent color among the populations surrounding Watts Bar Nuclear Plant, Tennessee, in 2010

Table B.55 Demographic composition of population, as sorted by distance from Watts Bar Nuclear Plant, Tennessee, in 1990, 2000, and 2010 *Source* (Kyne 2014)

	0–10	11–20	21–30	31–40	41–50	0–50	Outside
Year 1990							
Tract	5	16	37	68	52	178	3,288
Tract area (sq. mile)	417	772	1,533	2,545	1,188	6,455	95,115
Total population	17,803	61,485	135,601	210,773	190,986	616,648	10,738,744
White	17,358	58,597	130,706	204,156	162,222	573,039	8,078,649
Black	348	2,431	4,030	4,538	25,951	37,298	2,484,619
Asian	21	114	179	998	1,827	3,139	101,213
Native American	41	210	495	687	561	1,994	25,762
Others	32	134	191	388	449	1,194	48,461
Hispanic	99	220	802	1,160	1,640	3,921	128,545
Color	527	2,993	5,514	7,412	29,899	46,345	2,731,038
White (%)	97.50	95.30	96.39	96.86	84.94	92.93	75.23
Black (%)	1.95	3.95	2.97	2.15	13.59	6.05	23.14
Asian (%)	0.12	0.19	0.13	0.47	0.96	0.51	0.94
Native American (%)	0.23	0.34	0.37	0.33	0.29	0.32	0.24
Others (%)	0.18	0.22	0.14	0.18	0.24	0.19	0.45
Hispanic (%)	0.56	0.36	0.59	0.55	0.86	0.64	1.20
Color (%)	2.96	4.87	4.07	3.52	15.66	7.52	25.43
Female (%)	50.29	52.15	51.09	50.91	52.34	51.50	51.67
Old (65+ years) (%)	12.42	14.66	13.91	11.52	13.18	12.90	11.09
Kid (<5 years) (%)	6.41	7.48	7.27	7.77	7.76	7.59	8.81
Native-born (%)	99.43	99.53	99.38	98.99	98.17	98.89	97.90
Renter housing units (%)	15.80	22.71	20.52	20.82	28.32	23.18	30.92
Education (%)	6.29	10.84	10.33	14.24	21.23	14.99	18.02
Unemployment (%)	7.23	7.72	7.47	6.26	5.36	6.40	6.00
Poverty (%)	16.02	18.81	14.98	13.33	11.26	13.67	15.19
Mean household income ($)	43,804	43,813	47,294	54,053	58,609	52,697	57,100
Year 2000							
Total population	22,704	72,749	165,453	258,187	212,174	731,267	13,144,469
White	22,107	68,044	158,188	245,418	171,712	665,469	9,224,160
Black	323	2,868	3,782	5,971	32,436	45,380	3,226,594
Asian	9	272	686	1,516	3,024	5,507	226,113
Native American	71	233	444	681	755	2,184	37,045
Others	189	1,333	2,347	4,605	4,275	12,749	430,535
Hispanic	137	938	2,508	3,887	3,909	11,379	538,022
Color	644	5,116	8,852	14,852	42,370	71,834	4,165,552
White (%)	97.37	93.53	95.61	95.05	80.93	91.00	70.18

(continued)

Table B.55 (continued)

	0–10	11–20	21–30	31–40	41–50	0–50	Outside
Black (%)	1.42	3.94	2.29	2.31	15.29	6.21	24.55
Asian (%)	0.04	0.37	0.41	0.59	1.43	0.75	1.72
Native American (%)	0.31	0.32	0.27	0.26	0.36	0.30	0.28
Others (%)	0.83	1.83	1.42	1.78	2.01	1.74	3.28
Hispanic (%)	0.60	1.29	1.52	1.51	1.84	1.56	4.09
Color (%)	2.84	7.03	5.35	5.75	19.97	9.82	31.69
Female (%)	50.72	51.56	50.91	50.51	52.19	51.20	51.06
Old (65+ years) (%)	12.39	13.99	15.68	12.38	14.28	13.84	10.57
Kid (<5 years) (%)	8.04	7.45	6.94	7.44	7.27	7.30	8.41
Native-born (%)	99.46	98.82	98.40	98.15	96.89	97.95	94.51
Renter housing units (%)	15.75	21.95	19.01	20.28	28.15	22.32	29.20
College degree or higher (%)	8.63	10.85	13.89	18.84	24.46	18.22	22.56
Unemployment (%)	4.31	6.50	5.11	4.62	4.43	4.84	5.52
Poverty (%)	15.37	14.18	12.76	10.77	10.94	11.75	13.27
Mean household income ($)	49,660	51,163	56,858	62,857	65,471	60,696	66,652
Year 2010							
Total population	24,916	77,801	183,904	298,964	226,621	812,206	14,891,577
White	24,389	72,073	174,637	279,175	178,886	729,160	9,969,840
Black	191	3,168	4,182	9,915	36,386	53,842	3,856,603
Asian	0	522	1,045	3,276	4,772	9,615	382,465
Native American	101	135	376	872	614	2,098	36,373
Others	235	1,903	3,664	5,726	5,963	17,491	646,296
Hispanic	196	2,508	5,287	8,527	10,461	26,979	1,019,181
Color	712	7,387	12,975	25,893	55,226	102,193	5,469,325
White (%)	97.88	92.64	94.96	93.38	78.94	89.78	66.95
Black (%)	0.77	4.07	2.27	3.32	16.06	6.63	25.90
Asian (%)	0.00	0.67	0.57	1.10	2.11	1.18	2.57
Native American (%)	0.41	0.17	0.20	0.29	0.27	0.26	0.24
Others (%)	0.94	2.45	1.99	1.92	2.63	2.15	4.34
Hispanic (%)	0.79	3.22	2.87	2.85	4.62	3.32	6.84
Color (%)	2.86	9.49	7.06	8.66	24.37	12.58	36.73
Female (%)	51.03	52.15	49.95	50.72	51.83	51.00	51.19
Old (65+ years) (%)	15.04	15.54	18.69	14.46	15.62	15.86	11.13
Kid (<5 years) (%)	4.96	6.27	5.00	6.26	5.54	5.73	7.02
Native-born (%)	99.28	97.26	97.58	96.96	94.87	96.62	92.24

(continued)

Table B.55 (continued)

	0–10	11–20	21–30	31–40	41–50	0–50	Outside
Renter housing units (%)	20.19	21.88	19.57	20.88	28.02	22.71	27.97
College degree or higher (%)	9.74	13.80	16.17	22.12	27.14	20.95	25.63
Unemployment (%)	11.92	11.85	8.51	8.07	7.76	8.52	8.77
Poverty (%)	18.32	18.80	13.87	14.19	14.28	14.71	16.09
Mean household income ($)	48,760	47,342	54,341	62,930	62,752	59,019	63,819
Index	55	55	55	55	55	55	55

B.56 Comanche Peak Steam Electric Station, Texas

See Fig. B.56.
See Table B.56.

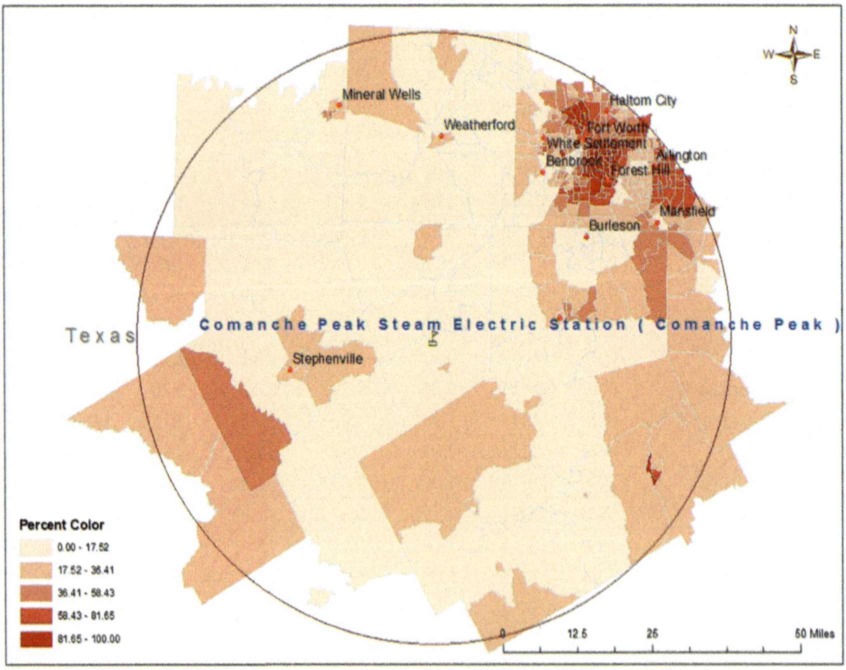

Fig. B.56 Spatial distribution of percent color among the populations surrounding Comanche Peak Steam Electric Station, Texas, in 2010

Table B.56 Demographic composition of population, as sorted by distance from Comanche Peak Steam Electric Station, Texas, in 1990, 2000, and 2010 *Source* (Kyne 2014)

	0–10	11–20	21–30	31–40	41–50	0–50	Outside
Year 1990							
Tract	6	10	28	120	104	268	4,997
Tract area (sq. mile)	356	1,003	1,427	2,308	992	6,086	262,510
Total population	16,383	27,568	87,665	413,835	336,153	881,604	16,104,907
White	15,526	26,606	81,737	335,381	232,870	692,120	12,095,405
Black	0	33	1,968	45,809	60,741	108,551	1,909,988
Asian	62	155	395	6,408	4,734	11,754	303,301
Native American	94	157	376	1,694	1,649	3,970	65,902
Others	703	615	3,197	24,539	36,153	65,207	1,730,315
Hispanic	1,275	1,614	6,661	44,466	59,795	113,811	4,180,319
Color	1,429	1,952	9,307	97,969	126,073	236,730	6,428,898
White (%)	94.77	96.51	93.24	81.04	69.28	78.51	75.10
Black (%)	0.00	0.12	2.24	11.07	18.07	12.31	11.86
Asian (%)	0.38	0.56	0.45	1.55	1.41	1.33	1.88
Native American (%)	0.57	0.57	0.43	0.41	0.49	0.45	0.41
Others (%)	4.29	2.23	3.65	5.93	10.75	7.40	10.74
Hispanic (%)	7.78	5.85	7.60	10.74	17.79	12.91	25.96
Color (%)	8.72	7.08	10.62	23.67	37.50	26.85	39.92
Female (%)	49.36	50.00	51.21	50.96	50.69	50.82	50.76
Old (65+ years) (%)	12.26	17.08	13.02	11.43	11.39	11.77	9.96
Kid (<5 years) (%)	9.85	8.24	8.81	9.37	10.41	9.68	9.80
Native-born (%)	96.67	96.92	97.15	94.59	92.17	94.03	90.86
Renter housing units (%)	15.94	17.71	25.30	32.17	29.23	29.53	34.14
Education (%)	16.77	13.24	14.91	22.89	13.38	18.12	20.44
Unemployment (%)	10.40	8.64	6.78	5.70	8.55	7.02	7.12
Poverty (%)	11.08	12.02	14.33	11.57	18.42	14.46	18.30
Mean household income ($)	63,899	57,368	54,192	63,538	53,324	58,706	58,452
Year 2000							
Total population	23,844	36,129	108,363	486,385	395,470	1,050,191	19,801,628
White	22,535	33,701	98,149	367,184	267,745	789,314	14,008,671
Black	20	53	2,036	53,905	57,331	113,345	2,272,209
Asian	45	218	808	8,460	8,021	17,552	550,840
Native American	261	380	666	2,873	2,598	6,778	106,977
Others	985	1,773	6,715	53,973	59,748	123,194	2,862,940
Hispanic	2,005	3,146	12,939	89,719	109,199	217,008	6,453,114
Color	2,470	4,094	17,582	161,044	179,682	364,872	9,559,410

(continued)

Table B.56 (continued)

	0–10	11–20	21–30	31–40	41–50	0–50	Outside
White (%)	94.51	93.28	90.57	75.49	67.70	75.16	70.75
Black (%)	0.08	0.15	1.88	11.08	14.50	10.79	11.47
Asian (%)	0.19	0.60	0.75	1.74	2.03	1.67	2.78
Native American (%)	1.09	1.05	0.61	0.59	0.66	0.65	0.54
Others (%)	4.13	4.91	6.20	11.10	15.11	11.73	14.46
Hispanic (%)	8.41	8.71	11.94	18.45	27.61	20.66	32.59
Color (%)	10.36	11.33	16.23	33.11	45.44	34.74	48.28
Female (%)	50.84	50.43	50.78	51.22	49.89	50.64	50.42
Old (65+ years) (%)	16.35	16.06	12.05	11.43	9.62	11.08	9.85
Kid (<5 years) (%)	7.15	7.48	8.33	8.70	9.63	8.93	9.30
Native-born (%)	96.64	95.96	94.60	90.44	85.71	89.42	85.92
Renter housing units (%)	16.43	16.57	24.35	32.26	28.56	29.09	33.00
College degree or higher (%)	22.38	18.02	17.70	23.45	16.00	19.90	23.42
Unemployment (%)	4.47	4.89	6.61	4.60	6.28	5.43	6.10
Poverty (%)	6.77	10.67	10.36	11.26	15.81	12.75	15.51
Mean household income ($)	74,906	63,012	62,529	68,180	62,038	65,396	67,919
Year 2010							
Total population	28,892	42,708	130,604	568,750	449,146	1,220,100	23,091,792
White	25,938	38,570	119,408	436,281	304,710	924,907	16,578,539
Black	162	191	2,761	70,101	57,885	131,100	2,733,566
Asian	178	199	1,223	11,676	10,447	23,723	900,113
Native American	165	408	815	2,765	2,330	6,483	113,197
Others	2,449	3,340	6,397	47,927	73,774	133,887	2,766,376
Hispanic	3,517	4,793	20,293	138,056	149,222	315,881	8,601,596
Color	4,191	6,125	26,419	228,601	225,170	490,506	12,534,673
White (%)	89.78	90.31	91.43	76.71	67.84	75.81	71.79
Black (%)	0.56	0.45	2.11	12.33	12.89	10.75	11.84
Asian (%)	0.62	0.47	0.94	2.05	2.33	1.94	3.90
Native American (%)	0.57	0.96	0.62	0.49	0.52	0.53	0.49
Others (%)	8.48	7.82	4.90	8.43	16.43	10.97	11.98
Hispanic (%)	12.17	11.22	15.54	24.27	33.22	25.89	37.25
Color (%)	14.51	14.34	20.23	40.19	50.13	40.20	54.28
Female (%)	52.09	49.17	51.30	51.33	49.60	50.63	50.39
Old (65+ years) (%)	20.27	16.92	13.02	11.36	9.42	11.23	10.09
Kid (<5 years) (%)	5.79	5.74	7.23	7.51	8.28	7.66	7.84
Native-born (%)	94.20	94.96	93.75	87.65	84.28	87.47	83.71

(continued)

Table B.56 (continued)

	0–10	11–20	21–30	31–40	41–50	0–50	Outside
Renter housing units (%)	17.45	18.80	24.59	30.32	27.61	27.95	31.05
College degree or higher (%)	28.15	20.40	19.56	25.62	18.57	22.30	25.95
Unemployment (%)	3.91	5.48	6.92	7.18	8.82	7.62	6.97
Poverty (%)	8.36	13.33	13.26	14.27	17.57	15.20	16.84
Mean household income ($)	75,416	71,102	67,592	68,754	62,768	66,798	68,801
Index	56	56	56	56	56	56	56

B.57 South Texas Project, Texas

See Fig. B.57.
See Table B.57.

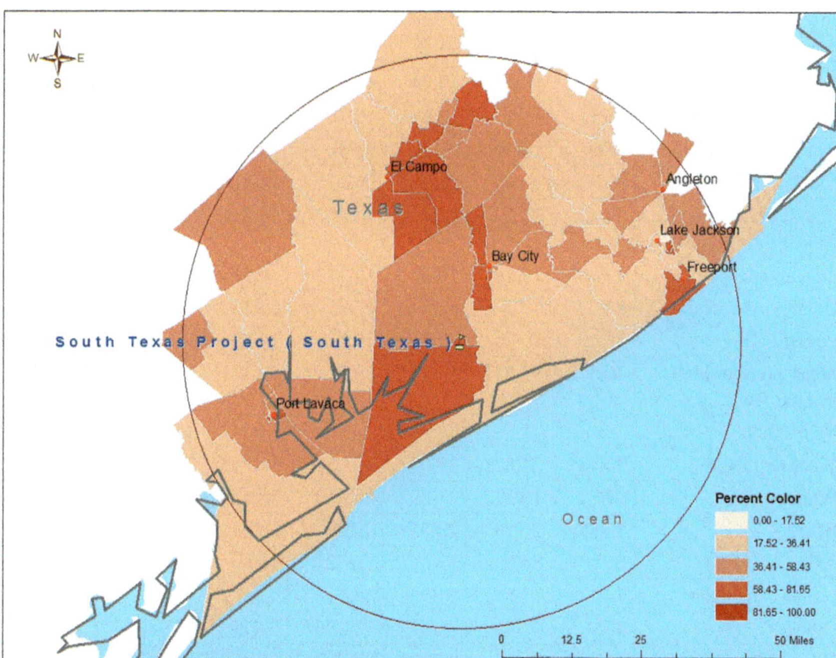

Fig. B.57 Spatial distribution of percent color among the populations surrounding South Texas Project, Texas, in 2010

Table B.57 Demographic composition of population, as sorted by distance from South Texas Project, Texas, in 1990, 2000, and 2010 *Source* (Kyne 2014)

	0–10	11–20	21–30	31–40	41–50	0–50	Outside
Year 1990							
Tract	1	9	7	23	12	52	5,213
Tract area (sq. mile)	25	1,484	1,003	1,428	725	4,663	263,933
Total population	3,233	25,996	28,373	98,056	43,455	199,113	16,787,398
White	1,171	19,716	22,227	77,477	32,043	152,634	12,634,891
Black	1,783	2,641	2,999	9,481	5,330	22,234	1,996,305
Asian	16	596	117	901	262	1,892	313,163
Native American	8	83	51	268	176	586	69,286
Others	251	2,967	2,977	9,932	5,646	21,773	1,773,749
Hispanic	347	6,735	6,038	21,018	9,892	44,030	4,250,100
Color	2,140	9,990	9,198	31,624	15,645	68,597	6,597,031
White (%)	36.22	75.84	78.34	79.01	73.74	76.66	75.26
Black (%)	55.15	10.16	10.57	9.67	12.27	11.17	11.89
Asian (%)	0.49	2.29	0.41	0.92	0.60	0.95	1.87
Native American (%)	0.25	0.32	0.18	0.27	0.41	0.29	0.41
Others (%)	7.76	11.41	10.49	10.13	12.99	10.93	10.57
Hispanic (%)	10.73	25.91	21.28	21.43	22.76	22.11	25.32
Color (%)	66.19	38.43	32.42	32.25	36.00	34.45	39.30
Female (%)	52.77	50.06	50.94	49.42	49.97	49.89	50.78
Old (65+ years) (%)	14.57	11.30	12.60	10.74	9.72	10.92	10.05
Kid (<5 years) (%)	9.90	10.39	9.42	9.73	10.49	9.94	9.79
Native-born (%)	97.03	94.66	95.98	95.97	94.21	95.43	90.97
Renter housing units (%)	29.45	23.79	22.79	27.00	27.57	26.04	33.99
Education (%)	13.40	12.41	10.02	15.17	14.14	13.82	20.40
Unemployment (%)	8.90	7.72	6.78	6.23	5.91	6.46	7.12
Poverty (%)	26.65	20.76	19.75	14.02	14.60	16.07	18.12
Mean household income ($)	46,372	52,637	50,393	59,729	59,121	57,074	58,481
Year 2000							
Total population	2,848	27,821	29,037	104,589	51,252	215,547	20,636,272
White	976	19,377	21,892	80,249	36,629	159,123	14,638,862
Black	1,489	2,718	2,655	9,230	5,561	21,653	2,363,901
Asian	0	867	123	1,372	482	2,844	565,548
Native American	7	122	210	442	173	954	112,801
Others	373	4,748	4,152	13,296	8,411	30,980	2,955,154
Hispanic	511	8,826	8,119	27,797	15,488	60,741	6,609,381
Color	2,049	12,727	11,180	39,626	22,160	87,742	9,836,540
White (%)	34.27	69.65	75.39	76.73	71.47	73.82	70.94
Black (%)	52.28	9.77	9.14	8.83	10.85	10.05	11.46

(continued)

Table B.57 (continued)

	0–10	11–20	21–30	31–40	41–50	0–50	Outside
Asian (%)	0.00	3.12	0.42	1.31	0.94	1.32	2.74
Native American (%)	0.25	0.44	0.72	0.42	0.34	0.44	0.55
Others (%)	13.10	17.07	14.30	12.71	16.41	14.37	14.32
Hispanic (%)	17.94	31.72	27.96	26.58	30.22	28.18	32.03
Color (%)	71.95	45.75	38.50	37.89	43.24	40.71	47.67
Female (%)	49.33	50.17	50.56	49.76	50.32	50.05	50.44
Old (65+ years) (%)	16.22	12.76	12.78	11.96	9.74	11.70	9.90
Kid (<5 years) (%)	7.09	9.19	8.31	8.78	9.50	8.92	9.29
Native-born (%)	94.98	89.36	93.70	93.14	90.63	92.16	86.03
Renter housing units (%)	27.52	24.47	22.17	24.85	28.60	25.29	32.89
College degree or higher (%)	12.51	12.96	9.90	16.89	15.46	15.04	23.32
Unemployment (%)	14.46	8.59	6.71	6.11	6.24	6.62	6.06
Poverty (%)	29.36	18.68	14.39	13.33	13.50	14.42	15.38
Mean household income ($)	47,534	52,902	55,810	64,295	62,168	60,886	67,862
Year 2010							
Total population	2,654	27,059	28,372	106,080	52,041	216,206	24,095,684
White	1,618	21,346	21,994	84,478	42,231	171,667	17,331,780
Black	843	2,935	2,658	9,621	4,945	21,002	2,843,664
Asian	9	754	306	1,450	306	2,825	921,011
Native American	0	50	127	299	136	612	119,068
Others	184	1,974	3,287	10,232	4,423	20,100	2,880,163
Hispanic	813	10,032	9,546	32,581	19,037	72,009	8,845,468
Color	1,802	13,972	12,872	45,184	24,634	98,464	12,926,715
White (%)	60.96	78.89	77.52	79.64	81.15	79.40	71.93
Black (%)	31.76	10.85	9.37	9.07	9.50	9.71	11.80
Asian (%)	0.34	2.79	1.08	1.37	0.59	1.31	3.82
Native American (%)	0.00	0.18	0.45	0.28	0.26	0.28	0.49
Others (%)	6.93	7.30	11.59	9.65	8.50	9.30	11.95
Hispanic (%)	30.63	37.07	33.65	30.71	36.58	33.31	36.71
Color (%)	67.90	51.64	45.37	42.59	47.34	45.54	53.65
Female (%)	54.60	49.15	50.10	50.08	51.48	50.36	50.40
Old (65+ years) (%)	14.81	13.72	14.12	12.78	12.13	12.94	10.12
Kid (<5 years) (%)	13.07	7.24	6.67	6.97	8.37	7.38	7.83
Native-born (%)	96.16	88.19	92.07	92.31	89.77	91.20	83.84
Renter housing units (%)	26.20	19.80	19.66	23.85	25.16	23.00	30.97

(continued)

Table B.57 (continued)

	0–10	11–20	21–30	31–40	41–50	0–50	Outside
College degree or higher (%)	22.02	14.14	13.02	17.47	15.92	16.13	25.85
Unemployment (%)	20.95	9.62	7.38	6.11	6.29	6.92	7.00
Poverty (%)	35.83	18.72	16.79	13.30	14.05	14.90	16.77
Mean household income ($)	58,182	57,343	63,317	65,116	61,833	62,983	68,751
Index	57	57	57	57	57	57	57

B.58 North Anna Power Station, Virginia

See Fig. B.58.
See Table B.58.

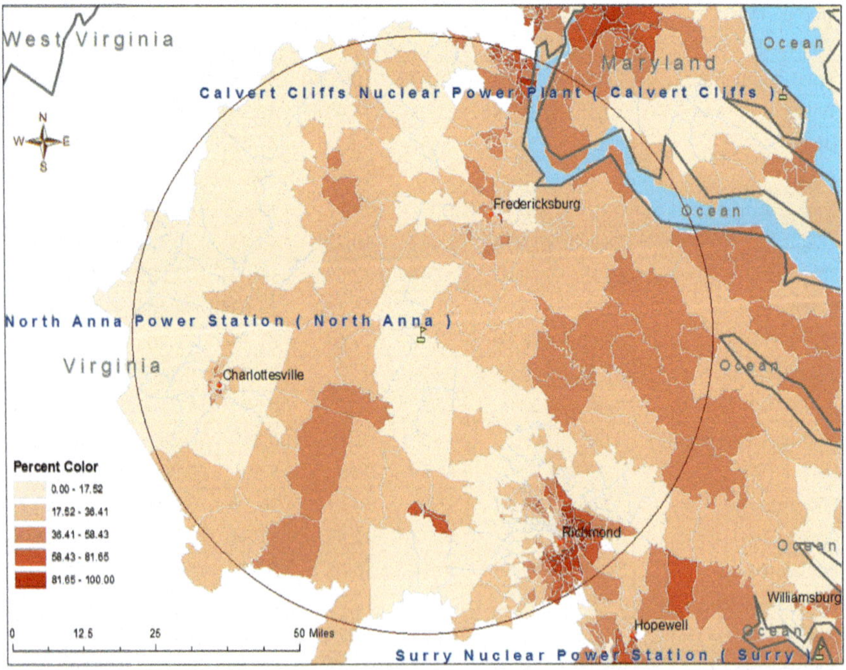

Fig. B.58 Spatial distribution of percent color among the populations surrounding North Anna Power Station, Virginia, in 2010

Table B.58 Demographic composition of population, as sorted by distance from North Anna Power Station, Virginia, in 1990, 2000, and 2010 *Source* (Kyne 2014)

	0–10	11–20	21–30	31–40	41–50	0–50	Outside
Year 1990							
Tract	6	26	55	134	143	364	2,949
Tract area (sq. mile)	363	1,037	1,730	1,502	1,549	6,181	49,000
Total population	14,747	64,514	147,361	402,522	455,975	1,085,119	9,883,689
White	11,623	51,398	120,330	335,257	292,911	811,519	7,378,020
Black	2,928	12,457	24,930	58,353	153,511	252,179	2,099,796
Asian	101	250	961	6,986	6,554	14,852	281,608
Native American	80	184	563	806	1,265	2,898	27,740
Others	21	237	561	1,128	1,726	3,673	96,481
Hispanic	70	563	1,567	4,522	5,427	12,149	263,207
Color	3,172	13,545	28,005	70,370	166,449	281,541	2,654,430
White (%)	78.82	79.67	81.66	83.29	64.24	74.79	74.65
Black (%)	19.85	19.31	16.92	14.50	33.67	23.24	21.25
Asian (%)	0.68	0.39	0.65	1.74	1.44	1.37	2.85
Native American (%)	0.54	0.29	0.38	0.20	0.28	0.27	0.28
Others (%)	0.14	0.37	0.38	0.28	0.38	0.34	0.98
Hispanic (%)	0.47	0.87	1.06	1.12	1.19	1.12	2.66
Color (%)	21.51	21.00	19.00	17.48	36.50	25.95	26.86
Female (%)	49.91	50.55	51.33	52.09	51.82	51.75	51.16
Old (65+ years) (%)	11.76	9.23	12.34	11.56	9.46	10.65	10.76
Kid (<5 years) (%)	8.79	9.75	8.43	8.21	9.56	8.91	8.69
Native-born (%)	98.69	98.92	98.22	96.64	97.55	97.40	93.96
Renter housing units (%)	11.23	16.68	26.57	33.05	34.61	31.63	31.53
Education (%)	7.51	14.42	19.05	31.19	23.46	25.03	25.38
Unemployment (%)	5.77	3.72	3.34	3.04	4.39	3.71	4.48
Poverty (%)	13.37	7.29	7.15	7.31	11.67	9.22	9.40
Mean household income ($)	57,543	65,484	67,508	74,276	65,708	69,120	73,231
Year 2000							
Total population	19,529	92,932	194,575	505,701	509,752	1,322,489	11,052,512
White	16,115	73,360	156,391	397,440	301,023	944,329	7,563,621
Black	2,964	16,117	30,418	78,514	181,818	309,831	2,542,420
Asian	118	837	2,224	14,738	9,363	27,280	444,435
Native American	30	357	681	1,602	1,819	4,489	33,556
Others	317	2,275	4,836	13,408	15,737	36,573	468,467
Hispanic	85	1,747	4,128	12,180	15,592	33,732	520,646
Color	3,471	20,635	40,365	114,911	215,072	394,454	3,729,566
White (%)	82.52	78.94	80.38	78.59	59.05	71.41	68.43
Black (%)	15.18	17.34	15.63	15.53	35.67	23.43	23.00

(continued)

Table B.58 (continued)

	0–10	11–20	21–30	31–40	41–50	0–50	Outside
Asian (%)	0.60	0.90	1.14	2.91	1.84	2.06	4.02
Native American (%)	0.15	0.38	0.35	0.32	0.36	0.34	0.30
Others (%)	1.62	2.45	2.49	2.65	3.09	2.77	4.24
Hispanic (%)	0.44	1.88	2.12	2.41	3.06	2.55	4.71
Color (%)	17.77	22.20	20.75	22.72	42.19	29.83	33.74
Female (%)	49.83	50.54	51.26	51.78	51.98	51.66	51.28
Old (65+ years) (%)	11.84	8.67	12.46	11.10	9.61	10.57	11.30
Kid (<5 years) (%)	7.56	8.79	7.58	7.91	8.42	8.12	7.91
Native-born (%)	98.52	97.71	96.89	94.29	95.50	95.44	90.70
Renter housing units (%)	11.12	14.81	25.17	31.32	32.33	29.39	29.75
College degree or higher (%)	12.00	18.76	26.63	37.69	27.29	30.36	30.31
Unemployment (%)	3.63	3.08	3.51	3.03	4.84	3.80	4.52
Poverty (%)	9.07	6.36	6.84	7.10	11.16	8.62	9.18
Mean household income ($)	63,854	74,573	78,375	81,903	71,855	76,783	80,212
Year 2010							
Total population	23,603	123,759	249,849	579,607	576,540	1,553,358	11,984,819
White	20,170	93,596	189,992	432,068	325,312	1,061,138	7,815,053
Black	2,667	23,261	40,418	97,102	204,309	367,757	2,829,628
Asian	25	1,734	7,227	24,751	17,741	51,478	678,333
Native American	74	389	809	1,727	2,273	5,272	35,882
Others	667	4,779	11,403	23,959	26,905	67,713	625,923
Hispanic	370	7,601	12,663	30,376	40,594	91,604	913,308
Color	3,577	34,485	65,145	164,562	275,434	543,203	4,664,333
White (%)	85.46	75.63	76.04	74.54	56.42	68.31	65.21
Black (%)	11.30	18.80	16.18	16.75	35.44	23.67	23.61
Asian (%)	0.11	1.40	2.89	4.27	3.08	3.31	5.66
Native American (%)	0.31	0.31	0.32	0.30	0.39	0.34	0.30
Others (%)	2.83	3.86	4.56	4.13	4.67	4.36	5.22
Hispanic (%)	1.57	6.14	5.07	5.24	7.04	5.90	7.62
Color (%)	15.15	27.86	26.07	28.39	47.77	34.97	38.92
Female (%)	50.72	50.50	51.24	50.98	52.29	51.47	51.19
Old (65+ years) (%)	13.62	8.90	12.76	11.54	9.67	10.86	12.00
Kid (<5 years) (%)	4.75	6.97	6.57	6.31	6.92	6.61	6.41
Native-born (%)	98.59	95.40	93.96	90.88	92.03	92.28	87.66

(continued)

Table B.58 (continued)

	0–10	11–20	21–30	31–40	41–50	0–50	Outside
Renter housing units (%)	12.97	17.75	23.10	29.25	30.68	27.67	27.95
College degree or higher (%)	17.62	25.36	31.31	41.09	31.09	34.23	34.66
Unemployment (%)	6.63	6.58	6.13	4.89	7.55	6.24	6.21
Poverty (%)	11.16	7.96	7.43	8.58	12.51	9.87	9.53
Mean household income ($)	75,634	80,880	87,353	87,536	75,335	82,290	86,787
Index	58	58	58	58	58	58	58

B.59 Surry Nuclear Power Station, Virginia

See Fig. B.59.
See Table B.59.

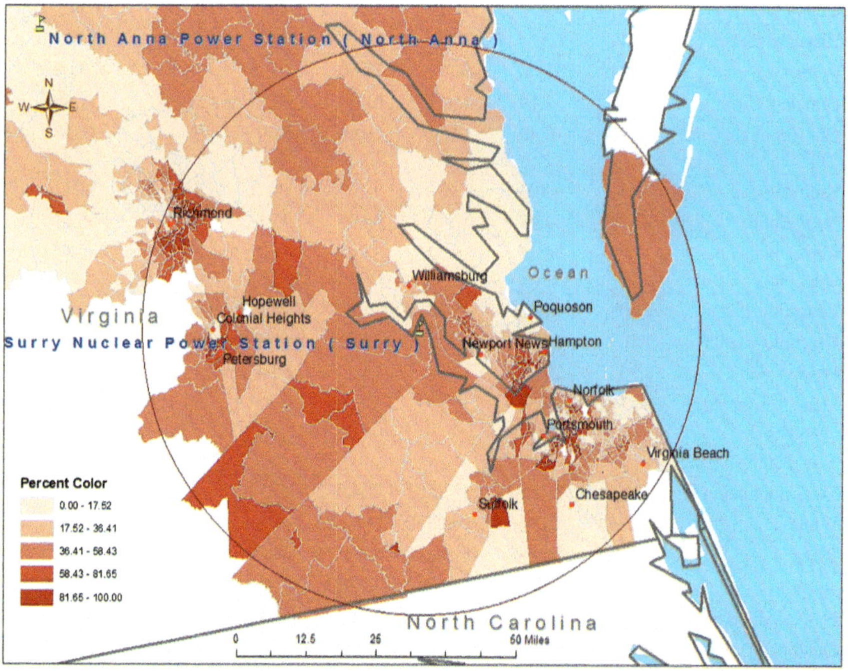

Fig. B.59 Spatial distribution of percent color among the populations surrounding Surry Nuclear Power Station, Virginia, in 2010

Table B.59 Demographic composition of population, as sorted by distance from Surry Nuclear Power Station, Virginia, in 1990, 2000, and 2010 *Source* (Kyne 2014)

	0–10	11–20	21–30	31–40	41–50	0–50	Outside
Year 1990							
Tract	29	73	93	212	43	450	3,652
Tract area (sq. mile)	356	892	1,551	1,797	1,003	5,599	90,995
Total population	108,530	252,904	321,598	700,649	140,317	1,523,998	11,291,965
White	77,393	175,672	214,999	434,235	108,717	1,011,016	8,793,509
Black	25,781	71,447	96,457	243,662	24,512	461,859	2,156,541
Asian	3,475	3,897	5,852	15,862	5,379	34,465	174,720
Native American	318	799	1,686	2,601	634	6,038	92,949
Others	1,575	1,078	2,622	4,287	1,069	10,631	74,215
Hispanic	3,403	3,822	7,214	13,015	3,484	30,938	193,446
Color	32,483	79,362	109,808	273,258	33,323	528,234	2,607,563
White (%)	71.31	69.46	66.85	61.98	77.48	66.34	77.87
Black (%)	23.75	28.25	29.99	34.78	17.47	30.31	19.10
Asian (%)	3.20	1.54	1.82	2.26	3.83	2.26	1.55
Native American (%)	0.29	0.32	0.52	0.37	0.45	0.40	0.82
Others (%)	1.45	0.43	0.82	0.61	0.76	0.70	0.66
Hispanic (%)	3.14	1.51	2.24	1.86	2.48	2.03	1.71
Color (%)	29.93	31.38	34.14	39.00	23.75	34.66	23.09
Female (%)	50.04	51.19	47.25	51.54	50.14	50.34	51.40
Old (65+ years) (%)	7.19	10.46	10.24	10.66	6.49	9.91	11.63
Kid (<5 years) (%)	9.81	9.40	8.93	9.76	10.71	9.61	8.21
Native-born (%)	94.86	97.19	97.09	96.91	95.73	96.74	96.66
Renter housing units (%)	38.79	36.15	36.33	36.41	24.77	35.51	28.95
Education (%)	26.08	20.02	18.38	16.82	18.71	18.48	21.10
Unemployment (%)	5.10	5.75	6.33	6.39	4.61	6.00	4.48
Poverty (%)	8.95	10.79	12.58	13.27	6.94	11.81	11.64
Mean household income ($)	64,169	59,585	56,975	57,769	66,543	59,126	61,416
Year 2000							
Total population	121,515	297,971	331,503	724,082	170,125	1,645,196	13,482,632
White	78,854	192,505	199,550	406,629	119,672	997,210	9,921,884
Black	33,064	89,683	111,240	273,671	34,826	542,484	2,575,678
Asian	3,586	5,976	6,728	18,159	8,417	42,866	332,097
Native American	604	1,157	2,032	3,293	619	7,705	115,645
Others	5,412	8,639	11,986	22,295	6,610	54,942	537,317
Hispanic	5,287	7,587	10,008	18,109	6,164	47,155	653,082
Color	44,981	108,939	135,586	325,179	53,621	668,306	3,846,659
White (%)	64.89	64.61	60.20	56.16	70.34	60.61	73.59

(continued)

Table B.59 (continued)

	0–10	11–20	21–30	31–40	41–50	0–50	Outside
Black (%)	27.21	30.10	33.56	37.80	20.47	32.97	19.10
Asian (%)	2.95	2.01	2.03	2.51	4.95	2.61	2.46
Native American (%)	0.50	0.39	0.61	0.45	0.36	0.47	0.86
Others (%)	4.45	2.90	3.62	3.08	3.89	3.34	3.99
Hispanic (%)	4.35	2.55	3.02	2.50	3.62	2.87	4.84
Color (%)	37.02	36.56	40.90	44.91	31.52	40.62	28.53
Female (%)	50.98	51.06	48.80	52.07	50.69	51.00	51.02
Old (65+ years) (%)	10.48	11.10	10.82	11.84	7.52	10.95	11.72
Kid (<5 years) (%)	8.29	8.01	7.66	8.27	8.74	8.15	7.92
Native-born (%)	94.52	96.18	96.12	96.03	94.03	95.76	93.10
Renter housing units (%)	33.37	35.12	33.98	34.94	22.90	33.52	27.74
College degree or higher (%)	30.09	25.04	21.36	20.05	23.56	22.31	26.15
Unemployment (%)	9.22	4.42	6.31	5.80	4.27	5.74	4.71
Poverty (%)	8.72	9.45	11.98	12.66	6.46	11.00	11.03
Mean household income ($)	72,811	64,749	62,916	62,622	73,177	64,827	70,393
Year 2010							
Total population	133,856	313,540	350,480	760,375	182,845	1,741,096	15,371,836
White	82,362	200,937	208,094	413,362	123,893	1,028,648	10,905,054
Black	38,403	93,141	118,805	294,501	38,901	583,751	2,929,341
Asian	5,491	8,059	8,781	23,183	11,486	57,000	565,423
Native American	263	928	1,784	2,682	617	6,274	124,693
Others	7,337	10,475	13,016	26,647	7,948	65,423	847,325
Hispanic	9,876	12,427	18,294	31,137	10,416	82,150	1,216,224
Color	57,482	120,163	152,312	365,185	65,643	760,785	5,072,013
White (%)	61.53	64.09	59.37	54.36	67.76	59.08	70.94
Black (%)	28.69	29.71	33.90	38.73	21.28	33.53	19.06
Asian (%)	4.10	2.57	2.51	3.05	6.28	3.27	3.68
Native American (%)	0.20	0.30	0.51	0.35	0.34	0.36	0.81
Others (%)	5.48	3.34	3.71	3.50	4.35	3.76	5.51
Hispanic (%)	7.38	3.96	5.22	4.09	5.70	4.72	7.91
Color (%)	42.94	38.32	43.46	48.03	35.90	43.70	33.00
Female (%)	49.85	51.91	49.69	51.50	50.47	50.98	51.11
Old (65+ years) (%)	12.04	12.97	11.04	12.20	9.37	11.79	12.31
Kid (<5 years) (%)	6.36	6.00	6.83	6.64	5.93	6.47	6.61
Native-born (%)	91.76	94.32	95.11	94.45	92.03	94.10	90.71

(continued)

Table B.59 (continued)

	0–10	11–20	21–30	31–40	41–50	0–50	Outside
Renter housing units (%)	31.95	32.15	32.06	33.09	21.26	31.47	27.22
College degree or higher (%)	32.47	29.21	24.82	23.56	27.69	25.94	30.06
Unemployment (%)	5.33	5.99	7.36	7.60	4.49	6.74	7.51
Poverty (%)	9.30	9.02	12.02	11.86	6.41	10.60	13.42
Mean household income ($)	76,794	72,330	67,347	66,378	80,390	69,885	71,291
Index	59	59	59	59	59	59	59

B.60 Columbia Generating Station, Washington

See Fig. B.60.
See Table B.60.

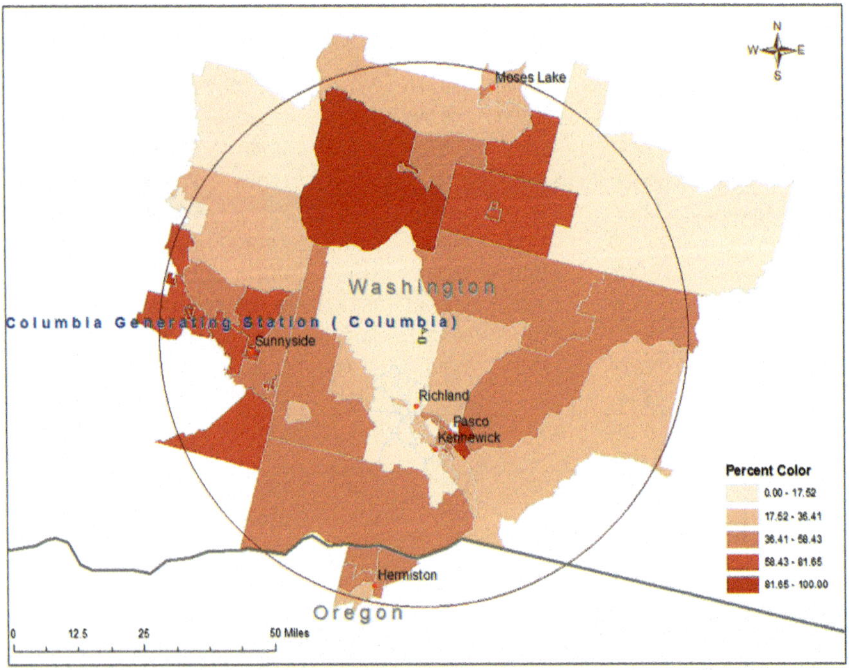

Fig. B.60 Spatial distribution of percent color among the populations surrounding Columbia Generating Station, Washington, in 2010

Table B.60 Demographic composition of population, as sorted by distance from Columbia Generating Station, Washington, in 1990, 2000, and 2010 *Source* (Kyne 2014)

	0–10	11–20	21–30	31–40	41–50	0–50	Outside
Year 1990							
Tract	1	31	22	8	9	71	2,221
Tract area (sq. mile)	209	1,136	1,432	1,479	525	4,782	164,895
Total population	3,530	80,090	83,319	28,897	38,595	234,431	7,474,582
White	3,030	70,657	66,091	17,605	33,563	190,946	6,760,395
Black	43	1,159	1,088	128	318	2,736	190,052
Asian	51	2,159	1,019	164	536	3,929	275,017
Native American	7	733	525	117	494	1,876	122,977
Others	397	5,376	14,604	10,879	3,691	34,947	126,184
Hispanic	810	8,135	18,057	13,558	5,302	45,862	270,747
Color	913	12,102	20,642	13,943	6,625	54,225	845,064
White (%)	85.84	88.22	79.32	60.92	86.96	81.45	90.45
Black (%)	1.22	1.45	1.31	0.44	0.82	1.17	2.54
Asian (%)	1.44	2.70	1.22	0.57	1.39	1.68	3.68
Native American (%)	0.20	0.92	0.63	0.40	1.28	0.80	1.65
Others (%)	11.25	6.71	17.53	37.65	9.56	14.91	1.69
Hispanic (%)	22.95	10.16	21.67	46.92	13.74	19.56	3.62
Color (%)	25.86	15.11	24.77	48.25	17.17	23.13	11.31
Female (%)	45.38	50.35	49.56	49.17	50.38	49.85	50.60
Old (65+ years) (%)	7.45	10.72	9.49	10.13	12.48	10.45	12.61
Kid (<5 years) (%)	12.52	9.71	11.34	13.36	9.93	10.82	8.82
Native-born (%)	85.33	93.30	89.70	79.77	93.41	90.25	94.13
Renter housing units (%)	23.65	34.90	34.51	36.26	33.88	34.57	34.34
Education (%)	18.59	25.49	14.58	9.45	12.73	17.64	22.19
Unemployment (%)	3.84	5.84	8.05	11.77	8.41	7.67	5.84
Poverty (%)	13.37	11.38	18.15	24.29	17.55	16.43	11.32
Mean household income ($)	68,778	62,309	53,379	45,794	46,521	54,816	60,293
Year 2000							
Total population	4,364	104,495	104,464	38,349	51,242	302,914	9,012,606
White	3,610	85,968	75,638	19,906	40,740	225,862	7,546,720
Black	20	1,226	1,088	95	860	3,289	234,795
Asian	7	2,998	1,189	252	684	5,130	444,306
Native American	0	494	936	202	294	1,926	132,807
Others	726	13,793	25,626	17,889	8,668	66,702	653,983
Hispanic	1,346	17,685	35,415	24,803	11,912	91,161	622,618
Color	1,403	24,533	39,916	25,431	14,360	105,643	1,703,621
White (%)	82.72	82.27	72.41	51.91	79.51	74.56	83.74

(continued)

Table B.60 (continued)

	0–10	11–20	21–30	31–40	41–50	0–50	Outside
Black (%)	0.46	1.17	1.04	0.25	1.68	1.09	2.61
Asian (%)	0.16	2.87	1.14	0.66	1.33	1.69	4.93
Native American (%)	0.00	0.47	0.90	0.53	0.57	0.64	1.47
Others (%)	16.64	13.20	24.53	46.65	16.92	22.02	7.26
Hispanic (%)	30.84	16.92	33.90	64.68	23.25	30.09	6.91
Color (%)	32.15	23.48	38.21	66.31	28.02	34.88	18.90
Female (%)	45.53	49.65	49.90	48.20	49.87	49.53	50.36
Old (65+ years) (%)	7.95	10.37	9.02	8.42	11.27	9.78	11.88
Kid (<5 years) (%)	8.68	9.21	10.70	13.10	9.92	10.33	7.87
Native-born (%)	85.15	88.28	83.12	67.04	89.25	83.93	90.51
Renter housing units (%)	23.36	29.09	31.09	34.76	31.04	30.64	32.88
College degree or higher (%)	14.80	28.64	15.37	10.32	14.14	19.40	26.97
Unemployment (%)	7.76	6.32	8.98	15.51	8.91	8.74	6.25
Poverty (%)	7.42	10.91	16.24	24.64	13.08	14.82	10.86
Mean household income ($)	78,951	73,296	61,975	50,840	57,875	64,507	70,554
Year 2010							
Total population	6,007	136,709	118,655	42,256	61,782	365,409	9,957,813
White	5,309	105,263	84,008	29,521	51,267	275,368	8,144,049
Black	29	2,314	1,514	161	741	4,759	291,553
Asian	74	4,332	2,004	121	660	7,191	633,904
Native American	0	1,274	1,297	321	874	3,766	152,058
Others	595	23,526	29,832	12,132	8,240	74,325	736,249
Hispanic	1,039	31,535	48,639	31,650	18,973	131,836	979,907
Color	1,211	41,078	54,415	32,270	22,168	151,142	2,350,377
White (%)	88.38	77.00	70.80	69.86	82.98	75.36	81.79
Black (%)	0.48	1.69	1.28	0.38	1.20	1.30	2.93
Asian (%)	1.23	3.17	1.69	0.29	1.07	1.97	6.37
Native American (%)	0.00	0.93	1.09	0.76	1.41	1.03	1.53
Others (%)	9.91	17.21	25.14	28.71	13.34	20.34	7.39
Hispanic (%)	17.30	23.07	40.99	74.90	30.71	36.08	9.84
Color (%)	20.16	30.05	45.86	76.37	35.88	41.36	23.60
Female (%)	44.18	49.87	49.52	48.56	47.32	49.08	50.35
Old (65+ years) (%)	9.59	10.46	9.92	8.02	10.97	10.07	12.57
Kid (<5 years) (%)	9.99	8.72	8.29	10.41	9.08	8.86	6.31
Native-born (%)	90.66	86.31	80.56	64.76	87.01	82.14	88.63
Renter housing units (%)	16.20	28.89	30.46	36.90	33.34	30.76	32.44

(continued)

Table B.60 (continued)

	0–10	11–20	21–30	31–40	41–50	0–50	Outside
College degree or higher (%)	17.45	28.27	16.75	7.84	14.78	20.04	30.46
Unemployment (%)	2.02	6.59	8.46	12.71	9.54	8.26	7.98
Poverty (%)	7.82	13.50	18.67	29.34	15.81	17.33	12.64
Mean household income ($)	79,791	72,614	59,086	47,739	58,236	63,491	70,963
Index	60	60	60	60	60	60	60

B.61 Point Beach Nuclear Plant, Wisconsin

See Fig. B.61.
See Table B.61.

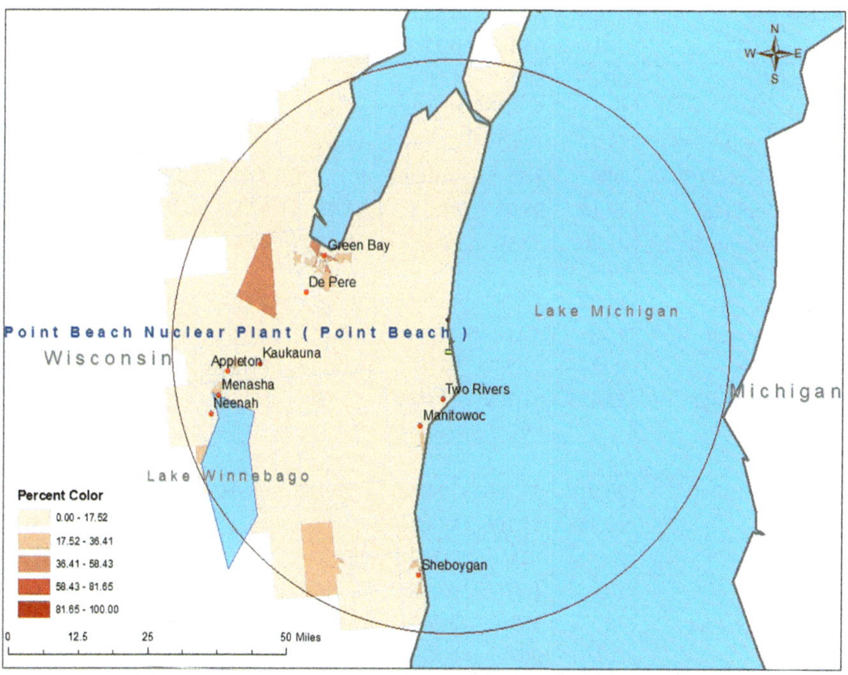

Fig. B.61 Spatial distribution of percent color among the populations surrounding Point Beach Nuclear Plant, Wisconsin, in 2010

Table B.61 Demographic composition of population, as sorted by distance from Point Beach Nuclear Plant, Wisconsin, in 1990, 2000, and 2010 *Source* (Kyne 2014)

	0–10	11–20	21–30	31–40	41–50	0–50	Outside
Year 1990							
Tract	5	13	48	30	18	114	1,295
Tract area (sq. mile)	192	1,154	1,689	712	1,403	5,150	60,346
Total population	20,905	51,500	173,433	109,702	62,752	418,292	4,473,476
White	20,582	50,155	167,146	106,176	61,258	405,317	4,108,993
Black	32	75	949	215	93	1,364	242,940
Asian	139	928	2,549	1,380	714	5,710	47,350
Native American	106	256	2,313	1,617	288	4,580	35,137
Others	45	88	473	315	397	1,318	39,044
Hispanic	86	314	1,341	870	938	3,549	84,057
Color	358	1,570	7,038	3,983	1,973	14,922	407,753
White (%)	98.45	97.39	96.37	96.79	97.62	96.90	91.85
Black (%)	0.15	0.15	0.55	0.20	0.15	0.33	5.43
Asian (%)	0.66	1.80	1.47	1.26	1.14	1.37	1.06
Native American (%)	0.51	0.50	1.33	1.47	0.46	1.09	0.79
Others (%)	0.22	0.17	0.27	0.29	0.63	0.32	0.87
Hispanic (%)	0.41	0.61	0.77	0.79	1.49	0.85	1.88
Color (%)	1.71	3.05	4.06	3.63	3.14	3.57	9.11
Female (%)	50.23	51.65	51.09	50.84	50.95	51.03	51.08
Old (65+ years) (%)	15.71	17.03	12.23	11.60	15.58	13.33	13.31
Kid (<5 years) (%)	8.89	8.55	9.36	9.31	8.65	9.12	8.93
Native-born (%)	99.15	98.05	98.40	98.08	97.86	98.23	97.45
Renter housing units (%)	21.80	27.10	32.07	28.36	23.84	28.65	29.59
Education (%)	11.08	12.78	16.20	15.88	12.64	14.87	18.01
Unemployment (%)	4.30	5.41	4.57	4.00	4.87	4.54	5.26
Poverty (%)	8.18	9.67	9.69	6.06	7.97	8.40	10.91
Mean household income ($)	53,475	51,709	57,095	64,344	53,847	57,584	57,762
Year 2000							
Total population	20,819	53,856	194,781	124,182	68,085	461,723	4,901,952
White	20,468	51,106	178,340	115,725	64,232	429,871	4,343,682
Black	2	123	2,339	603	416	3,483	296,872
Asian	103	1,276	4,066	2,924	1,470	9,839	74,815
Native American	43	387	3,205	2,213	283	6,131	43,530
Others	201	967	6,822	2,722	1,680	12,392	143,060
Hispanic	175	1,111	8,246	2,736	1,838	14,106	176,943
Color	444	3,251	19,591	9,649	4,611	37,546	638,480

(continued)

Table B.61 (continued)

	0–10	11–20	21–30	31–40	41–50	0–50	Outside
White (%)	98.31	94.89	91.56	93.19	94.34	93.10	88.61
Black (%)	0.01	0.23	1.20	0.49	0.61	0.75	6.06
Asian (%)	0.49	2.37	2.09	2.35	2.16	2.13	1.53
Native American (%)	0.21	0.72	1.65	1.78	0.42	1.33	0.89
Others (%)	0.97	1.80	3.50	2.19	2.47	2.68	2.92
Hispanic (%)	0.84	2.06	4.23	2.20	2.70	3.06	3.61
Color (%)	2.13	6.04	10.06	7.77	6.77	8.13	13.03
Female (%)	49.51	51.19	50.15	50.61	50.61	50.43	50.64
Old (65+ years) (%)	16.06	16.24	11.79	11.45	15.47	12.95	13.11
Kid (<5 years) (%)	6.40	7.20	8.14	8.24	7.00	7.81	7.71
Native-born (%)	98.84	97.06	95.73	97.09	96.56	96.51	96.38
Renter housing units (%)	18.83	25.88	32.49	27.21	24.44	28.41	28.34
College degree or higher (%)	13.40	16.27	20.25	21.36	15.58	19.04	22.74
Unemployment (%)	3.16	4.18	3.91	3.01	2.99	3.53	4.79
Poverty (%)	5.57	7.15	7.29	5.02	5.70	6.34	8.88
Mean household income ($)	61,514	61,446	65,898	72,787	63,011	66,550	67,158
Year 2010							
Total population	19,752	53,456	202,115	140,059	70,207	485,589	5,152,358
White	18,850	50,204	179,187	127,270	65,421	440,932	4,473,619
Black	279	289	3,674	1,711	709	6,662	342,000
Asian	252	1,565	5,466	3,887	2,364	13,534	110,461
Native American	129	244	3,599	3,289	255	7,516	41,673
Others	242	1,154	10,189	3,902	1,458	16,945	184,605
Hispanic	221	2,018	15,574	4,811	2,683	25,307	285,242
Color	1,071	4,608	31,504	14,984	6,438	58,605	843,917
White (%)	95.43	93.92	88.66	90.87	93.18	90.80	86.83
Black (%)	1.41	0.54	1.82	1.22	1.01	1.37	6.64
Asian (%)	1.28	2.93	2.70	2.78	3.37	2.79	2.14
Native American (%)	0.65	0.46	1.78	2.35	0.36	1.55	0.81
Others (%)	1.23	2.16	5.04	2.79	2.08	3.49	3.58
Hispanic (%)	1.12	3.78	7.71	3.43	3.82	5.21	5.54
Color (%)	5.42	8.62	15.59	10.70	9.17	12.07	16.38
Female (%)	51.27	50.78	50.53	50.04	50.69	50.47	50.40
Old (65+ years) (%)	19.15	16.83	11.85	12.32	15.85	13.41	13.39
Kid (<5 years) (%)	4.35	6.04	6.86	6.24	6.13	6.38	6.29
Native-born (%)	98.31	96.44	94.19	96.40	96.26	95.54	95.41

(continued)

Table B.61 (continued)

	0–10	11–20	21–30	31–40	41–50	0–50	Outside
Renter housing units (%)	17.23	24.82	30.54	25.50	24.06	26.86	26.75
College degree or higher (%)	15.05	18.25	22.56	24.91	20.59	22.13	26.14
Unemployment (%)	6.47	7.90	6.28	5.42	5.99	6.17	6.70
Poverty (%)	9.06	9.16	10.86	7.53	9.51	9.43	11.83
Mean household income ($)	58,942	58,520	61,510	71,688	57,426	63,316	65,460
Index	61	61	61	61	61	61	61

Reference

Kyne D (2014) Environmental justice issues in communities hosting US nuclear power plants (Order No. 3617964). Available from ProQuest Dissertations & Theses Global. (1528556250). Retrieved from http://ezhost.utrgv.edu:2048/login?url=http://search.proquest.com.ezhost.utrgv.edu:2048/docview/1528556250?accountid=7119

Index

D. Kyne, *Nuclear Power Plant Emergencies in the USA*,
DOI 10.1007/978-3-319-50343-1

The manufacturer's authorised representative in the EU is Springer Nature Customer Service Centre GmbH, Europaplatz 3, 69115 Heidelberg, Germany. If you have any concerns regarding our products, please contact ProductSafety@springernature.com

Printed and bound by CPI Group (UK) Ltd, Croydon, CR0 4YY
01/10/2026
02225860-0010